Lecture Notes in Computer Science   882

Edited by G. Goos, J. Hartmanis and J. van Leeuwen

Advisory Board: W. Brauer   D. Gries   J. Stoer

D. Hutchison   A. Danthine
H. Leopold   G. Coulson (Eds.)

Multimedia Transport and Teleservices

International COST 237 Workshop
Vienna, Austria, November 13-15, 1994
Proceedings

Springer-Verlag
Berlin Heidelberg New York
London Paris Tokyo
Hong Kong Barcelona
Budapest

Series Editors

Gerhard Goos
Universität Karlsruhe
Vincenz-Priessnitz-Straße 3, D-76128 Karlsruhe, Germany

Juris Hartmanis
Department of Computer Science, Cornell University
4130 Upson Hall, Ithaka, NY 14853, USA

Jan van Leeuwen
Department of Computer Science, Utrecht University
Padualaan 14, 3584 CH Utrecht, The Netherlands

Volume Editors

David Hutchison
Geoff Coulson
Computing Department, Lancaster University
Lancaster LA1 4YR, United Kingdom

Andre Danthine
Institut Montefiore, University of Liège
B-4000 Liège, Belgium

Helmut Leopold
Broadband Communications Division, Alcatel Austria
Scheydgasse 41, A-1211 Vienna, Austria

CR Subject Classification (1991): H.4.3, C.2, I.7.2, B.4.1

ISBN 3-540-58759-4 Springer-Verlag Berlin Heidelberg New York

CIP data applied for

This work is subject to copyright. All rights are reserved, whether the whole or part of the material is concerned, specifically the rights of translation, reprinting, re-use of illustrations, recitation, broadcasting, reproduction on microfilms or in any other way, and storage in data banks. Duplication of this publication or parts thereof is permitted only under the provisions of the German Copyright Law of September 9, 1965, in its current version, and permission for use must always be obtained from Springer-Verlag. Violations are liable for prosecution under the German Copyright Law.

© Springer-Verlag Berlin Heidelberg 1994
Printed in Germany

Typesetting: Camera-ready by author
SPIN: 10479316   45/3140-543210 - Printed on acid-free paper

Preface

Although many distributed multimedia applications now exist as pilot projects on local networks, these prototypes have yet to be translated into realistic applications running over large scale heterogeneous high-speed networks. To help bring about this important transition, a number of initiatives such as the COST 237 Multimedia Telecommunications Services project in Europe and the Multimedia Communications Forum in the US have recently been established. These groups identify a *lack of generic system support* as the primary technological factor holding back the deployment of realistic, large scale, distributed multimedia applications. There are two basic technologies required to make feasible such support: an appropriate *transport service* for communications needs, and a suitable set of generic multimedia teleservices to provide a framework for application development.

It is now accepted that significant enhancements to existing transport services are needed to adequately support large scale distributed multimedia applications. In particular, the transport service must be extended to support quality of service configurability and multicast/multipeer connectivity, and must be supported by a variety of high-speed network types. The area of multimedia teleservices is equally crucial. Generic high level services, such as multimedia enhanced email, conferencing frameworks and shared application frameworks, are necessary to ease the evolution from present day pilot applications to commercial inter-operable products. The present workshop addresses both of these technological areas with particular attention paid to the *integration* of the two. The emphasis of the workshop is on service and architectural aspects of distributed multimedia application support from the transport layer upwards.

In total, the call for papers resulted in 46 papers being received from 14 countries. Following the review procedure, the program committee selected 18 papers to be presented at the workshop and 6 papers to be presented at a poster/demo session. The workshop's keynote address will be given by Roland Hüber, director of the European Commission's RACE/ACTS research programmes, who will no doubt provide us with a stimulating opening to the workshop! His talk is entitled "Multimedia communication - making use of all senses and resources". The workshop proper is organised as a series of sessions and an ongoing poster display. The six sessions cover the following subject areas: teleservices, technological support for teleservices, quality of service and synchronisation, multipeer communication, broadband network transport and variable bit rate video transport.

COST 237 Multimedia Transport and Teleservices is organised by the CEC COST 237 Multimedia Telecommunications Services Project and hosted by Alcatel Austria AG. Many people have worked hard in preparation for the workshop - the COST 237 Steering Committee in planning the event in the first place, the Program Committee in reviewing and selecting the technical papers and posters and, of course, the local Organising Committee. I hope you will all join me in thanking Alcatel Austria AG for agreeing to host our workshop in the beautiful city of Vienna, and in a special vote of thanks to Helmut Leopold who has done all the *real* work as Chair of the local Organising Committee.

September 1994   David Hutchison, Conference Chairman

Program Committee

David Hutchison, Lancaster University, UK (Chair)
Jon Crowcroft, UCL, UK
Andre Danthine, University of Liege, Belgium
Michel Diaz, LAAS/CNRS, France
Christophe Diot, INRIA, France
Domenico Ferrari, University of California, USA
Serge Fdida, University of Paris, France
Gary Herman, Bellcore, USA
Andrew Lister, Queensland University, Australia
Craig Partridge, BBN, USA
Joe Pasquale, UCSD, USA
Steve Pink, SICS, Sweden
Bernhard Plattner, ETH Zürich, Switzerland
Radu Popescu-Zeletin, GMD-Fokus, Germany
Otto Spaniol, Aachen University, Germany
Jean-Bernard Stefani, CNET, Paris, France
Ralf Steinmetz, IBM ENC, Germany
Harmen van As, IBM Zürich Research, Switzerland
Giorgio Ventre, University of Napoli, Italy

Steering Committee

Andre Danthine, University of Liege, Belgium (Chair)
Christophe Diot, INRIA, France
Andonis Galetsas, CEC, Brussels
David Hutchison, Lancaster University, UK
Svend Jager, JYDSK, Denmark
Helmut Leopold, Alcatel Austria
Vassili Loumos, NTUA, Greece
Radu Popescu-Zeletin, GMD-Fokus, Germany
Melanie Pralong, Swiss Telecom, Switzerland
Sandor Stefler, PKI, Hungary
Giorgio Ventre, University of Napoli, Italy

Organising Committee

Helmut Leopold, Alcatel Austria (Chair)
Georg Blechinger, Moser+Blechinger PR, Austria
Geoff Coulson, Lancaster University, UK
Franz Edler, Alcatel Austria
Gerhard Weiss, communications service, Austria
Eike Wolf, Alcatel Austria
Gabriela Würth, Alcatel Austria

Contents

Session A: Teleservices: Multimedia Mail, Archiving and Retrieving
Chairperson: Radu Popescu-Zeletin, GMD-Fokus (D)

Towards a Complete Multimedia Mail: Use of MHEG in Standard
Messaging Systems 1
B. Kervella, V. Gay, E. Horlait, University of Paris (F)

A Mail-Based Teleservice Architecture for Archiving and Retrieving
Dynamically Composable Multimedia Documents 14
*H. Thimm, GMD-IPSI (D), K. Röhr, GMD-Fokus (D), Th. C. Rakow,
GMD-IPSI (D)*

The Global Store Server - A Multimedia Teleservice Component 35
C. Blum, L. Neumann, Fraunhofer Institute for Computer Graphics (D)

Session B: Teleservice Support
Chairperson: Geoff Coulson, University of Lancaster (UK)

From Broadband Network Services to a Distributed Multimedia
Support-Environment 47
H. Leopold, K. Frimpong-Ansah, N. Singer, Alcatel Austria AG (A)

Managing Shared Ephemeral Teleconferencing State: Policy and
Mechanism 69
*S. Shenker, XEROX (USA), A. Weinrib, Bellcore (USA), E. Schooler,
USC/ISI (USA)*

Computational Components for Synchronous Cooperation on
Multimedia Information 89
C. Loge, V. Gay, E. Horlait, University of Paris (F)

A Binding Architecture for Multimedia Networks 103
*A.A. Lazar, Columbia University (USA), S.K. Bhonsle, K.S. Lim,
National University of Singapore*

Session C: Quality of Service and Synchronization
Chairperson: Andre Danthine, University of Liege (B)

Implementation of an End-to-End Quality of Service Management Scheme   124
L. Fedaoui, University of Paris (F), A. Seneviratne, University of Technology, Sydney (AUS), E. Horlait, University of Paris (F)

A Formal Description Technique Supporting Expression of Quality of Service and Media Synchronisation   145
H. Bowman, University of Kent (UK), L. Blair, G.S. Blair, A.G. Chetwynd, Lancaster University (UK)

On the Synchronization Mechanisms for Multimedia Integrated Services Networks   168
W. Yen, I. F. Akyildiz, Georgia Insitute of Technology (USA)

Session D: Multipeer Communication
Chairperson: Georgio Ventre, University of Napoli (I)

Efficient Support for Multiparty Communication   185
C. Szyperski, International Computer Science Institute, Berkley (USA), G. Ventre, University of Napoli (I)

QoS Negotiation for Multicast Communications   199
L. Mathy, O. Bonaventure, University of Liege (B)

Support for High-Performance Multipoint Multimedia Services   219
G. Carle, J. Schiller, C. Schmidt, University of Karlsruhe (D)

Session E: Broadband Network Transport Issues
Chairperson: Serge Fdida, University of Paris (F)

Providing Support for Data Transfer in a New Networking Environment   241
R. Schatzmayr, R. Popescu-Zeletin, Technical University of Berlin (D)

Congestion Avoidance for Video over IP Networks   256
T. Sakatani, NTT Human Interface Labs (J)

Network Layer Scaling: Congestion Control in Multimedia Communication with Heterogeneous Networks and Receivers   274
H. Wittig, J. Winckler, J. Sandvoss, IBM ENC (D)

Session F: Variable Bit Rate Video Coding Transport
Chairperson: Christophe Diot, INRIA (F)

Resource Requirements for VBR MPEG Traffic in Interactive Applications   294
M. Hamdi, P. Rolin, Y. Duboc, M. Ferry, ENST de Bretagne (F)

Transmission of MPEG2 Applications over ATM Networks   310
T. Andrade, A.P. Alves, INESC (P)

Poster Session

Protocols for Multimedia Conferencing - An Introduction to the ITU-T T.120 series   322
W.J. Clark, BT Labs (UK)

A Platform for Multimedia Telecooperation Bridging Endsystem Heterogeneity   334
G. Dermler, University of Stuttgart (D), T. Gutekunst, Swiss Federal Institute of Technology (CH), E. Ostrowski, Technical University of Berlin (D), N. Pires, INTERSIS, T. Schmidt, M. Weber, Siemens AG (D) H. Wolf, University of Ulm (D)

A Scheme for Multimedia and Hypermedia Synchronisation   340
N. B. Pronios, T. Bozios, INTRACOM S.A. (GR)

Integration of Existing Applications into a Conference System   346
D. Riexinger, K. Werner, IBM ENC (D)

The CIO Multimedia Communication Platform   356
A. Rozek, P. Christ, Stuttgart University Computer Center (D)

Broadband Multimedia and Collaboration Tools. Idea Project   368
M.A. Blanco, R. Montero, TELEFONICA I+D (E), F. Almerico, TECNATION (I), G. Venuti, CSELT (I), P. Cremonese, FINSIEL

Authors Index   379

Towards a Complete Multimedia Mail: Use of MHEG in Standard Messaging Systems

Brigitte Kervella, Valérie Gay and Eric Horlait
Université Pierre et Marie CURIE,
Institut Blaise PASCAL, Laboratoire MASI
4, place Jussieu,
75252 PARIS Cedex 05, France.
Tel: (+33 1) 44.27.71.28 - Fax: (+33.1.) 44.27.62.86
E-mail: kervella, gay or horlait@masi.ibp.fr

Abstract. This paper addresses multimedia aspects in messaging services. It highlights the functionalities that are missing to provide what we call a complete multimedia mail service and it gives standard solutions to integrate those additional functionalities in standard messaging systems.

A complete multimedia messaging system does not provide only means to exchange simple multimedia messages having a reasonable size. It should first enable the transfer of multimedia information, directly or by references (e.g. for huge message). In case it uses a reference, files may be retrieved in connected mode. This is, in fact, already included in most messaging systems. Second, it should also be open to the introduction of new media types. This requires the introduction of functionalities to prevent the fact some workstations cannot support those media types. Third, it should enable the composition, the exchange and the presentation of formatted multimedia messages.

MIME and X.400-based solutions are presented for the two first issues. For the third issue, we have chosen MHEG as a basis. This future standard is promising and this paper presents how to implement MHEG functionalities in messaging systems.

Keywords. Multimedia electronic mail, MHEG, MIME, X.400, Standards.

1 Introduction

Nowadays, there is an increasing number of means to handle multimedia information (e.g. video disk, multimedia workstations). This evolution has transformed the universally accepted electronic mail into the multimedia electronic mail. To be complete, multimedia messaging should propose the transfer of messages containing text, graphics, still or animated picture, audio, video or facsimile. It should also give the opportunity to send messages containing formatted multimedia and hypermedia information. In this paper, when using the term multimedia, we refer to multimedia and hypermedia. The terms in italic are defined in the different standards.

Multimedia information needs to be displayed by appropriate equipment software and hardware (high definition screen, audio interface, microphone, loudspeakers, video interface, electronic card, etc.). It represents an important cost of equipment. Nowadays, parts of these elements are already existing on modern workstations [1].

The main condition for the development and the wide use of the multimedia electronic messaging is the standardization. In fact, the user's working environments are so various with heterogeneous hardware and software that distributed multimedia applications need to be standardized to be able to interoperate.

The two major standards of messaging systems are IPMS[1] [2,3,4,5] (X.420) and MIME[2] [6] standards. In spite of the gap between these two standards, some efforts have been made to enable message exchange between them. In the first one, even if it is not really a multimedia messaging, some facilities are provided to extend the standard in order to exchange different media. It is usually used in the industrial world. The latter is an extension of Internet mail and is mainly used by universities.

The MHEG[3] [7] standard is applicable to all kinds of multimedia applications that need to exchange formatted information. The MHEG format enables the expression of spatio-temporal and conditional relations between entities (as well as interactivity). It suits to real-time and it gives a final form representation of information without additional processing needed to restructure the information before its presentation. This standard is therefore an interesting candidate to be integrated in complete multimedia mail for the formatted multimedia part.

This paper describes, in section 2, the functionalities required to build complete multimedia messaging services. It presents the standard data type handled by standard messaging system and their possible extensions. It describes the processing of unknown data types. Finally, it indicates the need for tools to compose, exchange and present formatted multimedia messages and it proposes MHEG as a candidate for the provision of those functionalities. Section 3 is dedicated to the integration of MHEG in standard messaging systems. It summarizes the support that the MHEG standard provides. After this introduction, it details the integration of this standard to the messaging system. Then it describes the way to compose MHEG multimedia object, their incorporation in an electronic message and their presentation. The conclusion discusses the advantages of using MHEG for complete messaging services.

2 Towards a complete multimedia electronic mail

This section outlines some requirements and shows how to provide them with respect to the data type in multimedia message, the processing of unknown data type and the multimedia message formatting. They are based on a general study of the multimedia electronic mail, existing standards and prototypes, presented in [8].

---

[1]IPMS: InterPersonal Messaging System
[2]MIME: Multipurpose Internet Mail Extensions
[3]MHEG: Multimedia/Hypermedia Expert Group

2.1 Media types in multimedia messages

A message may contain different kinds of media. It is separated in bodyparts, each of them containing a particular media type. A simple multimedia message contains bodypart types like text, audio, video and pictures and may include other multimedia messages. Unlike text messages, the size of multimedia messages may range from some Kbytes to some Mbytes. Even with advanced compression techniques, a video sequence might be too large for a message transport system and for the storage capacities of the receivers' workstations. To avoid this problem, a possible solution would be to include references to large bodyparts within the message instead of the content itself. This way, the recipient may decide or not to retrieve, in connected mode, the referenced media.

The MIME messaging provides all these bodypart types and some others like *multipart* type for message containing bodyparts of various types and *application* type for data which does not fit to the previous ones. But it does not propose the synchronization of its components. For huge messages, the MIME messaging enables to reference this message using the *message* type and *external-body* subtype. After having received the reference, the recipient may retrieve the message using file systems, FTP[4] anonymous or other mechanisms.

The IPMS media types are *text, facsimile, teletex, videotex, encrypted, message, mixed, bilaterally agreed, nationally defined* or *externally-defined*. The VMGS[5] or X.440 standard [9] provides an extended bodypart *voice*. It was previously a basic bodypart. The sound quality is not good enough to call it an audio bodypart. X.420 messaging does not provide video and picture media types either, but the MHEG capabilities may be used to provide them.

The X.420 standard does not provide the opportunity to reference a data, but an external reference mechanism was introduced in MMMS[6] [10]. This messaging system is based on the X.420 standard and its reference extension uses the *externally-defined* bodypart. The sender may replace a huge bodypart by a reference. In this case, only the external reference is transmitted in the message via the X.420 system. The data structure of an external reference is based on the international standard for DORs[7]. A DOR contains information about the type and location of the information, the communication service to be used, as well as quality of service parameters. The message is placed in a local or global server and will be transfer to the recipient on demand using a data transfer service (e.g. RDT[8]).

---

[4]FTP: File Transfer Protocol
[5]VMGS: Voice MessaginG System
[6]MMMS: MultiMedia Mail System
[7]DOR: Distinguished Object Reference
[8]RDT: Referenced Data Transfer

2.2 Processing of unknown media types

The evolution of multimedia applications goes fast therefore new data types may not be processable by some workstations. It is important to foresee the arrival of unknown media types. The non-restitution of one media type can in fact remove the meaning of the message. For example, when two persons that do not know each other make an appointment, they may send their photos through multimedia electronic mail. If the photo restitution is not possible on the receiver workstation, the message has no more meaning.

Two alternatives are possible to process these unknown media types. Either the messaging system sends a *probe* to the recipient to test if a certain message can be accepted or the message is send to the recipient anyway and the receiver messaging system must warn the users that a data type has not been presented.

The first alternative is possible in an X.420 messaging. The *probe* takes into account the message size and the data types contained in it and the message is sent if the receiver messaging system notifies that the message could be accepted.

The other solution is applicable to both messaging systems MIME and X.420. It is to foresee a mechanism which uses an *alternative* bodypart when a message, or some parts of it, cannot be presented. This *alternative* bodypart may be a simple text which tells the recipient what part of the message cannot be processed and its type. It can be of the following form:

```
Sorry !
A bodypart
cannot be presented.
Its type was <bodypart type>
```

The recipient messaging system has also to notify to the sender what part of the message has not been presented and why.

2.3 Scenario for multimedia messages

Most multimedia messaging systems send and present the bodyparts sequentially, without any synchronization or scenario linking these bodyparts. It is the major issue in existing multimedia messaging systems. The next section presents a solution based on the MHEG standard to enable the sending of multimedia messages containing spatial and temporal relationship between the different media.

3 MHEG & multimedia messaging systems

3.1 General overview of MHEG standard

The objective of the MHEG standard is to ease the development of multimedia applications in open environments by ensuring the cross-platform portability of elementary units of information called multimedia and hypermedia objects (MHEG objects). It defines the representation and encoding of MHEG objects to be interchanged within or across applications or services, by any means of interchange. These objects, encoded using ASN.1[9] (part I) or SGML[10] (part II) will provide a common basis for other standards and for many applications which will be developed. The specification addresses the need for minimal resource terminals and makes use of other standards for text, image, video and other objects. MHEG accepts all kind of monomedia or data formats taking into account the resources of the workstations.

The various levels of interchange are shown in figure 1. The MHEG objects embed the non MHEG data (*content data*) (e.g. MPEG) and use other protocol elements (*OPE*) (e.g. ASN.1) for the object interchange. The application used may use a script interchange standard (e.g. C++) at the lower level.

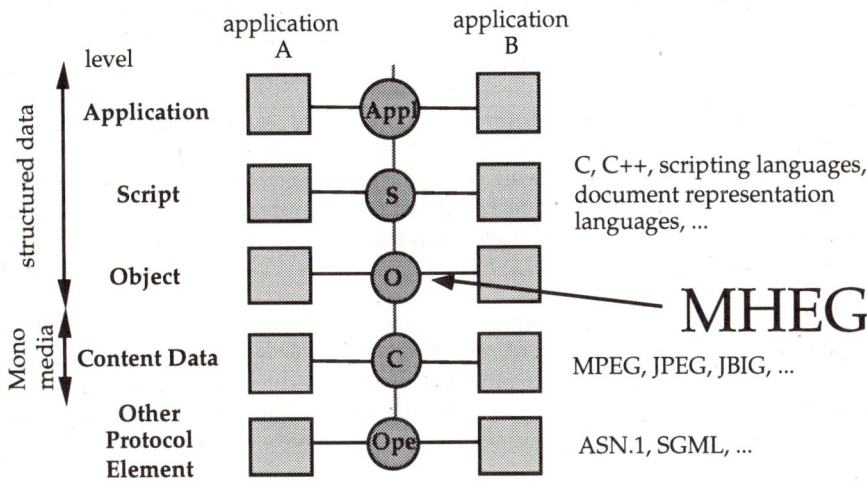

Fig.1. Data interchange services model

The MHEG standard specification is object oriented. The classes defined can be used to specify the objects containing monomedia information, the relations between objects, the dynamic behaviour of objects and the information required to optimize the real-time handling of objects. Two fictive classes have been defined to help the understanding of the standard. They are the *component* and *behaviour* classes.

---

[9]ASN.1: Abstract Syntax Notation number One
[10]SGML: Standard Generalized Markup Language

The first one contains the *content* and *composite* classes. The *content* class contains or references monomedia information and a set of parameters needed for the presentation. The *composite* class defines a structure to specify relationships between objects. These relationships allow the specification of time sequencing, positioning in space and logical interaction between the components.

For presentation reuse purpose, a clear separation has to be made between the MHEG interchange object which contains the original reusable structural information (*component* object) and a specific presentation of the original *component* object. A specific presentation of the original *content* object is called *presentable* (or *runtime content*) and a specific presentation of the original *composite* object is called *tree* (or *runtime composite*).

The latter one groups the classes concerning the behaviour of objects. The *action* class specifies the set of actions to be applied on a certain object. The *link* class defines the connection of one object with one or more objects. Combined with the use of actions, the links define spatio-temporal and conditional relationships between objects. The *script* class enables the encapsulation of an external piece of software performing a script in the MHEG formalism.

A clear separation is also made between the *script* object, which contains the original script data, and the *script instance* (or *runtime script*) that corresponds to a specific activation of the original script data. In the actual version, MHEG does not describe the script language and how to use it. A third part of the standard is studied to support scripting languages and should be an International Standard in 1996.

The other classes are *container* and *descriptor* classes. The *descriptor* class provides the information needed to ease the presentation of objects. The *container* class is a new class which has been created to avoid confusion between two different topics that are the *container* and the composition facilities previously included in the *composite* class. So the *container* class enables to interchange a set of objects without any link between them.

The scope of the classes defined in this standard enable their use in a wide range of applications and domains. It is recognized that certain applications may require specific functionalities not directly provided by classes. Therefore extensibility of the standard is provided by new object attributes, new actions to be applied on the objects, added conditions for triggering links and media types supported.

A MHEG tool kit devoted to the presentation of multimedia applications is developed within CCETT[11] projects to support the development of MHEG standard. Figure 2 presents the components of the tool kit for the presentation: the *MHEG engine*, the presentation server, the *MHEG class library* (MCL[12]), the *ASN.1 parser/formater* and an object manager *OCAM*[13]. A MHEG editor is available to compose the object.

---

[11]CCETT: Centre Commun d'Etudes de Télédiffusion et Télécommunications
[12]MCL: MHEG Class Library
[13]OCAM: Object & Content Access Management

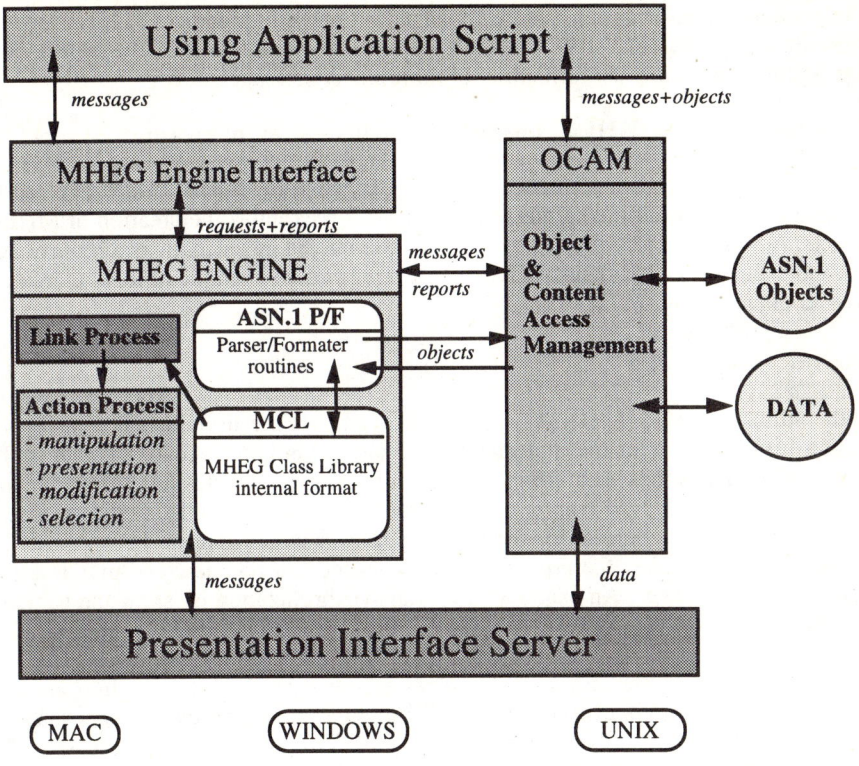

Fig. 2. MHEG components for the presentation

The *MHEG engine* manages the interpretation and the presentation of MHEG objects under the control of the application using it. The presentation server, called *ARAMIS*, manages the sessions, windows and presents monomedia contents.

The *MCL* is a portable library of classes implementing the hierarchy of MHEG objects. It offers a representation of the MHEG objects and *runtime* objects with their behaviour to present them in easy way. The *MCL* purpose is mainly to reduce the development time of *MHEG engines*. The *ASN.1 parser/formater* provides functions for the management of inputs and outputs from files. These files contain the description of MHEG objects in ASN.1 format to the internal structure of MHEG objects in *MCL*, and vice versa. It is used by the *MHEG engine* and the MHEG editor. The *OCAM* gives the access to monomedia content or ASN.1 objects file to the customer application.

The editor called *ANIMA* has a graphic interface which offers a *layout view* corresponding to the spatial composition, a *timeline view* representing the behaviour of objects in time and a *library view* displaying every object and data being edited. The interface of *ANIMA* is user friendly and does not require to be familiar with MHEG concepts. Figure 3 shows what the *ANIMA* editor looks like.

Complementary information can be found in the MHEG document [11]. After this overview of the MHEG standard, the next section integrates it in the multimedia messaging system, for the formatting of multimedia messages.

3.2 Creation of a MHEG object for inclusion in a message

The possibility to create formatted messages is the main functionality required to create a complete multimedia message. This section explains the creation of MHEG objects. A MHEG object enables to position monomedia in space, to synchronize and link them and to express their behaviour. Then it gives an example of how to create a MHEG object using *ANIMA*.

Spatial positioning

In the editor *ANIMA*, the *layout view* window allows the user to make easily the composition in space of the multimedia message. This latter is called *scene* in the editor and corresponds to a *composite* object. Monomedia desired for the *scene* composition are retrieved from a data base or created by specialized editors. These *content* or *composite* objects are shown in the *library view* window. To compose the message, the user takes the *content* object from the *library view* and puts it at the right place in the *scene*. An example of spatial synchronization is "show the name of product 2 cm above the image".

In the 1994 version, the various media types accepted by the MHEG editor are still pictures, video, drawing, text, audio and button. These *content* objects may be modified by a specific editor and copied. Their size can be reduced or increased. The overlay of content is possible using instructions like "move to back". This editor also provides usual instructions for text content like "change font", "change size", "center", "justify" or "align" the text.

Temporal positioning

The third window of the editor, *time view*, is in charge of temporal positioning of media objects. It is a grid with a time scale on the top. Its unit of time can be changed by a function proposed in the menu. The synchronization of elements is easily done using the grid. Each element in the *composite* object may have a limited presentation in time or not. An example of temporal synchronization is "show the text 2 seconds after the end of the video".

Conditional synchronization

There is another type of synchronization: the conditional synchronization. A current state of the object in the presentation may trigger a reflex action on another object, for example: "when the audio ends, ask the question". This synchronization type can be expressed by a link defined between objects and it would be provided by a function in the editor menu. The basic behaviour of each object and *runtime* objects, e.g. volume, is part of the object or part of the *runtime* object itself. The actions associated with the separated *link* object may in turn modify the basic behaviour of the other objects or of the *runtime* objects, for example to increase a volume.

The link contains a condition, a set of parallel actions and targets to which they are applied. When the link condition is satisfied, the link is fired, i.e. the set of parallel actions is processed on the targets. The overall link condition is expressed as a trigger accompanied by optional additional conditions. The trigger and additional conditions must be satisfied simultaneously in order to fire the link. The parallel targeted action set is made up of groups of serial targeted actions. The groups are to be processed in parallel but it is not a real parallel processing, the MHEG engine does not required it. Each group of serial targeted actions contains reference to the targets of the actions and a list of *action* objects.

Behaviour

The actions enable to express or change the behaviour of an object or a *runtime* object. These actions are specific to the target type (object or *runtime* object). The *prepare* action makes an object available to be process by the *MHEG engine*. The *destroy* action has an inverse effect. Other actions are associated to a class.

Some actions enable to describe the *runtime* object behaviour. *Set speed*, for example, changes the speed presentation of the presentation instance. *Set opacity* action gives the opportunity for the presentation instance to appear or disappear gradually. *Set box* action presents a box to show the perimeter of the visible size. *Set scrollability* action specifies whether the presentation instance is presented with a scroll mechanism or not. Other operations are defined in the standard.

Creation of a MHEG object using ANIMA

This section presents the creation of a composite object. Figure 3 shows how a user may create an object and what the ANIMA windows look like.

This composite object contains 3 *content* objects that are 2 video sequences (Video1 and Video2) and a graphic (Graphic1). They can be seen in the *library view* window called MHEGApp1:1 in the figure 3. In the *layout view* (called MHEGApp1:2 in figure 3), the spatial composition of the object is shown. The Video1 *content* which is a filmed speech of a lady is positioned in the top left part of the window. The graphic *content* which is positioned on the right part represents a map. The Video2 *content* is positioned in the lower part of the window. It is a video sequence which gives a short presentation of a town by night. The *time view* window gives the temporal presentation of the *composite* object. MHEGApp1:3 window indicates that the speech (Video1) is played first. The Graphic1 is presented after a delay of 10 seconds and its duration is of about one minute. Then Video2 is played and the speech stops. The total presentation duration of the *composite* is of about 2 minutes. The reference mark placed on the *time view* window at the date=1 minute is needed in the following.

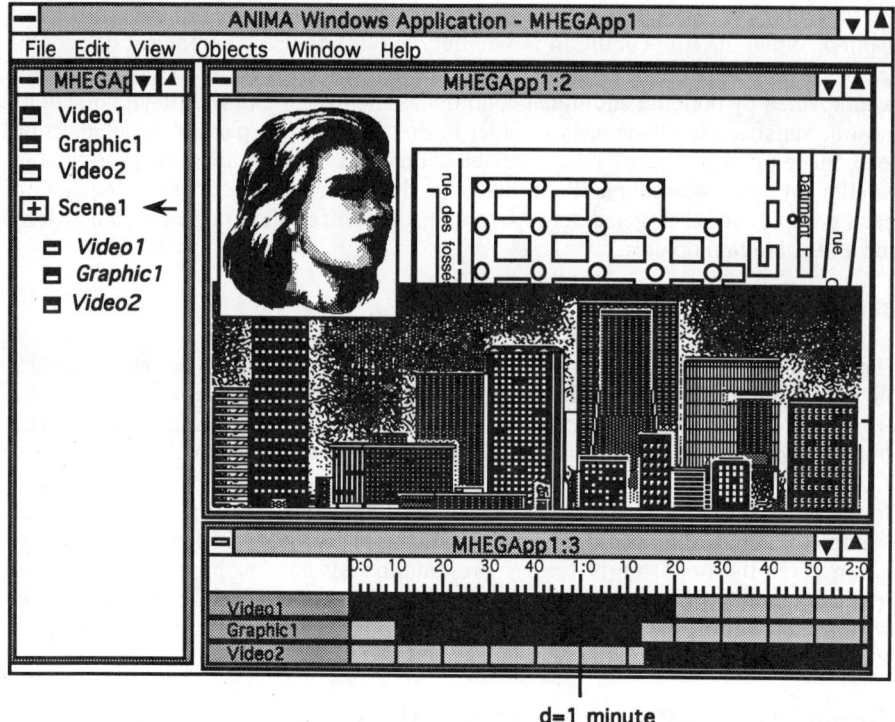

Fig.3. An example of a MHEG object

3.3 Integration of MHEG objects in X.420 and MIME

The user of the multimedia messaging has the possibility to design either a message with the bodypart types of the messaging system or an elaborated message containing spatial and temporal synchronization using a MHEG object. A multimedia message integrating the MHEG bodypart may also contain classical bodyparts. An example is given in figure 4. This section indicates how MHEG objects are integrated in a MIME or X.420 message.

All MHEG object characteristics described using the editor are coded in ASN.1 by the *ASN.1 parser/formater*. In the future, they may be coded in SGML. To integrate this object in a particular bodypart we may create a *MHEG* bodypart with an encoded type set to ASN.1.

This example of multimedia message sent by Susan is an answer message to Elena. It contains 2 classical bodyparts and a *MHEG* bodypart. The two classical bodyparts are the message previously sent by the recipient and a text. The text bodypart is a short answer and the *MHEG* bodypart is an additional answer with complete multimedia explanations.

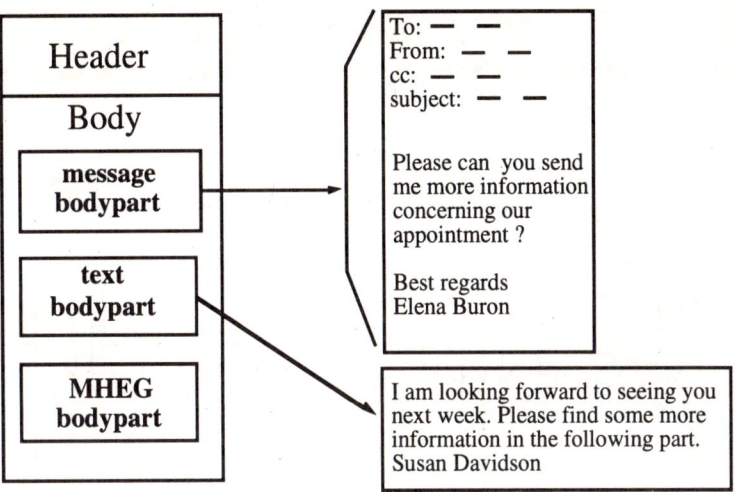

Fig. 4. An example of an extended message

For its integration in MIME, we have the possibility to use the subtype in the content types. This enables the extension of MIME standard to MHEG data types. The *application* content type is to be used for data which do not fit in any of the other categories and particularly for data to be processed by application programs. The *MHEG* bodypart must be processed by the *parser/formater* before its presentation to the recipient. A new subtype called *MHEG* is therefore created to integrate the bodypart MHEG. This new subtype must be registered with Internet assigned numbers authority (IANA) before its use. Another solution is to use a private subtype called *X-MHEG* but it does not fit with our objective to have a multimedia messaging based on standard.

For its integration to X.420, some extended bodyparts have been added [6] to the IPMS standard in the *externally-defined* bodypart using a macro. They are the *file transfer* and the *voice*. We can define a new bodypart type for MHEG using the same macro.

The MHEG object can now be precisely integrated in the message in a *MHEG* bodypart. If the object size is huge, we put its reference instead. As described in section 2.1, this mechanism already exists in the MIME standard and is the purpose of an extension for X.420 standard. Once the message is integrated, the transfer can be done as for any other message. During the transfer, to avoid a double coding (with possible incompatibility), it seems necessary to prohibit the implicit conversion.

At the message reception, the *MHEG* bodypart is presented using the presentation server *ARAMIS* and other bodyparts with the messaging system. Before being presented, the MHEG bodypart must be decoded by the *ASN.1 parser/formater*. Then, the MHEG object is available for the *MHEG engine* and may be presented using *ARAMIS*. Figure 5 corresponds to the screen seen by the recipient of the multimedia message described in figures 3 and 4.

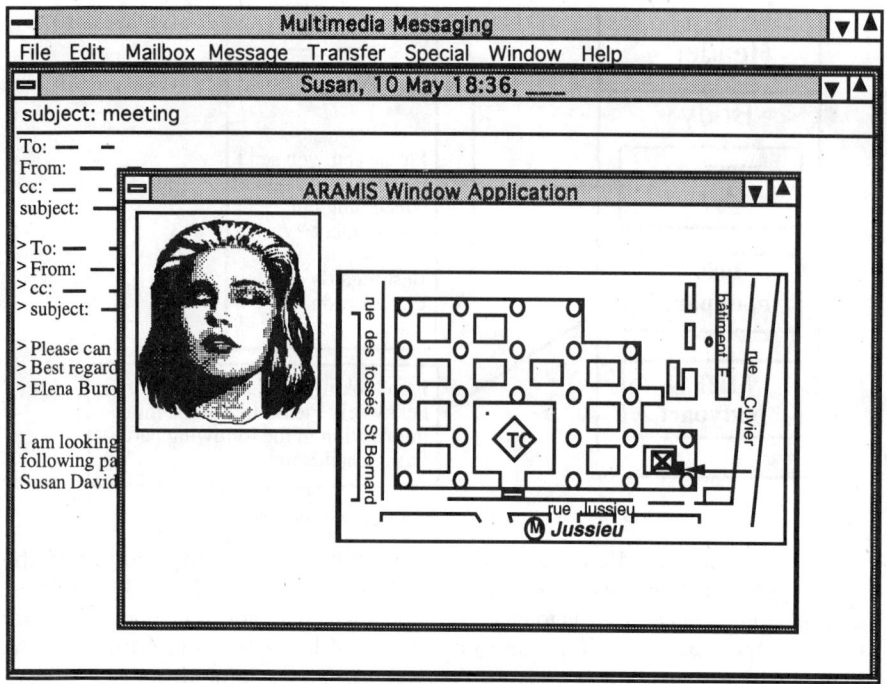

Fig. 5. Multimedia message presentation at the date "d=1 minute"

The multimedia message of our example is displayed by the recipient messaging system like the other messages except for the *MHEG* bodypart. Its presentation is made, using the presentation server *ARAMIS,* in a new window somewhere on the workstation screen. Figure 5 shows the presentation of the *MHEG* bodypart at the date "d=1 minute" as indicated by the reference mark in figure 3.

4 Conclusion

The main particularity of this paper is its proposal to integrate the MHEG functionalities in standard messaging system to design a standard complete multimedia messaging system. The advantages of using MHEG to build our complete messaging services are numerous. First, it is a future standard. It should be a Draft International Standard in 1994. A second advantage is the range of applications it may be included in. This will ease the integration of the complete multimedia mail in distributed multimedia environment together with other multimedia applications. Finally, in the CCETT project, most of the MHEG tool kit components are now available. We are now studying the way to implement the complete multimedia messaging system we have presented in this paper and the MHEG tool kit will be of great interest.

5 References

1. Huitema C., 'The Challenge of Multimedia Mail', Computer Networks and ISDN System, Vol. 17.
2. ISO 10021 and CCITT X.400, Series of Recommendation - Message Handling Systems - Information Processing Systems - Text Communication - MOTIS, October 1991.
3. ISO 10021(7) and CCITT X.420, Message Handling Systems - Information Processing Systems - Text Communication - MOTIS - Interpersonal Messaging System ISO/CCITT, October 1991.
4. ISO 10021(5) and CCITT X.413, Message Handling Systems - Information Processing Systems - Text Communication - MOTIS - Message Store: Abstract Service Definition, October 1990.
5. ISO 10021(4) and CCITT X.411, Message Handling Systems - Information Processing Systems - Text Communication - MOTIS - Message Transfer System: Abstract Service Definition and Procedures, October 1991.
6. Borenstein N., 'MIME: Mechanisms for specifying and Describing the Format of Internet Message Bodies', Network Working Group, RFC 1521, September 1993.
7. MHEG, Coded Representation of Multimedia and Hypermedia Information Objects, ISO/IEC JTC1/SC29/WG12, March 1993.
8. Gay V., Kervella B. and Horlait E., 'Conception of Multimedia Electronic Mail Based on Standards', 4th Workshop on Future Trends of Distributed Computing Systems (FT-DCS'93), IEEE Computer Society Press, Lisboa, Portugal, September 1993.
9. CCITT X.440, Message Handling Systems: Voice Messaging System ISO/CCITT, October 1991.
10. Hoepner P. et al, 'Technical Documentation on Functional Components and Interfaces', WP4.1, R2060/PTT/CIO/DS/P/004/b1, September 1993.
11. Picquet P. and Bertrand F., 'MHEG Toolkit and CCETT experiments', MHEG93/627 SC29/WG12 N111, November 1993.

A Mail-Based Teleservice Architecture for Archiving and Retrieving Dynamically Composable Multimedia Documents*

Heiko Thimm[1], Katja Röhr[2], and Thomas C. Rakow[1]

[1] GMD - Integrated Publication and Information Systems Institute (IPSI)
Dolivostraße 15, D-64293 Darmstadt, Germany
e-mail: {thimm, rakow}@darmstadt.gmd.de

[2] GMD - Research Institute for Open Communication Systems (FOKUS)
Hardenbergplatz 2, D-10623 Berlin, Germany
e-mail: roehr@fokus.berlin.gmd.d400.de

Abstract. In this paper, a teleservice for archiving and retrieving multimedia documents using public networks is described. This teleservice encourages a broad range of commercially applicable multimedia archiving applications suitable for an asynchronous access mechanism. It is based on an integrated architecture comprising stand alone archive clients and a multimedia archive server which is realized using a database management system. Archive clients access the archive server via an extended X.400 Multimedia Mail Teleservice. This teleservice reflects the specific requirements of dealing with multimedia documents in a networked environment by supporting a global reference mechanism. The archive server can dynamically compose new versions from an original archived multimedia document including extractions of subsequences of continuous data streams, coding and quality transformations. Thus, users can retrieve multimedia documents that explicitly reflect his individual workstation environment, information needs, and preferences. Since this feature allows to determine in advance the data volume to be retrieved and transmitted, users have more control over their service charge. The concept, architecture, and functionality of the teleservice as well as a sample instantiation of the proposed architecture showing an application called multimedia calendar of events are described. A comprehensive discussion of this prototypical implementation provides our experiences.

1 Introduction

In the near future, many users on possibly heterogeneous platforms will have access to multimedia mail allowing them to interchange multimedia documents over public networks. Besides its usage for conventional interpersonal messaging, this technology can also be exploited for innovative, commercial multimedia applications. It is especially suited for domains that can cope with the mail delay, i.e. which do not require immediate delivery of requested multimedia data, like, e.g., a product offering service, a virtual travel agency, subscription service for multimedia documents, or cooperative authoring of multimedia documents.

In this paper, we introduce an archiving and retrieval teleservice for multimedia documents which employs such a multimedia mail implementation. Connectionless,

---

* This work is partially granted by DeTeBerkom GmbH, Berlin, as project "Globally Accessible Multimedia Archives (GAMMA)" within the BERKOM II initiative.

asynchronous multimedia mail is applied as a means for interchanging multimedia documents between *archive clients* and an *active multimedia archive server*. Since the special requirements of multimedia data within a networked environment are reflected in the underlying mail implementation as well as in the archiving components, it extends those functionality provided by usual electronic document archives. One feature is that searching for documents is possible by descriptive search criteria addressing document contents as well as multimedia specific data. This allows, for example, to select documents which do not contain video clips longer than 1 minute. Another feature is support for dynamic document composition. It allows to retrieve other versions of originally archived multimedia documents which are dynamically created by the archive. In other words, a copy of the original document is tailored to meet the user's individual information needs, preferences, workstation environment and cost restrictions (charge of the teleservice provider, network and storage costs). Furthermore, it is possible to retrieve only a description of a document's structure which can be helpful for a user who has to decide if it is worthwhile to retrieve the document completely. The uncertainty with respect to the returned amount of data and the resulting costs faced in formulating queries is less critical. Users can control which of the following types of query result they want to receive: (1) a report of the number of matching documents, (2) for each matching document some application specific descriptive information, or (3) complete documents.

The *archive access protocol* defined between archive clients and the server is using the multimedia mail service developed at GMD-FOKUS which is based on the principles of CCITT Recommendation X.400 Message Handling System [3]. The client/server-communication employs both, the *store-and-forward mechanism* inherent to electronic mail and the *referenced object access* (ROA) mechanism which complements electronic mail submission and delivery in case of high-volume multimedia information content such as video.

In our prototypical implementation of the proposed teleservice, the archive server which is described in-depth in [28] is based on the VODAK *database management system* (DBMS) [18] which is a research-prototype developed at GMD-IPSI. Our standalone archive clients provide functionality for the authoring and presentation of the multimedia documents as well as the management of the user access to the archive server. Except for its asynchrony, for the teleservice users, the utilization of the mail service is transparent.

The remainder of this paper is organized as follows. We conclude this section with a review of related work. In section 2, we introduce the multimedia mail implementation relevant for us. The architecture of our teleservice is described in section 3, while in section 4, we explain its functionality. A sample instantiation of the proposed architecture is outlined in section 5. In section 6, we evaluate our approach and discuss the experiences made by a sample implementation. We conclude the paper with a review of issues that have been out of scope, and an outline of our current state of implementation as well as future work.

Related Work

To our knowledge, the integration of a multimedia database system with multimedia electronic mail for the realization of a multimedia archiving teleservice is a new approach not reported in any previous work. Research in both areas from which technologies are applied have started almost simultaneously. Information about multimedia mail can be found, e.g., in [2, 7, 22] while research on archives for documents (mostly office documents) is reported, e.g., in [5, 30]. With respect to multimedia database systems, a research field which is still developing, first approaches were database systems for specialized data such as spatial databases [20, 26] and pictorial databases [27]. Spatial databases were attractive because of the fact that the semantics of the objects and operations could be clearly defined. One of the first multimedia efforts in managing multimedia data was the *Multimedia Information Manager* in the ORION object-oriented database system, developed at MCC [31]. The integration of the new data types is accomplished through a set of definitions of class hierarchies and a message passing protocol not only for the multimedia capture, storage, and presentation devices, but also for the captured and stored multimedia objects. This way, a high degree of flexibility is achieved since new storage or presentation devices are easily included by providing the corresponding classes as subclasses of the existing classes. Another project is MINOS, in which the *Multimedia Object Presentation Manager* was developed at the University of Crete [4]. Synchronization mechanisms for distributed multimedia systems are addressed in [25]. An approach for an integration of multimedia into a distributed office application environment based on a multimedia database is described in [24].

2 Multimedia Mail

Research within the area of multimedia mail has been going on for almost 10 years [2, 19, 22]. While the early approaches have been very restrictive and support for continuous data was not fully provided, today's approaches are less restrictive and more powerful.

Within the BERKOM project, a Multimedia-Mail Teleservice for public networks is under development [7, 19]. It is based on the principles of CCITT Recommendation X.400 *Message Handling System* (MHS) [3] and provides an extended *Interpersonal Messaging Service* running in an open heterogeneous environment on a series of different hardware and software platforms. Implementations may include access to de facto standard Internet Mail SMTP [21] and competing multimedia mail standard MIME [1].

In terms of X.400, the body of a message consists of a sequence of *bodyparts* (Fig. 1). A bodypart contains data of a certain information type. Due to the fact that the information types supported by X.400 do not meet the requirements of an extensive multimedia mail service, the BERKOM project makes use of the extension mechanism of X.400 (88), the *externally-defined bodypart type*. This mechanism allows to integrate additional (commonly used) multimedia and hypermedia information representations as message bodyparts. Graphics, video, and audio as well as multimedia/hypermedia documents may now be included in a message. A special LINK bodypart contains information about how contents of bodyparts are related. The only content types which are mapped onto standardized bodypart types are TEXT and MESSAGE.

The *Message Transfer System* (MTS) is based on a *store-and-forward mechanism*. Most of the existing nodes between originator and recipient in a X.400 MTS have message size restrictions. Large messages, one of the characteristics of multimedia information conveyed through electronic mailsystems, may lead to congestion in *Message Transfer Agent* (MTA) implementations as well as may go beyond storage capaci-

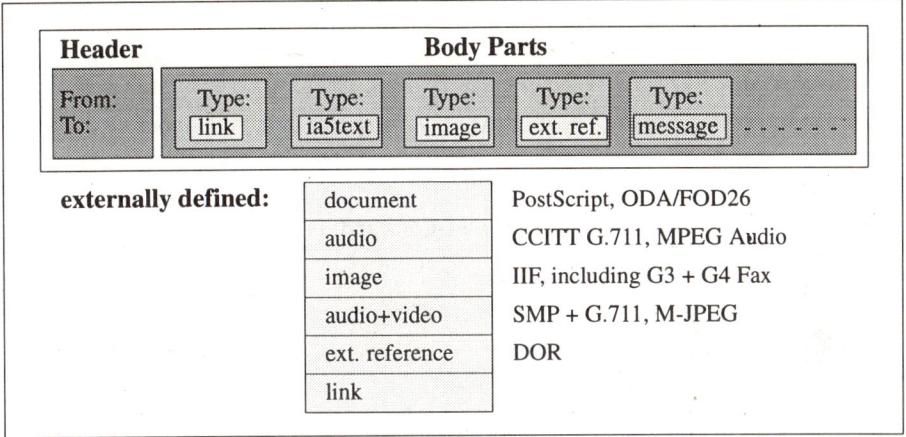

Fig. 1. The BERKOM Multimedia Message Structure

ties at the recipients' sites. As a consequence, a powerful *reference mechanism* is needed, which makes it possible to include a reference to mono or multimedia data within a message instead of including the data itself. The data is then stored in a specific, remotely accessible store (Fig. 2).

The *external references* are based on the universal reference mechanism *Distinguished Object Reference* (DOR) [14], defined as part of the *Referenced Object Access* (ROA) model, described in the international standard *Distributed Office Application Model* (DOAM[13]). The DOR standard describes a data structure and a coded representation of external references that specifies the store, the access method(s), the type of the data object, a local reference to the data object and quality-of-service (QoS) parameters. It allows data objects to be localized and distinguished globally.

The recipient of a message is free to resolve the reference using communication protocols which provide direct access to the stored data object. In the BERKOM project, a non-realtime service and protocol, the *Referenced Data Transfer* (RDT) [16, 23] has been implemented to retrieve the referenced data object from the store. In case a client wants to view message bodyparts such as audio and video without prior local storage, a *realtime* protocol must be used between the store and the client's site. There are a number of activities going on regarding the design and implementation of such realtime protocols, in particular for the transport/network layer, where the transport user may specifiy QoS parameters (e.g. throughput and transit delay) to be fulfilled by the transport connection. The integration of such realtime protocols within the Multimedia-Mail Teleservice is under development.

As mentioned above, the LINK bodypart of a multimedia mail describes the relationships between the mail's different content parts. The BERKOM-profile uses the concept of *links* between two parts of information to allow structuring for the whole range from simple annotations to complex references between bodyparts. This scheme may also lead to hypertext/hypermedia-like messages as required for the Multimedia Archive Teleservice.

For separation of archiving specific information from the remainder of the message body, an externally defined bodypart type called ARCHIVE is introduced. Multi-

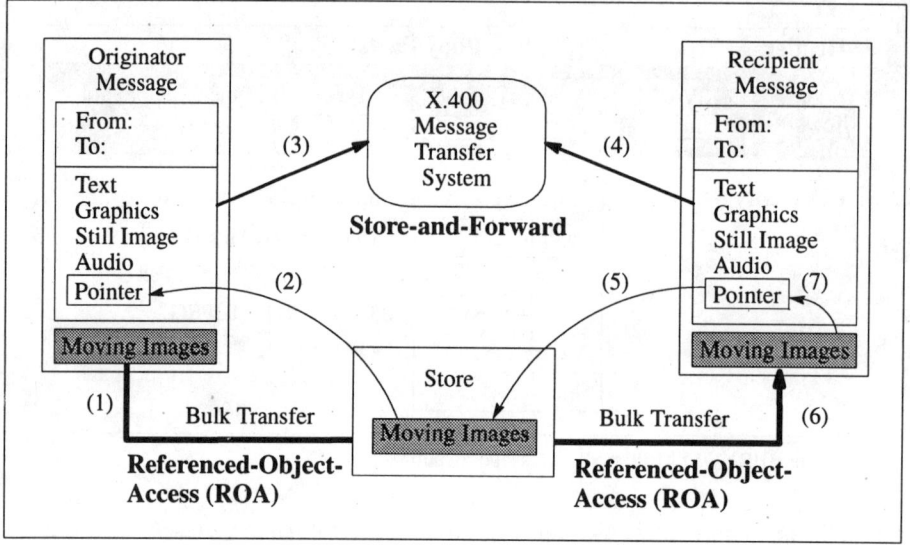

Fig. 2. Two exchange mechanisms: Store-and-Forward plus Referenced-Object-Access

media mails exchanged between archiving recipients always provide a bodypart of this type which can be accompanied by a document contained in the remaining bodyparts. Its contents is only completely interpretable by the receiver specified in the mail header. From the perspective of the mailsystem, however, an ARCHIVE bodypart is a document component as well which is treated in the usual way.

Figure 3 provides an illustration of these aspects. Besides the header which contains mailing specific information, the shown sample mail consists of nine bodyparts. The LINK bodypart defines a mapping of the remaining eigth body parts into a hypermedia document structure. In general, one branch directly below the root embodies the actual multimedia document (of arbitrary size and structure) to be communicated (in Fig. 3 this is the left branch). The other branch exclusively comprises the ARCHIVE body part containing archiving specific information. This can be perceived as an archive order with an attached hypermedia document.

3 Architecture

Figure 4 shows the central components of our architecture which is based on a *client-server* approach. The archive server and each client are directly connected to a *Multimedia Mail User Agent* (MMM-UA). Each MMM-UA has access to a X.400 *Message Transfer Agent* (MTA) and an *External Reference Manager* (XRM) [23] in order to support the two interchange mechanisms explained in section 2. In contrast to the server's database which contains *completely* structured multimedia documents for an undefined period of time, the global store only contains *document parts* which are automatically removed after the *guaranteed access time* specified in the DOR value is exceeded. In our prototypical implementation, the multimedia documents are accessed via the *data manipulation language* (DML) provided by the DBMS used for the realization of the archive server. Alternatively, a standardized language such as, e.g., SQL [8], or "maila-

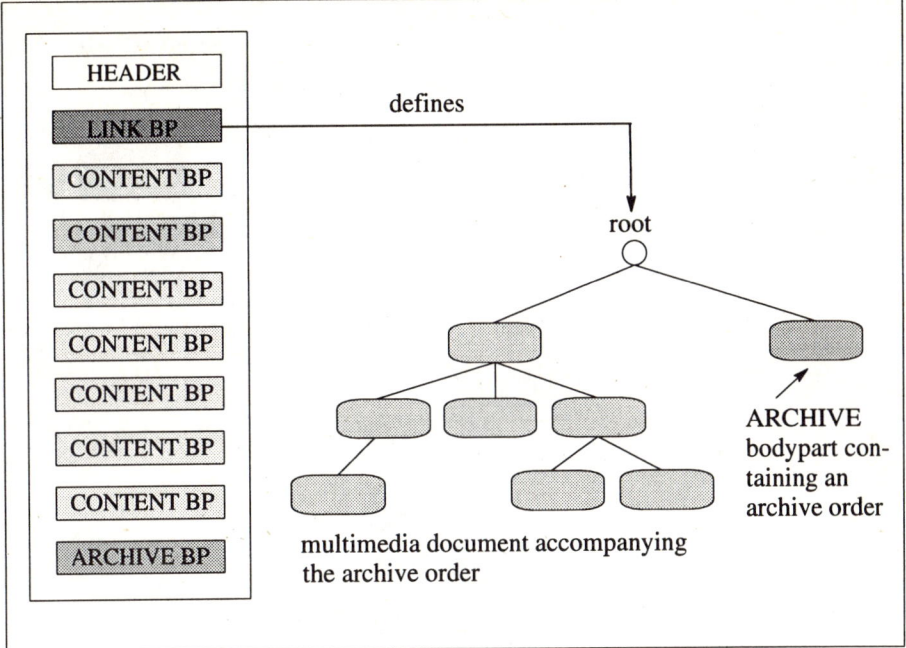

Fig. 3. Structure of a sample archiving multimedia mail

dopted" standards for synchronous client-server access, e.g. DFR [15], or RDA [10], could also be supported (respectively those subsets of these standards that specify the query functionality). However, the deficiencies in currently available versions of such languages with respect to a realization of the required archiving functionality have to be compensated by more sophisticated client functionality. The provided *descriptive access* to the documents includes access to a set of *application specific search attributes* that serve as predefined search criteria for the client (attribute-based retrieval of multimedia documents). Thus, in the database, every archived document is associated with an application specific set of attribute/value-pairs.

3.1 Archive Access Protocol

The archive access protocol, illustrated in Fig. 5, consists of rules that precisely define how an archive client can communicate with the server and vice versa.

Archive orders are issued to the archive server in separate *order mails*, i.e. a mail may only contain a *single* individual order (Note that for archiving, the order is accompanied by a multimedia document). Each order which is provided in the mail's bodypart of type ARCHIVE includes an *order type identifier*, the client's own *order identifier* necessary for the management of its individual orders, and the *actual order*. The server returns *1..n order result mails*, that each include in the ARCHIVE mail bodypart an *order result type identifier*, the *total number* of order result mails, and the mails *own order result number*. In the possibly remaining bodyparts a multimedia document can be contained. Multiple order result mails belonging to the same order are returned whenever

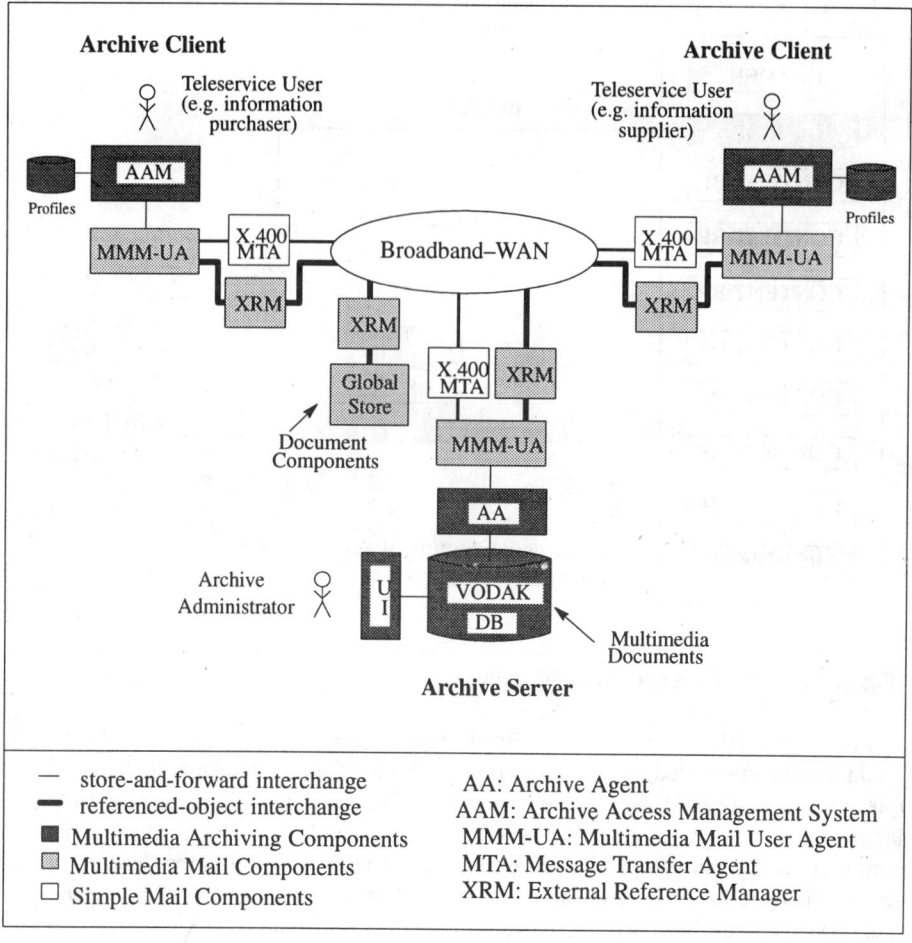

Fig. 4. Architecture of a multimedia archiving teleservice

more than one document has to be delivered to the client separating the documents from each other.

For the delivery of documents to the server, optionally, the ROA mechanism can be exploited for relevant, i.e., high volume document components. If specified by the user, the server automatically fetches the raw data from the global store (With respect to Fig. 5, path (1) shows the deposition of the bulk data on the global store, while (2) shows the DOR resolution optionally performed by the archive). Likewise, if specified by the user, document components not already residing on a global store are deposited at the global store by the archive. This is performed prior to the store-and-forward interchange of the document containing the DOR (With respect to Fig. 5, path (3) shows the deposition of the bulk data by the server, while (4) shows the DOR resolution at presentation time).

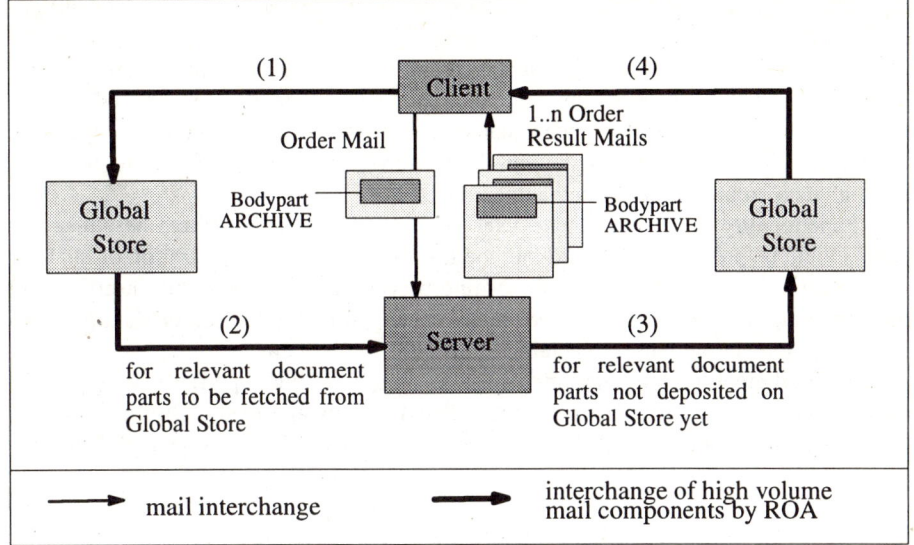

Fig. 5. Archive access protocol

3.2 Archive Client

Archive clients are standalone *Archive Access Management Systems* (AAM) which manage the user access to the archive server and make the utilization of the underlying mail system transparent to the teleservice users. Its functionality directly supports multimedia document authoring and/or the import of multimedia documents created by other authoring environments. The concrete *archive service profile* and the kind of user (see section 4) determine the functionality to be supported.

For the user's convenience, in a concrete implementation, a high level graphical user interface allows *direct manipulative* creation of archive orders. After its specification, the AAM encapsulates an archive order in a multimedia mail which in turn is sent to the server. Result mails are dispatched, unpacked and their availability is notified to the user. Orders as well as results are persistently stored on the filesystem for reuse in other sessions. It is imagined that issued orders are collected in an archive order list while an order's results are collected in an order result list. Both lists should be part of the user interface (see section 5 for an example). The AAM makes sure that only those documents are retrieved, that are currently not available at the client's site by cross-checking the involved document identifiers.

3.3 Archive Server

The archive server consists of a database and a so-called *Archive Agent* (AA for short). Information about the multimedia documents and the document themselves are stored in the database. This allows to support a descriptive access to stored documents based on search criteria that address the document contents or multimedia specific properties, respectively. The AA manages the connection to the mailsystem (sending and receiving). It maps statements of the archive access protocol into corresponding database operations.

When an archive order mail is received and the originator has the permissions required for the order execution, the archive order is interpreted, i.e. it is mapped into database operations. Its result is encapsulated in *1..n* single multimedia mails separating the multimedia documents to be delivered. This separation is necessary due to the possibly high data volume of multimedia documents.

Functionality and an adequate user interface for typical archive administrator tasks should be provided. It should include functions for the administration of the teleservice users, the management of the billing process, and for the direct insertion and manipulation of multimedia data. For maintenance purposes, the server should be dynamically connect- and disconnectable to/from the mailsystem.

In an efficient implementation, the server should be able to partially recognize redundant data e.g. by comparison of received and already archived DORs. In case of a match, the already archived DOR database instance can be shared by several document instances.

4 Access to Dynamically Composable Multimedia Documents

In general, three different *service profiles* for an archiving teleservice can be differentiated: (1) *Document Pool* – filing and retrieval of documents, (2) *Information Service* - in contrast to the former here it is also possible to retrieve only document parts and to modify archived documents, (3) *Work Flow Management* - support for computer supported cooperative work. The functionality provided to the users who can be differentiated in *information purchasers* (only read access to multimedia documents), and *information suppliers* (write access as well) depends on the service profile.

In the following, we first summarize the functionality provided by the teleservice introduced in this paper that provides the notion of an *archive order*. Then we provide our motivation and solution for dynamic document composition. Finally, we discuss the notion of *profiles* which provide flexibility to control the dynamic document composition process and the parameterization of archive orders.

4.1 Archive Orders

Our proposed teleservice provides an information archive profile which functionality is accessible by issuing *typed archive orders*. Five different order types are supported which execution semantics and results are manipulated by *type specific order parameters*. In 4.3 we explain the origin of these order parameters. In Table 1, the different order types and their corresponding parameters are summarized.

Note that the search criteria of a SEARCH archive order instance can address the documents contents as well as the structure inclusively multimedia specific data. This allows, e.g., to exclude those documents which contain videos longer than 1 minute.

Order Type	Order Parameters	High level pseudo code description of the algorithm which executes the order and generates result mails
INSERT: An accompanying multimedia document is inserted into the archive.	**boolean** P_{INS1}; P_{INS1} is called DOR resolution request parameter;	... insert document in archive; **if** (P_{INS1} == true) **then** **for all** DORs of document store bulk data in server as well; // data is fetched from store and stored in // the server as well **endif**; prepare order completion notification mail; send prepared mail; ...
SEARCH: Selection of documents based on an accompanying descriptive search query. (Note that search criteria can address the document contents as well as the document structure inclusively multimedia specific data).	**integer** P_{SEAR1}, P_{SEAR2}; <u>constraint:</u> P_{SEAR2} >= P_{SEAR1}; P_{SEAR1} is called search result constraint paramter 1; P_{SEAR2} is called search result constraint paramter 2; **boolean** P_{SEAR3}; P_{SEAR3} is called DOR creation request parameter;	... n:= EvaluateQuery(thequery); //n is number of matching documents **case** (n>P_{SEAR2}): prepare a search result notification mail providing n; (P_{SEAR1}< n<=P_{SEAR2}): **for all** matching documents prepare a mail containing the application specific attribute values of the document; (n<=P_{SEAR1}): **for all** matching documents { make a copy of the document; **if** (P_{SEAR3} ==true) **then** substitute each high volume document component by a DOR; // deposit raw data on store if not // already there **endif** <u>enforce document constraints</u>; //dynamic document composition prepare a mail containing the document; }; **endcase**; send prepared mails; ...

READ:
The document(s) with the given document identifier(s) is(are) retrieved.
	boolean
	P_{READ1},
	P_{READ2},
	P_{READ3},
	P_{READ4};
	
	constraint:
	only one of
	P_{READ1},
	P_{READ2},
	P_{READ3} can be true;
	
	P_{READ1} is called read result parameter 1;
	P_{READ2} is called read result parameter 2;
	P_{READ3} is called read result parameter 3;
	P_{READ4} is called DOR creation request parameter;
		...
		for all documents to be read
		{
		case
		  (P_{READ1} == true):
		    prepare a mail containing a description of the document structure;
		  (P_{READ2} == true):
		    prepare a mail containing the application specific attributes of document;
		  (P_{READ3} == true):
		    {
		    make a copy of the document;
		    if (P_{READ4}==true)
		      then substitute each high volume document component by a DOR;
		      // deposit raw data on store if not
		      // already there
		    endif;
		    <u>enforce document constraints</u>;
		    //dynamic document composition
		    prepare a mail containing the document;
		    };
		endcase;
		};
		send prepared mails;
		...

UPDATE: The application specific attributes of the document with the given document identifier are updated.		... update attribute values of given document; prepare order completion notification mail; send prepared mail; ...
REMOVE: The document with the given document identifier is removed.		... remove document with given document identifier; prepare order completion notification mail; send prepared mail; ...

Table 1. Summary of supported archive orders

4.2 Dynamic Document Composition

A multimedia information archive which only allows to retrieve documents as originally archived faces the users with the problem that the retrieved information might not fully reflect the user's individual workstation environment, or his preferences with respect to structure, or contents of documents, respectively. Thus, e.g., it can happen that some components of retrieved multimedia documents cannot be presented since the required presentation facilities (hardware/software) are lacking, or the documents include detailed kind of information despite the user is only interested in overview kind of information. Within the context of a networked environment for commercial utilization, this is even more problematic since the users' final charge, more or less, depend on the retrieved data volume. To allow users to accomplish a minimum of such retrieved "missmatching" information, our teleservice provides functionality for dynamic document composition available for SEARCH and READ archive orders. It allows to retrieve multimedia documents that more properly fit to the individual workstation environment, preferences with respect to structure and contents of documents, as well as service cost restrictions than the original one. The individual restrictions are expressed in *document constraints* which are derived from the different *profiles* which are discussed in the next subsection. The document constraints are enforced by the archive server resulting in a new document (constraint driven dynamic document composition). Note that the original document is not altered.

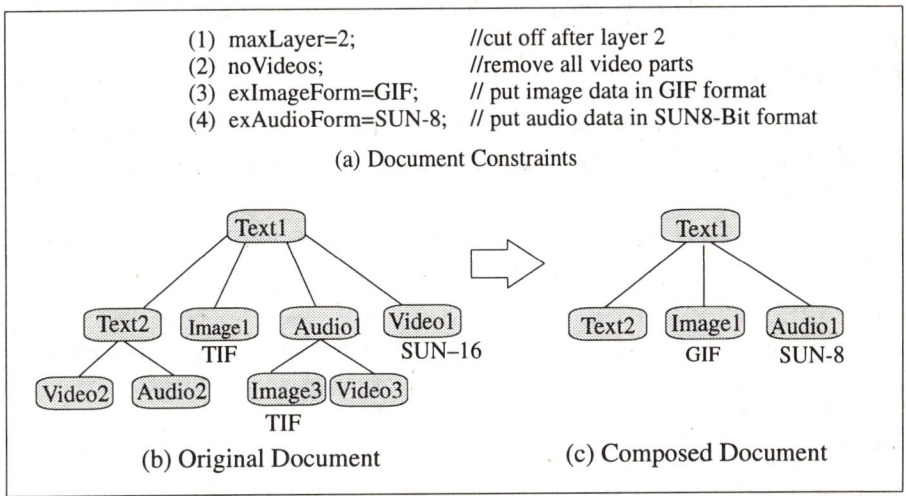

Fig. 6. Sample dynamic document composition

The types of *document composition functions* called by the constraint enforcement can be categorized as follows (Note the execution of such a function can include that the raw data is fetched from the global store):

Functions that alterate the document structure: Arbitrary document parts are removed. For example, constraint (1) of Fig. 6 (a) cuts off all document parts below layer 2 since, e.g., they provide detailed information which are not relevant for a user who is only interested in overview information, and constraint (2) projects out all video parts which, e.g., a user on a workstation with low performance does not want to receive.

Functions that perform coding transformations: Specific document parts are transformed from one coding into another. For example, constraint (3) of Fig. 6 (a) transforms relevant images contained in the multimedia document into GIF images since, e.g., the user's workstation environment only provides a GIF viewer.

Functions that perform quality transformations: Specific document parts are transformed form one level of quality into another. For example, constraint (4) of Fig. 6 (a) transforms relevant audio parts into SUN 8 Bit audios, since, e.g., the receiver does not have a SUN Sparc®10 workstation which is required to present SUN16 Bit audios.

Functions that extract subsequences out of a continuous data stream: Specific subsequences of a continuous data document component are extracted based on (a) *temporal constraints*, or, for video, based on (b) *video data specific constraints*. An example for (a) is the extraction of the first 20 seconds of an audio document component. An example for (b) is the extraction of only the first and last scene. More sophisticated examples can be mentioned such as the extraction of only every tenth video frame to generate some kind of summary.

4.3 Profiles

The input data for the generation of archive orders performed by clients are (Fig. 7):

- interactively specified order data such as, e.g., the search criteria for a concrete SEARCH order, and
- order data such as, e.g., the parameter values for the SEARCH order and the document constraints which are derived from so-called profiles.

For SEARCH and READ orders, in the first step of the order execution, the relevant archived documents are evaluated. In the next step, the document constraints contained in these types of orders are enforced. As a result, there are individualized copies of the relevant archived documents to be returned to the client. Within this context, as Fig. 7 illustrates, the profiles are a means to determine the final individualized documents.

In the following, we describe the different profiles supported by our teleservice.

Technical Profile: The individual hardware and software environment of the client is described in the technical profile. It is assumed that this profile is specified and updated by the person who conducts the installation of the archive client software.

Default Global Preference Profile: This profile comprises a set of default document constraints that have to be fulfilled by every document to be delivered to the client. It also contains default values for the order parameters. For its interactive specification, the client user interface provides a specification mask from which the user can select predefined options.

Order Specific Profiles: The default global preference profile is ignored if the user interactively, in addition to the other order data, specifies an order specific preference profile. The client user interface provides a corresponding specification mask from which predefined options can be selected. This feature is available for INSERT, SEARCH, and READ orders. Concerning concrete INSERT orders, it allows to manipulate the value of the supported parameter. Concerning

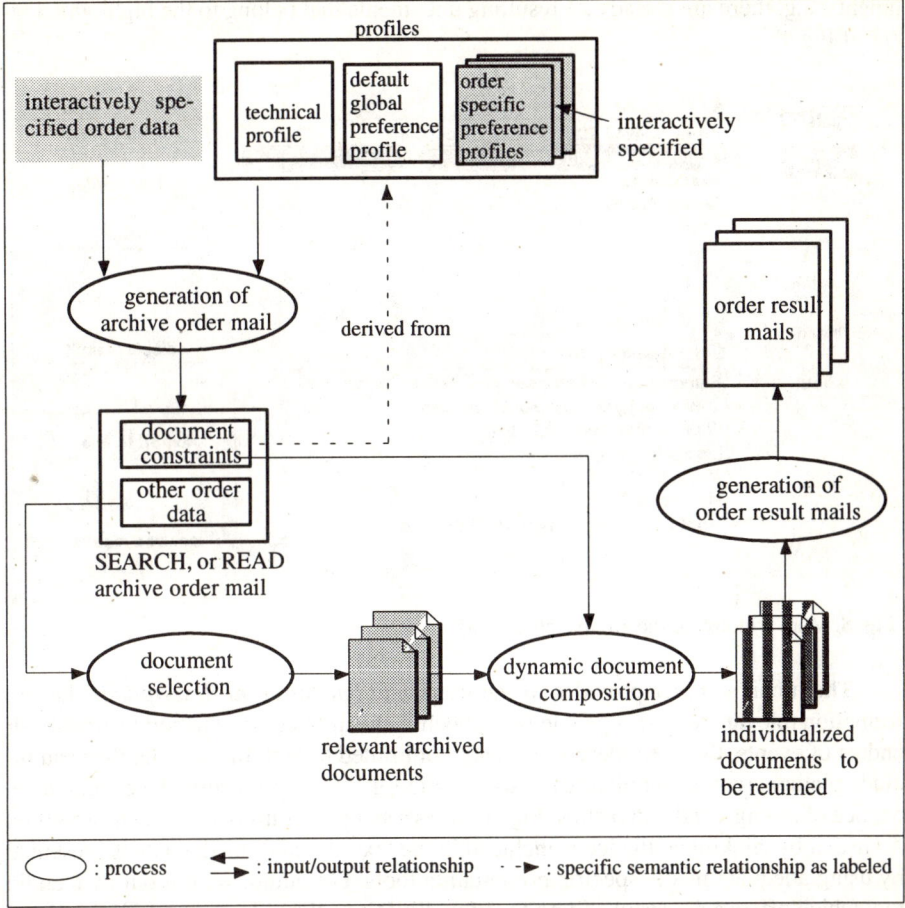

Fig. 7. The relationship between the profiles and the results of SEARCH and READ archive orders

concrete SEARCH and READ orders, in addition to the opportunity to manipulate the parameter values, this allows to specify document constraints to be enforced only for this individual order instead of the ones defined by the default global preference profile.

5 Sample Application

We built a concrete prototype that demonstrates an instantiation of our proposed teleservice architecture. This prototype shows a *multimedia calendar of events* (CoE) application which fits very well to the characteristics of applications encouraged by our proposed teleservice.

Figure 8 provides a screendump of the main window of our archive client's user interface. In the upper list, the user's orders are collected, while in the lower one, the result of a selected order is viewed. Note, a result can comprise a set of multimedia documents, e.g. there are already 21 resulting documents that belong to the highlighted order of Fig. 8.

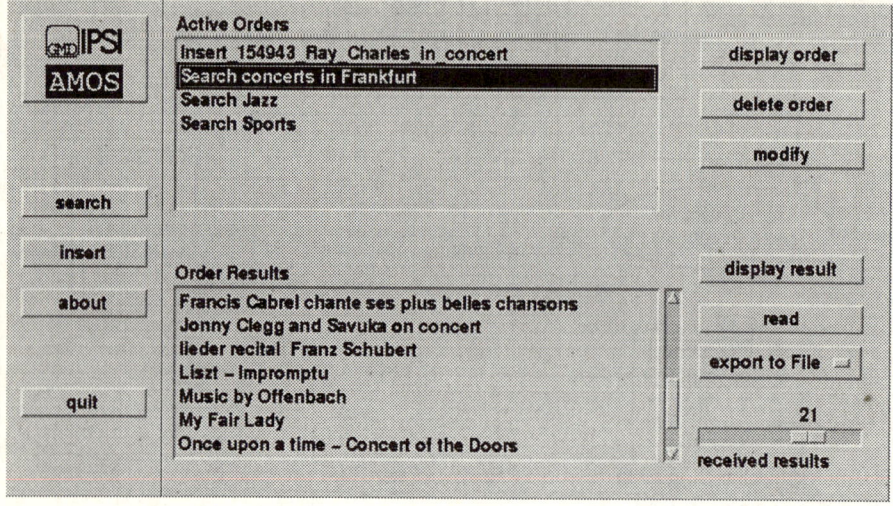

Fig. 8. Main window of the archive client [28].

The archive server provides to the users *multimedia event descriptions*, i.e. information relevant to attract people for a specific event. In contrast to conventional calendars of events, the event descriptions are not limited to text. In addition, they can include pictures (e.g. some pictures characteristic for the event), audio pieces (e.g. a sequence of a song), and video clips. Fig. 9 shows a sample multimedia event description. As usual, by clicking at the icons included in the text, the multimedia data is presented by using adequate media specific presentation tools. For audio, we use our own audio tool which offers a powerful editing functionality. Note, the video presentation initiated by clicking at the icon with the label *DOR* involves the ROA mechanism. The window at the bottom shows the CoE-specific attributes and corresponding individual values.

6 Discussion

In this section, we evaluate our concept and prototypical implementation of the proposed teleservice. The relevant aspects are differentiated and interrelated to each other. Conceptual as well as technical issues are addressed.

6.1 Applicability of Teleservice

The described teleservice architecture is highly generic. The customization to a concrete multimedia archiving application requires only the definition of the application specific search attributes for the multimedia documents to be dealt with. The client-server protocol remains the same. However, support for multimedia documents created by an external authoring tool is lacking, i.e. the documents have to be created via the clients own editing functionality. This should be complemented by functionality to interchange (in either way) multimedia documents between external authoring environ-

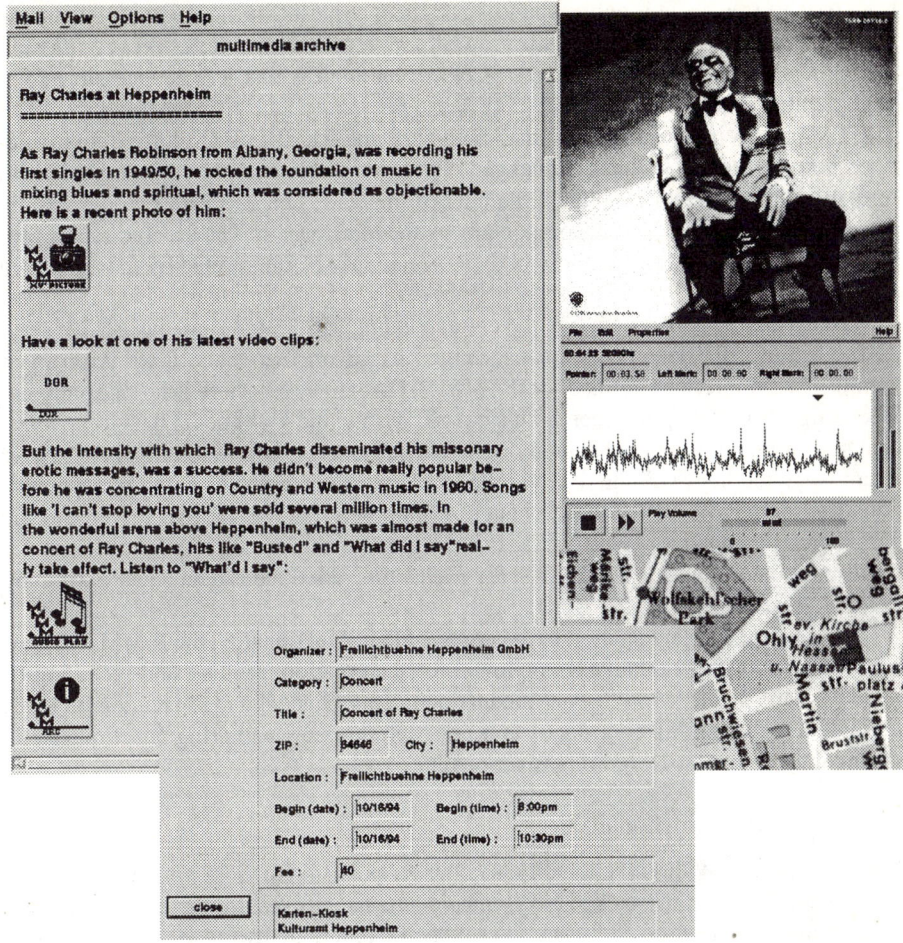

Fig. 9. Sample Multimedia Event Description [28]

ments and the client based on standardized formats like, e.g., SGML/HyTime [11, 12], or "HyperODA" [9].

6.2 Utilization of Multimedia Mail

The incorporated mail system's two interchange mechanisms, asynchronous store-and-forward for complete documents and (possibly synchronous) referenced-object-access for high volume document parts, provide to the archiving components adequate means for the interchange of multimedia documents. A consequence of using electronic mail is that there is no guaranteed *response time*. This is a drawback of the proposed archiving teleservice implementation, especially if used for kiosk information systems. However, on the other hand, by using electronic mail, the archiving teleservice's *availability* for the users is decoupled from the server's availability since mails are buffered in the mail system. Hence, mails can be issued at any time regardless if the receiver (client or server) is available. Thus, in contrast to usual client/server architectures where the server has a passive role, i.e. it is only responding to client requests, the mail service enables an *active server*. This allows the realization of active archive services, like, e.g., a subscription service. Another advantage of electronic mail is that it is is accessible for a large number of users.

By introducing the additional mail bodypart type ARCHIVE (X.400 externally defined bodypart type), in mails exchanged within our teleservice, archiving specific information are separated from the remainder of the mail. The ARCHIVE bodyparts are considered as usual document components by the mail system. The archive access protocol statements contained in these bodyparts can only be interpreted by the addressed receiver.

Between the mail system (i.e. the corresponding MMM-UA) and the archive server, the multimedia documents are exchanged based on files since there is no main memory based interface defined for the MMM-UA. However, there is no reason to neglect a tighter coupling between both systems, despite the multimedia data to be interchanged might not fit into main memory in one portion. A possible solution is to strictly interchange the high-volume data via the ROA mechanism and to employ a synchronous transport protocol for the exchange of data between the archive server and the global store, respectively, vise versa. For the DBMS, this would require support of a continuous data exchange mechanism for exchanging continuous data on a "portion by portion" basis.

To ease the implementation of the archive client functionality, the MMM-UA's inherent mail tool functionality is employed for the creation, sending and receiving of mails, and the presentation of multimedia documents. Since the mail creation includes the authoring of multimedia documents, our teleservice's authoring functionality is identical to the mail tool's editing functionality.

6.3 Teleservice Functionality

Our teleservice provides functionality which is an extension of those normally provided by conventional, non-networked electronic document archives. Multimedia documents are encouraged to be used within archives. Specific functionality for dealing with multimedia data in a networked environment such as functionality to control the contents of search result mails, and functinality for dynamic document composition are explicitly supported. The multimedia data handling capabilities (multimedia mail bodyparts, two data interchange mechanisms) of the underlying multimedia mail service are fully exploited. The application itself can control the interchange mechanism to be applied. Flexibility to control order parameters and document composition constraints is given by means of profiles. The idea to use within this context a cost axis [17] to show what kind of alterations of the overall service utilization costs are implied by which profile modifications seems to be also advantageous for our teleservice.

6.4 Using an Object-Oriented DBMS

For the realization of the described teleservice functionality, the archive server must be based on a differentiated and semantically rich database model. Especially, the structure of the multimedia documents must not be hidden in order to support dynamic document composition and descriptive queries containing multimedia specific search arguments. These requirements can be best met with an object-oriented DBMS like VO-

DAK. In fact, it turned out that with VODAK's object-oriented data definition language VML [18] which is also a programming language an adequate database schema can be defined. The contents of the multimedia documents are treated as database instances in order to fully support the shown dynamic document composition functionality. For audio data, we applied VML's built-in AUDIO datatype that provides built-in methods for performing such transformations. Since other built-in multimedia datatypes are not available yet, for other document parts, the required transformation methods were implemented in the corresponding object types. An in-depth description of the server's modeling is given in [28]. This shows that for multimedia applications, DBMS which explicitly support multimedia data by adequate datatypes are demanded.

The archive server's processing includes a mapping of archive access protocol statements into corresponding database operations. Due to the fact that VODAK offers the powerful object-oriented query language VQL [6], even complex archive orders can be almost completely mapped into regular query statements. These statements have just to be submitted to VODAK's query interpreter, i.e. further processing steps need not be encoded outside the database schema (Recall that an object-oriented database besides the actual data objects provides methods which operate on these objects). Thus, in this aspect, our first prototypical implementation could be achieved rather quickly.

There are different alternatives for the organization of the database transactions. An object-oriented DBMS like VODAK allows the implementation of advanced transaction concepts. Currently we investigate our approach to balance the more parallelism against the more overhead for such an advanced transaction concept.

7 Conclusion and Future Work

We described and discussed the architecture and functionality of a teleservice for archiving and retrieving dynamically composable multimedia documents. Its architecture is based on the integration of a standardized multimedia mail system with standalone archive clients and an archive server which is realized using an object-oriented DBMS. By supporting dynamic document composition, users can retrieve multimedia documents that fit much better as the original archived one to their individual workstation environment as well as preferences with respect to structure and contents of the multimedia documents. Thus, a better control of the service costs (teleservice provider charge, network and storage costs) is possible for the users and it is likely that the charge more accurately reflects the value of the information gained by using the teleservice.

Not all issues involved within an archiving teleservice for multimedia documents have been addressed. Especially, authentification and security mechanisms required to prevent unauthorized utilization and access to the documents have not been addressed. However, we believe that the mail system should deal with these issues. We also did not consider the billing aspect which is important for commercial utilization of the teleservice. However, we are convinced that even sophisticated billing models which allow that each information supplier can define his own charging policies for his documents (cost sharing, e.g. retrieval of short information is toll free while retrieval of complete documents is charged based on data volume, applied interchange mechanism, ... etc.; refund for ticket order; ... etc.) can be implemented in a straight forward manner by using the modelling power of an object-oriented DBMS.

The current state of our prototypical implementation includes the described archiving functionality almost completely. Dynamic composition of multimedia documents is partially supported and the ROA interchange between the archive server and the global store in either way are still not realized yet. A future goal of our work is an upgrade of the archiving teleservice's profile by functionality for work flow management to support the distributed authoring, composition, and management of multimedia documents, i.e. support for *asynchronous* cooperative work [29]. As sample application, we will use the CoE application for which the event descriptions should be cooperatively created following a formal specification of sequential and parallel steps performed by several information suppliers.

Acknowledgement

Thanks to K. Hofrichter, P. Hoepner, E. Moeller, H. Pusch, and G. Schürmann of GMD-FOKUS for providing comprehensive support with respect to the multimedia mail implementation. We would also like to thank S. Jakob, and A. Ozimek of GMD-IPSI who helped us by implementing the prototype. Thanks again to E. Moeller for helpful comments on an earlier version of this paper.

References

1. Borenstein, H., Freed, N.: "MIME (Multipurpose Internet Mail Extensions): Mechanisms for Specifying and Describing the Format of Internet Message Bodies", RFC 1341, Bellcore, Innosoft, June 1992
2. Borenstein, N.S.: "Multimedia Mail From the Bottom Up or Teaching Dumb Mailers to Sing", *Conf. Proc. USENIX - Winter '92*, San Francisco, CA, 1992, pp. 79-89
3. CCITT Recommendation X.400 series: 1988, Data Communication Networks, Message Handling Systems, Blue Book
4. Christodoulakis S., Ho F. and Theodoridou, M.: "The multimedia Object Presentation Manager of MINOS: A Symmetric Approach", *Proc. Int. Conf. on Management of Data*, Washington D.C., USA, May 1986, pp. 295-310
5. Clifton, H., Garcia-Molina, H., Hagmann, R.: "The Design of a Document Database", *Proc. of the ACM Conference on Document Processing Systems*, Santa Fe, New Mexico, USA, Dec. 1988, pp. 125-134
6. Fischer, G.: "Updates in Object-Oriented Database Systems by Method Calls Queries", *Proc. of 3rd ERCIM Database Research Group Workshop*, Pisa, Italy, Sept. 1992, pp. 69-76
7. Hofrichter, K., Moeller E., Scheller, A., Schürmann, G.: "The BERKOM Multimedia Mail Teleservice", *Proc. of the Fourth Workshop on Future Trends of Distributed Computing Systems*, Lisbon, Portugal, September 1993, IEEE Computer Society Press, Los Alamitos, California, USA, 1993, pp. 23-30
8. ISO/IEC, Database Language SQL2 and SQL3, international commitee document, JTCI/SC21, WG3 DBL SEL-3b, April 1990
9. ISO/IEC, Information Technology - Open Document Architecture (ODA) and Interchange Format - Temporal Relationships and Non-linear Structures, DIFF 8613/14, ITU–T Draft Rec. T.424, 1993
10. ISO/IEC, Information Processing Systems - Open Systems Interconnection - Remote Database Access (RDA), Part 1: Generic Model, Service, and Protocol, 1991

11. ISO/IEC Information Processing, Hypermedia Time-based Structuring Language (HyTime), International Standard 10744, 1992
12. ISO/IEC, Information Processing - Text and Office Systems - Standardized Generalized Markup Language (SGML), International Standard 8879, 1986
13. ISO/IEC, Information Technologie – Text and office systems – Distributed Office Applications Model (DOAM), Part 1: General Model, International Standard 10031, 1991
14. ISO/IEC, Information Technologie - Text and office systems - Distributed Office Applications Model (DOAM), Part 2: Distinguished-object-reference and associated procedures (DOR), International Standard 10031, 1991
15. ISO/IEC, Information Technology - Text and office systems - Document Filing and Retrieval (DFR) - Part 1 and Part 2, International Standard 10166, 1991
16. ISO/IEC, Information Technology – Text and Office Systems – Referenced Data Transfer (RDT) – Part 1:Abstract Service Definition, Part 2: Protocol Specification, International Standard 10740, 1993
17. Kalkbrenner, G., Pirkmayer, T., van Dornik, A., Hofmann, P.: "Quality of Service in Distributed Hypermedia-Systems", *Proc. of the RPODP '94*, Second Int. Workshop on Principles of Document Processing, Damstadt, Germany, April 1994
18. Klas, W., Aberer, K., Neuhold, E.J.: "Object-Oriented Modeling for Hypermedia Systems using the VODAK Modeling Language (VML)", *Object-Oriented Database Management Systems*, NATO ASI Series, Springer Verlag Berlin, August 1993
19. Moeller, E., Neumann, L., Schürmann, G., Thomas, S., Weber, R., Wolf, F.: "The BERKOM Multimedia-Mail Teleservice", *Computer Communications*, forthcoming
20. Orenstein, J., Manola, F.: "PROBE Spatial Data Modeling and Query processing in an Image Database Application", *IEEE Trans. Software Eng. 14*, No. 5, 1988, pp. 611-629
21. Postel, B.: "Simple Mail Transfer Protocol", August 1982, Internet RFC–821
22. Postel, B., et al.: "An Experimental Multimedia Mail System", *ACM TOIS*, Vol. 6, No. 1, January 1988, pp. 63-81
23. Pusch, H.: "Design and implementation of a global reference mechanism for arbitrary data objects", *Computer Standards & Interfaces*, 17(1994), forthcomming
24. Rückert, J., Paul, B.: "Integrating Multimedia into the Distributed Office Application Environment", *Datenbanksysteme in Büro, Technik und Wissenschaft*, GI-Fachtagung Braunschweig, Germany, March 1993, Springer Verlag Berlin, pp. 181-188
25. Steinmetz, R.: "Synchronization Properties in Multimedia Systems", *IEEE Journal on Selected Areas in Communications*, Vol.8 No. 3, April 1990, pp. 401-412
26. Stonebraker, M., Rowe, L.: "The Design of POSTGRES", *Proc. ACM SIGMOD*, Washington D.C., USA, May 1986, pp. 340-355
27. Tamura, H. and Yokoya, N.: "Database Systems: A Survey", *Pattern Recognition*, Vol. 17 No. 1 1984.
28. Thimm, H., Rakow, T.C.: "A DBMS-Based Multimedia Archiving Teleservice Incorporating Mail", *Proc. of the 1st Int. Conf. on Applications of Databases*, Vadstena, Sweden, June 1994, pp. 281-298
29. Thimm, H.: "A Multimedia Enhanced CSCW Teleservice for Wide Area Cooperative Authoring of Multimedia Documents", Position Paper for the ACM CSCW '94 Workshop on Distributed Systems, Multimedia, and Infrastructure Support in CSCW, Chapel Hill, NC, USA, October 1994

30. Tsichritzis, D.C., Christodoulakis, S., Lee, Vandenbroek, J.: "A Multimedia Office Filing System", in Tsichritzis, editor, *Office Automation*, Springer Verlag Berlin, 1985, pp. 43-65
31. Woelk, D., Kim, W.: "Multimedia Information Management in an Object-Oriented Database System", *Proc. of the 13th VLDB Conference*, Brighton, 1987, pp. 319-329

The Global Store Server — A Multimedia Teleservice Component

Christof Blum and Luc Neumann

Fraunhofer Institute for Computer Graphics (IGD)
Wilhelminenstr 7, D–64283 Darmstadt, Germany
e-mail: blum@igd.fhg.de, neumann@igd.fhg.de

Key Words
Multimedia, Global Store, Teleservices, Communications, Electronic Mail.

Abstract. The Global Store Server is part of the Multimedia Mail architecture which is currently being realized within the BERKOM Project Multimedia Teleservices, funded by the German Telekom and managed by DeTeBerkom.
This paper focuses on the various motivations behind the usage of and design criteria for the development of a Global Store Server. A comparison to other multimedia mail activities coping with bulky messages is made.
The paper illustrates the service interface and the internal architecture of the Global Store Server, including system administration, security management, and accounting. Furthermore, it focuses on specific design goals related to its use in a more generalized teleservice approach, including OSI management concepts.

1 Introduction

Today's distributed administration scenarios require powerful means for the transmission and distribution of structured documents (hereafter called *messages*). Asynchronous (or "offline") transmission services such as electronic mail are involved as well as isochronous communication services for computer–supported cooperative work (CSCW).

Regarding electronic mail, up to now only text messages are predominantly used, whereas provisions for the use of messages of more complex structure are still limited to isolated communities. However, emerging multimedia-capable messaging systems [16] [19] [7], are beginning to overcome the obstacles, as shown in this paper.

Compared to traditional text messages, multimedia messages that encompass bulky message parts (e.g., high resolution still images and A/V streams) occupy storage space which is higher in orders of magnitude, even if state–of–the–art data compression (JPEG, M–JPEG, MPEG) is employed for the reduction of redundancies and irrelevancies.

Moreover, multimedia messages often contain *continuous media* (CM), such as audio annotations, moving image sequences, and A/V data streams, thus requiring realtime presentation (and realtime capture) which could easily exceed the technical capabilities of the components that participate in the communication process (i.e., originator, receiver(s) and transmission instances).

1.1 The Demand for a Global Storage Component

On the way from the originator to a receiver, messages are typically being relayed between an arbitrary number of mail servers, or *Message Transfer Agents* (MTAs), as they are called in the X.400 terminology [4]. Unfortunately, this store–and–forward technique has a serious limitation regarding the transfer of high–volume multimedia data: the limited storage capacity of the MTAs.[1] Moreover, when the transition from one protocol stack to another is required, the same obstacle may occur at the respective gateway.

As shown in Figure 1, the introduction of a *global store* provides a solution to overcome this limitation. Bulky message parts are now being extracted from the message body at the sender's side. Only the external references to the extracted parts need to be sent via the store–and–forward message handling service.

It should be noted that such a global storage component may not be mixed up with *message stores* (MS), as they are now part of the X.400 specification: In contrast to a message store, the global store discussed here is used to download the message *bodies* (or parts of structured message bodies) "only". Hence, the global store is neither aware of any header information needed for addressing and forwarding, nor will there be global store access protocols comparable to the P7 message store protocol in X.400('88). Moreover, documents could be downloaded to the global store which have never been *messages* in the sense of the electronic mail scenario, if the service element for global store access is accessible separately to a multimedia mail user agent.

Besides the original reason for introducing this central storage component, its incorporation into an conventional messaging scenario offers the potential for additional features and benefits, it raises new questions, and may lead to other side effects or drawbacks which will be discussed below.

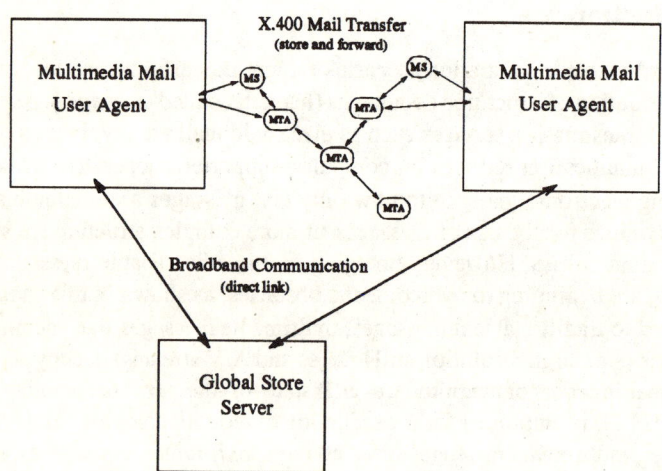

Fig. 1. Global Store Server within MMM scenario

---

[1] In today's messaging systems, the message size is often limited to approx. 100 Kbytes.

Figure 1 shows the Global Store Server as a component of the BERKOM Multimedia Mail (MMM) scenario.

1.2 The BERKOM MMM Teleservice

Within the BERKOM[2] Project a *Multimedia Mail* (MMM) teleservice based on the CCITT Recommendation X.400('88) has been specified and a prototype has been implemented by a mixed industry/research consortium coordinated by DeTeBerkom and consisting of[3] Danet GmbH (Darmstadt), Deutsches Forschungsnetz (DFN Berlin), Digital Equipment (CEC Karlsruhe), Fraunhofer Institute for Computer Graphics (IGD Darmstadt and Rostock), GMD (FOKUS Berlin), IBM (ENC Heidelberg), Liebing & Ullfors (Berlin), and Siemens AG & Siemens Nixdorf (Munich) [19]. The main goal has been to provide MMM user agents on a variety of different hardware platforms, such as SUN, IBM, DEC, HP, and NeXt workstations, as well as PCs.

Two key components of the BERKOM MMM effort may be pointed out:

- The specification of a multimedia document structure based on the use of externally defined X.400('88) body part types, hereafter called X.400('88-BERKOM). For a description of the supported representation types and the relation to other emerging multimedia message structures, refer to [17].
- the Global Store Server; the requirements for this component and the resulting architecture are described below.

At the CeBIT '94, the BERKOM MMM system was presented. Interoperability of the different X.400('88) MTAs[4] was shown as well as coexistence of several Global Store Servers (two were located at the CeBIT fare, one was located remotely in Berlin).

2 Related Work

In the following, major activities are mentioned which aimed at extending the functionality of electronic mail to allow for the exchange of multimedia messages in such a way that bilateral agreements between message parties about proprietary message formats are no longer required.

2.1 Internet MIME

The *Multipurpose Internet Mail Extensions* (MIME) [16] are based on the Internet SMTP [20]. Like the BERKOM MMM system, MIME supports the handling of external references. The access to the extracted message parts is done via FTP. On the one hand, no special protocol elements needed to be developed using the FTP approach, but on the other hand, no special support for downloading the message and creating the external reference is offered to the originator. He has to find a general purpose FTP site or provide and manage its own site for this purpose.

---

[2] BERliner KOMmunikationssystem
[3] in alphabetical order
[4] The following products were used: ISODE pp, Digital's MAILbus400, SIEMENS' ISOCOR/ISOPLEX 800, and CoConet.

Due to the importance of SMTP, which can be regarded as the worldwide de-facto messaging standard in the academic community, a gateway between X.400('88-BERKOM) and MIME is being realized by DFN in cooperation with GMD and IGD [2]. The implementation is based on the X.400('88)–MIME gateway which is included in the release 1.0 of the *ISODE* Consortium software[5].

2.2 RACE CIO

The RACE II project R2060 *Coordination, Implementation and Operation of Multimedia* (CIO) develops a common communication platform which is based on different network technologies, and realizes multimedia teleservices on top of it. One specified and developed teleservice is a prototype multimedia mail system (MMMS) [7] [1].

The MMMS message structure differs only in minor points from the X.400('88-BERKOM) message structure. For example, the representation of images differs. The external reference mechanism is identical in both services. However, in the current CIO realization the storage component for bulky message parts is co–located with the originator's site. So it may not be regarded as a remote component for both, originator and recipients. As a consequence the originator has the responsibility to provide worldwide access and storage space for the extracted message parts.

2.3 RACE Eurobridge

The activities within the RACE II project R2008, EuroBridge, focus on the definition of a multimedia mail application, called EMMA[6] [10] [15]. Whether the system will be based on Internet/MIME or X.400('88), has not yet been decided.

The approach is to concentrate the intelligence within the user agent as opposed to the message transfer system in order to facilitate the development of a tool which does not have to wait for the widespread deployment of advanced messaging protocols. The storage manager is intended to provide location transparency. In contrast to the BERKOM MMM approach, the specification of *local* message archiving and management components is not regarded as being local matter. Their functionality will be specified within the project.

2.4 BERKOM MMC

Besides the MMM service, a second teleservice, called *Multimedia Collaboration* (MMC) is currently under development by the BERKOM project. It allows several users to share applications and to participate in audiovisual conferences.

In principle a global store facility is suitable as an MMC component as well, acting as a source for A/V documents that will be presented simultaneously to all conference participants However, technical details concerning the interworking between both teleservices and the required protocol modifications are still under discussion.

---

[5] *ISO Development Environment*
[6] *EuroBridge Multimedia Mail Application*

3 The BERKOM MMM Global Store Server

3.1 Requirements

Looking from a user's perspective, a Global Store Server (GSS) should offer the ability to store data objects (especially high volume data) at a worldwide accessible location for a certain charge, depending on the size of the data and the intended "expiration date" of the object at the Global Store.[7] Therefore the creation of unique references to data objects to be stored via the GSS is required. These references will be handed out to the provider of a stored data object. Anybody aware of such a reference (e.g., somebody who received it from the originator via a multimedia mail message) should be able to obtain[8] the data. A variety of different user demands hold for this data transfer between Global Store and receiver, depending on the data object's type, size, and on the characteristics of the receiver's platform.

The requirements for the GSS management and the maintenance of the stored data objects are addressed in section 5. They may call for additional service elements (e.g., for the subscription of users) but these will not affect the key service elements that result from the requirements above.

3.2 Design Principles

For the provision of unique external references, the functional model of *Referenced Object Access* (ROA) and the universal reference mechanism *Distinguished Object Reference* (DOR) [12], both described within the *Distributed Office Application Model* (DOAM) [11], are employed for the MMM-Teleservice. The DOAM is based on a client-server concept and defines guidelines for the design of ROA services and protocols, using DORs to identify data objects uniquely. The Global Store Server is accessible via such ROA services and protocols and performs *consume*- and *produce*-operations to resolve and generate DORs.

3.3 Functional Description

The ROA service elements used within the MMM scenario are called *Referenced Data Transfer Service Element* (RDTSE) and *External Reference Produce Service Element* (ERPSE). The Referenced Data Transfer (RDT) standard [14] defines the RDTSE and a corresponding protocol, in order to retrieve a data object, or to negotiate the "expiration date" of a data object referenced by a DOR. Such a standard doesn't exist for the ERPSE. Hence, a new service element and protocol has been specified[9], that allow the storage of a data object, whose DOR is created remotely on the storage site [19]. Based on these ROA services and protocols the following access functionalities are provided by the Global Store Server (see Fig. 2).

---

[7] It is quite obvious that data objects cannot be stored infinitely, thus requiring an expiration date attribute.

[8] We refrain from using the term "retrieve the data" since it is not our intent to offer a database or an archival service.

[9] The realization of the ERP service and protocol has been performed by the MMM project partner GMD–FOKUS.

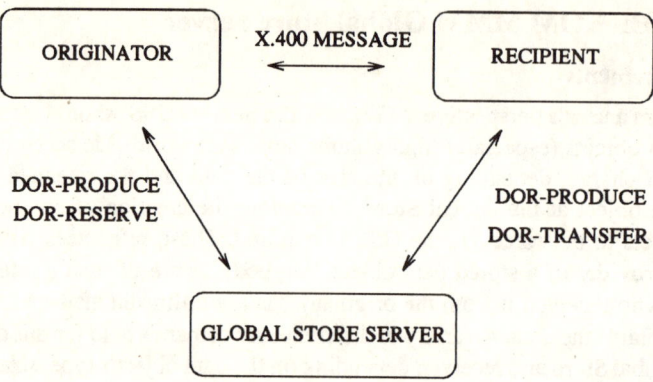

Fig. 2. Global Store service definition

- DOR–PRODUCE
  The originator stores a data object at the GSS and can specify a duration for the lifetime of this data object. The GSS creates a new DOR and sends it back to the originator. This DOR contains information about the used communication services, the type and the location of the data object, and QoS–parameters i.e. expiration date and the mode of usage.
- DOR–RESERVE
  The originator can reserve some storage space for a specific duration of time and obtains a DOR pointing to the reserved storage space.
- DOR–TRANSFER
  This operation allows an owner of a DOR to request the GSS to return the data object associated with the DOR which is supplied in the invocation.
- DOR–EXTEND
  The owner of a DOR can request the GSS to change the QoS of the DOR. In other words, if possible, the data object will be available longer (as defined by its DOR–PRODUCE operation).

Thus, the Global Store Server must be able to resolve, create and change the DORs. Due to the fact that the usage of the GSS causes costs, additional management functions are necessary to enable the maintenance of user accounts. This means the GSS must be able to identify all users of the GSS, and must use information concerning the user's identity to make access decisions and to calculate the costs. Therefore at least such functionalities as *Subscribe–User, Unsubscribe–User* are required to maintain user authentication data.

4 Global Store Server Architecture

As shown in Fig. 3, the Global Store Server consists of the three functional components: the *Reference Manager* (RM), the *Global Store Manager* (GSM), and the *Global Store* (GS). The Global Store represents the kernel of the Global Store Server and provides

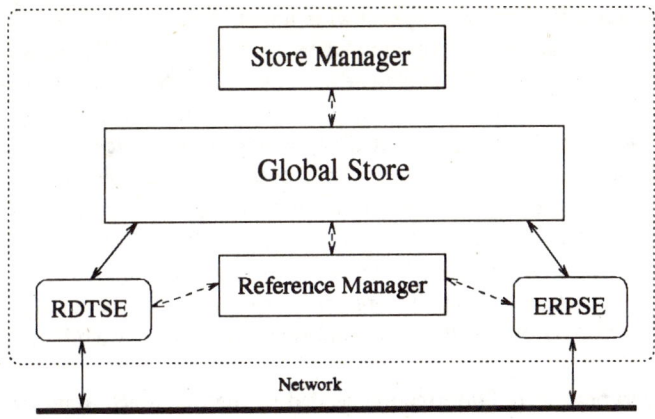

Fig. 3. Global Store Server architecture

two kind of interfaces; one for access purposes and one for management purposes. The access interface of the Global Store component has been designed to provide the access facilities necessary for the Global Store Server access via the service entities. The Global Store Manager uses the management interface to administer the resources and user accounts of the Global Store. Moreover, the Global Store provides information needed by the External Reference Manager to resolve and create DORs.

The behavior of the Global Store Server, as described in section 3.3, has following implications to the functionality of the different components:

- *Reference Manager.*
  The Reference Manager prepares the access to the Global Store according to the abstract operation requested by a client. The requirements of the client are verified according to whether they can be guaranteed or not. Therefore the Reference Manager makes use of specific functionalities for the DOR management which is provided by the Global Store component.
- *Global Store Manager.*
  In principle the Global Store Manager is responsible for typical management tasks, such as accounting and registration of users, and justification of the resources and costs. So the store administrator may for instance for reasons of limited storage capacity restrict the period of time for data storage. Additionally the deletion of data objects is within the responsibility of the store manager, since the deletion takes effect in regard to the deletion criteria (i.e. the date up to which the data object is stored) defined in the reference. Hence, the originator and all recipients are aware of this criteria. If one of the recipients or the originator is enabled to perform any deletion request, then the data object value may be deleted without knowledge of the recipients.
- *Global Store.*
  It is obvious that the Global Store component primarily controls the data objects on the mass storage. Consequently the Global Store is responsible for some addressing information of the DOR, regarding the location within the addressed storage. The

identifier for this location is systemwide unique and independent of the physical location of the data or the path to the data. Additionally the Global Store specifies the application service elements to be used by a client for data object value access. Furthermore the Global Store supports the negotiation of how long a reference should be valid. If the requested time cannot be guaranteed, the Global Store provides information about the supported possibilities. These possibilities may be defined by the system administrator via the Store Manager.

As already stated, the usage of the resources of the Global Store Server is subject to fees. For this purpose the Global Store provides a minimal system access control. This means that an account is kept for every registered client in order to make identification possible and to provide information to calculate a bill for the produced costs.

In order to provide the information needed by the Reference Manager, the Global Store contains an internal DOR management. Therefore each data object conists of a data-object value and attributes, describing data-object related data, such as reference, encoding, time of change, etc.

5 Global Store Management

In principle the Global Store Server can be seen as a storage teleservice used by the MMM users. The service provider needs management functions for the organization, monitoring and control of such a service.

The following subsections illustrate the management of the Global Store Server as it is realized within the MMM teleservice.[10] These management functionalities can roughly be classified into four groups:

- system administration
- accounting management
- security management
- fault management

5.1 System Management

The system management allows to create, to initialize, and to modify the system configuration. It gives the operator the flexibility to adapt the system to the different requirements of the large number of individuals, which are sharing the resources of the system.

Within the MMM-Teleservice, the Global Store Server offers a shared multimedia storage system for all MMM-participants. Its system management is primarily oriented by the requirements of the used access protocols of the DOAM family. In other words, the operator has to take care of the amount of storage capacity promised for a specific period of time for each user and must find a trade-off between costs and available storage capacity.

---

[10] Many of the solutions were motivated by the underlying goal of keeping the management overhead for the MMM prototype as small as possible.

5.2 Accounting Management

The services offered by the GSS are subject to service charges. Therefore, functionalities like the negotiation of tariffs, recording and billing of services are necessary. The accounting strategy of the GSS, including the registration of the user and the calculation of the charge, is realized as follows:

If the user wants to delete his account, he must give an order to the GSM. The GS then closes the account and returns the bill of costs to the GSM. The forwarding of the bill of costs to the user (e.g. via email), confirms the closing. After a fixed time period the GSM is invoking the GS to calculate the accumulated costs. The calculation is based on a function over the allocated storage space and the duration and time of the GSS usage. The current policy is that the charge has to be paid for every *produce* and *extend* operation on the GS, i.e., the originator of a message which is being placed on the GS is being charged whereas the retrieval of this data object is free of any service charge. This may of course be reconfigured if a different service charge policy is desired.

5.3 Security Management

To offer a secure service, the handling of security aspects concerning identification, authentication, authorization, and data access control have to be taken into consideration.

The system access control of the GSS is a two-step process of identification and authentication [18]. Therefore, users are registered by sending a subscribe request (e.g. via e-mail), that enfolds a desired account name, e-mail and postal address. A unique login identifier assigned by the GSM, based on the desired account name, is used to register the user in the GS access control list. This registration is confirmed by forwarding the login identifier to the user, and from now on the user is authorized to store data objects on GS, or to extend the fidelity time of stored data objects.

Within the MMM Teleservice, the data access control is performed by the usage of DORs. In other words, by sending out the DOR via an MMM message, the originator automatically enables the recipient(s) of this DOR to gain access to the referenced data object.

5.4 Monitoring

Both passive and active monitoring functionality is provided for the GSS. An audit trail mechanism logs all access activities and an active warning mechanism informs the system administrator about critical system states. Both features support the detection, localization, and correction of errors.

5.5 OSI Management

Vendors of telecommunication systems, network services, and application components offer a variety of different management systems. Due to the incompatibility of these systems, the combined operation of multiple services and applications turns out to be expensive and difficult. The future trend towards open network provision may amplify this problem. As a consequence, standardized management concepts need to be taken into consideration:

- *OSI Systems Management*.
  The ITU-T Recommendation X.700 [6] provides an object–oriented information model and a functional model. The former is based on *Managed Objects* (MO) which represent an abstract management–oriented view of the properties of a real or logical resource. An MO is characterized by a set of attributes that is visible at its *Management Object Boundary* and by the specified management operations including *notification* operations. The functional model describes the interactions between the system manager, called *Managing Application* and an agent, called *Management Agent*. The latter communicates with the MOs and reports the MOs notifications back to the manager. The communication between manager and management agent is defined by the *Common Management Information Service* (CMIS) and the respective *Common Management Information Protocol* (CMIP).
- *Telecommunications Management Network* (TMN).
  A TMN [5] provides the control over a telecommunication network including all applications which run on that network. It contains an interface to the telecommunication network for the exchange of control messages and management operations. The following main functional components are defined: *Operations System Functions* (OSF) for the system management; *Network Element Functions* (NEF) for the interface between the TMN and the network; *Mediation Functions* (MF) for the conversion and filtering of NEF data, and *Workstation Functions* (WSF) for the interaction between a user and the OSFs. The communication protocols which are used between the OSF, MF, and NEF components are based on CMIS/CMIP. Hence the underlying information model is OSI Management–conformant.

Within the BERKOM Project *BERMAN*, a management platform was developed and initial proposals for the management of the MMM and MMC teleservices on the basis of X.700 were outlined [8].

Some of these proposals are directed to the management of the MMM system as a whole, e.g., the use of the X.500 Directory Service [3] for the negotiation of UA capabilities in order to limit data format conversions. This seems to go beyond the original scope of X.400. One may object furthermore that this goes beyond the scope of applicability of the management standard as well.

The other proposals which are made in [8], [9] are directed in particular to the Global Store Server and offer. They are grouped into the following categories:

a) *Service Provision*
b) *Maintenance*
c) *Performance monitoring*
d) *Security control*
e) *Accounting*
f) *Customer Administration*

This list is suitable to further structure and extend the existing management functions of the Global Store Server according to the X.700 guidelines, in order to gain more flexibility for the usage of the Global Store Server within a telecommunication enviroment. This extended flexibility is reflected by the capability of the management across administrative domains and by the introduction of the role of the *customer*. In other words, the coexistence of different service offerings on a single Global Store Server can now be managed remotely. Three roles are distinguished in this scenario: the service *provider*, one or more *customers*, and the *users*. The users will be authorized by their customer to use the service and the customers are responsible for the payment of all costs caused by their users. Therefore additional management capabilities for the handling of users are made available to the customer by the service provider. Finally, the user may choose between a multitude of similar services.

By comparing the list items with the management functions described above, it turns out that the categories b) and c) correspond to the set of system administration functions, described in section 5.1, category d) and parts of category a) correspond to the set of security management functions, described in section 5.3, and category e) corresponds to the set of accounting functions, described in section 5.2. One might further subdivide category a) into *announcement* and *configuration*, whereby the former represents an extension of the hitherto described functionality. It reflects the orientation towards *open network provision*. To sum up, the generic components of the BERKOM Management Platform can easily be mapped on existing functions and allows the integration of the Global Store Server in an open multi-service enviroment.

6 Conclusion and Perspectives

We have shown the demands and requirements for a global store server as an essential component within multimedia teleservices. Its benefits, design principles, and current solutions were highlighted. An outlook towards the role of OSI-conformant service management and the use of value-added services for temporary deposition of bulky data in a global network, based on the global store, was given.

Future perspectives may include the extension of the Global Store functionality by retrieval and archival functions, e.g., based on the *Document Filing and Retrieval* (DFR) Standard [13], the introduction of security features, and last but not least the realization of realtime capabilities.

Acknowledgments

The authors would like to thank the members of the BERKOM Multimedia Mail and BERMAN working groups for fruitful and open-minded discussions and their colleagues at IGD, especially Uwe Schneider and Rüdiger Strack for valuable suggestions and contributions.

This work belongs in parts to the above-mentioned BERKOM Project *Multimedia Mail* (MMM), funded by the DeTeBerkom GmbH, Berlin.

References

1. M.P.P. Baveco, G.H. Kruithof, L.J. Teunissen, and P. Hoepner. MultiMedia Mail using OSI — Bridging the gap between what exists and what is required. In Spaniol O. [21], pages 293–298.

2. M. Bogen et al. X.400(88–BERKOM) – X.400(88-MIME) Gateway Specification. Technical report, DFN–Verein, Berlin, October 1993.
3. CCITT. *Data Communication Networks Directory, Recommendation X.500*. Blue Book. CCITT, 1988.
4. CCITT. *Message Handling Systems and Service Overview, Recommendation X.400*. Blue Book. CCITT, 1988.
5. CCITT. *Principles for a Telecommunications Management Network (TMN), Draft Recommendation M.3010*. CCITT, 1991.
6. CCITT. *Open Systems Interconnection – Management Framework – Systems Management and Structure of Management Information, Recommendation X.700*. CCITT, 1992.
7. RACE Project R2060 (CIO). Report containing Technical Documentation on Functional Components and Interfaces. Deliverable R2060/PTT/CIO/DS/P/004/b1, CEC, September 1993.
8. A. Dittrich et al. Szenarien und Anforderungsanalyse für das Management multimedialer Teledienste. Technical report, DeTeBerkom, Berlin, March 1993.
9. A. Dittrich et al. Konzepte für das Management des Global Store. Technical report, GMD FOKUS, Berlin, March 1994.
10. RACE Project R2008 (EuroBridge). The Specification and Implementation of EuroBridge Multimedia Mail Applications. Deliverable R2008/SSE/WP2/I/042/A, CEC, November 1993.
11. *ISO/IEC DIS 10031-1: Information Technology — Text and Office Systems — Distributed-Office-Application model (DOAM) — Part 1: General Model*. ISO/IEC, 1991.
12. *ISO/IEC DIS 10031-2: Information Technology — Text and Office Systems — Distributed-Office-Application model (DOAM) — Part 2: Distinguished Object Reference (DOR) and associated procedures*. ISO/IEC, 1991.
13. *ISO/IEC DIS 10166: Information Technology — Text and Office Systems — Document Filing and Retrieval (DFR)*. ISO/IEC, 1991.
14. *ISO/IEC IS 10740: Information Technology — Text and Office Systems — Referenced Data Transfer (RDT)*. ISO/IEC, 1992.
15. K. Lenssen, A. Spinner, and S. Hand. Towards the Messaging System of Tomorrow. In Spaniol O. [21], pages 287–292.
16. *MIME (Multipurpose Internet Mail Extensions)*, Request for Comments (RFC) 1341. Internet Network Working Group, June 1992.
17. E. Moeller et al. The BERKOM Multimedia–Mail Teleservice. *Computer Communications*. to be published.
18. D. Russel and G.T. Gangemi Sr. *Computer Security Basics*. O'Reilly & Associates, Inc., Sebastopol, CA, 1992.
19. G. Schürmann, E. Moeller, A. Scheller, G.R. Hofmann, and Rückert J., editors. *The BERKOM Multimedia–Mail Teleservice, Release 2.2*, Berlin, May 1993. DeTeBerkom.
20. *Simple Mail Transfer Protocol*, Request for Comments (RFC) 821. Internet Network Working Group, August 1982.
21. Williams F. Spaniol O. editor. *Broadband Islands: Towards Integration*, Athens, Greece, June 1993. Elsevier Science Publishers B.V.

From Broadband Network Services to a Distributed Multimedia Support-Environment

Helmut Leopold, Kwaku Frimpong-Ansah and Nikolaus Singer

Alcatel Austria AG, Broadband Communications Department,
Scheydgasse 41, A-1211 Vienna, Austria
e-mail: [leopold,frimpong,singer]@aut.alcatel.at

Abstract. We are on the brink of another major step in the age of "information", stimulated by: Firstly multimedia computing characterized by a wide range of potential applications that combine information sources such as voice, graphics, hi-fi quality audio and video. Secondly advances in network technology which have made it feasible to build high-speed networks operating at hundreds of Mbit/s over long distances. Consequently distributed processing in the new information age will have to support the Quality of Service (QoS) requirements of the new multimedia applications and be able to exploit in full the increasing capabilities of the new broadband networks.

The most likely candidate for broadband communications is B-ISDN based on ATM technology. The current status of development of ATM encompasses generic bearer services which fulfill on the one hand the major needs of current uses like LAN-to-LAN interconnection, transmission of digital signals with constant bit rates, and provides on the other hand interworking with other communication technologies like Frame Relay, ISDN, DQDB, etc. The next step in the development of broadband communications is the specification, standardization and validation of appropriate services, which suitably support the evolving distributed multimedia applications.

This paper provides an overview of the state-of-the-art of broadband services and applications as currently discussed in various bodies like ITU-T, ETSI and the ATM Forum. A unification of the terminology is undertaken and a basic framework is described in order to provide an understanding of the concepts. It is important to note that the activities of standardization bodies are just targeted towards service specifications, but will not consider the realization of these services. The Telecommunications Information Networking Architecture Consortium (TINA-C) follows an approach which tackle both issues, the service specifications on the one hand and the realization of these services supported by an appropriate Support Environment (SE) on the other hand. It is discussed and suggested that the models and concepts of TINA-C should be universally adopted for the development of broadband telecommunication services.

Keywords: Broadband Communications, Multimedia Communications, Broadband Network Services, Broadband Telecommunication Services (Bearer and Teleservices), Open Distributed Processing, TINA-C

1. Introduction

Recently, multimedia computing has emerged as a major area of research motivated by the wide range of potential applications made feasible by combining information sources such as voice, graphics, hi-fi quality audio and video. At the same time advances in network technology have made it feasible to build high-speed networks operating at some hundred Mbit/s over long distances. Example of such new multiservice networks are high bandwidth MANs, such as DQDB and FDDI and finally the ATM technology which will be the basis for the future B-ISDN. Thus the evolution of distributed computing is being influenced by two major factors: the new distributed multimedia applications requirements and by the increasing capability of high performance multiservice networks.

This technology push is influencing, and acting as a catalyst for the specification of new distributed multimedia applications. These new applications places stringent requirements on the communication infrastructure. Some of these requirements, which are identified in [24] and summarized in [28], are: minimum latency, guaranteed bandwidth, real-time transmission, bounded jitter, Quality of Service (QoS) guarantees, orchestration (i.e. synchronization) within single information streams and between multiple streams, multipeer communication support, media and application specific protocols, error recovery strategies etc.

In [27] it was clearly worked out, that some of the application requirements, have been tackled in the standardization of broadband. However, much remains to be done. What is required is to establish an integrated architecture which encompasses ATM, AAL and higher layers and to provide a complete set of B-ISDN services which will meet the needs of distributed multimedia applications [26].

In parallel to these evolving areas of multimedia computing on the one hand and high performance multiservice networks on the other hand, a new standardization area is becoming increasingly important: Open Distributed Processing (ODP) [20]. Besides other issues, ODP deals with the problem of information exchange within heterogeneous distributed systems in general.

The integration or harmonization of B-ISDN, the OSI Reference Model (OSI-RM), the ODP architecture and the emerging multimedia end-user service requirements is identified as an essential issue. This has led to an important area of research in order to develop an appropriate distributed Multimedia Support Environment (MM-SE). Outputs of these activities are for example: investigation of modern multimedia applications [33]; identification of the distributed multimedia requirements on the communication systems of tomorrow [24,28]; the relationship between the B-ISDN RM and the lower layers of the OSI-RM [32]; the understanding of the basic differences and commonalities of ODP and OSI [25].

The concern of this paper is to provide an overview of the state-of-the-art of broadband services as currently discussed in various bodies like ITU-T, ETSI and ATM Forum. A unification of the terminology is undertaken and a basic framework is described in order to provide an understanding of the "service concept". It is important to note that the activities of standardization bodies are just targeted towards service specifications, but will not specify any implementation issues. The

Telecommunications Information Networking Architecture Consortium (TINA-C) follows an approach which tackle both, the service specifications on the one hand and the realization of these services by an appropriate Support Environment (SE) on the other hand. The activities of TINA-C are seen as an important input for the development of broadband telecommunication services and are thus presented here.

This paper is structured as follows. The basic "service terminology" is presented in Sect. 2 and the state-of-the-art broadband network services are summarized in Sect. 3. Then we introduce, in Sect. 4 the set of broadband telecommunication services which are currently under investigations in various bodies. A justification for proposing the TINA-C framework architecture is provided in Sect. 5 and a general TINA-C overview is following in Sect. 6. The conclusion, which summarizes the activities which have to take place in the future, is presented in Sect. 7.

2. From Services to Applications - a terminology discussion

The requirements of the new distributed multimedia communication environment are expediting the convergence of the separate worlds of telecommunication and information technology. It is clear that standards have a very important role to play in development of distributed multimedia systems, and, to allow useful standards to emerge, it is essential that experimental designs are implemented as soon as possible. This will help to achieve some verification of various technological approaches and to provide appropriate demonstrations to convince potential customers who will use the new communication technology in the future.

For the reasons above, many people with different expertise and backgrounds are building, prototyping and discussing the same issues. Thus it is a very important prerequisite to have first of all a common understanding of the terminology centered around the "service issues" such as telecommunication service, tele-service, bearer service, supplementary service, layer service and application.

A first very important point which may be a potential source of misunderstanding between people with different background is the understanding of the terms "service" and "application". Very basically, a service is a commodity offered at the interface between two potential role-partners. The roles are those of a service provider and a service user. An application is the result of construction; it is the conglomeration of objects, relationships and modelling, which are brought together for some specific purpose. A purpose for the construction of an application could be again the provision of services. To be more practical, in the field of telecommunication we have to distinguish between two basic service types:

1. *Telecommunication services* as defined by ITU-T (I.210) which makes a distinction between *bearer services* which provide the capability for the transmission of information between User-Network Interfaces (UNIs) and *tele-services* which provide, in addition to the network bearer functions, a complete capability, including terminal equipment functions, for communication between users according to defined protocols.

Telecommunication services are offered to users at a reference point in a network configuration and specify the whole protocol stack plus additional information such as QoS attributes as defined in I.140. Note that additionally to these basic services, so called *supplementary services* are used to extent the functionality of the bearer and tele-services respectively[1].

2. *Layer services* are services of a specific protocol layer to the next higher layer in a protocol architecture as defined by ITU-T (X.200; OSI-RM) for example.

Using the OSI-RM as a reference architecture, these two basic service types are related as follows: the bearer services are defined up to OSI layer 3 and tele-services are defined up to OSI layer 7.

To summarize, from a customer's point of view, a "service" means something which is offered by a service provider to his customers in order to satisfy a specific telecommunication requirement. "Application" means "what a customer uses the service for"[2]. In this context it is important to note, that there is no relationship to the OSI application layer 7.

3. Todays Broadband Network Services

B-ISDN with its ATM technology, which is progressing in the standardization bodies will be the most important candidate of the broadband multiservice networks of tomorrow. Currently there are a lot of activities all over the world to establish pilot networks, field trials and regular public network services using the new broadband technology. In Europe there is a big effort influenced by EURESCOM to establish an ATM based pan-European broadband network as basis for further development of the technology itself and to offer broadband services to potential customers on a wide range.

One of the main achievements of the ATM technology is, that an ATM network works on a semantic and time transparent basis. That means that the network does not differentiate between various information characteristics, nor does it have to deal directly with the end-user time characteristics. Thus the basic ATM network, i.e. the ATM layer of the B-ISDN RM, offers a very generic bearer service which is independent of any media characteristics (i.e. ATM bearer service).

Since end-users may not be satisfied with an ATM bearer service only, we have to consider additional protocols above the ATM layer. The purpose of the ATM Adaptation Layer (AAL) on top of the ATM layer and situated outside the pure ATM network, is to provide the necessary functions which are not provided by the ATM layer to offer more specific media dependent services. The various layer services offered by the AAL protocols, standardized in I.363, are summarized below.

---

[1] E.g. "Dall Diversion" is a supplementary service to the tele-service "telephony".
[2] For example, the tele-service "High Resolution Image Retrieval" can be used for the application "Medical Imaging".

3.1. ATM Adaptation Layer Services

Currently there are 4 types of adaptation protocols progressing through ITU-T, ETSI and ATM Forum, which are called AAL-1, AAL-2, AAL-3/4 and AAL-5. Each AAL type is foreseen to support specific communication requirements like Connection Oriented (CO) and Connectionless (CL), real-time issues, error control mechanisms, Variable Bit Rate (VBR) and Constant Bit Rate (CBR) signal characteristics, etc. We will first summarize the most important issues around these protocol classes and then provide descriptions of the realized B-ISDN services, which are based on the AAL services.

The *AAL-1* [10,12] offers a service which accepts Service Data Units (SDUs) at a fixed clock rate for transmission over the network and delivers them at the same clock rate in a CO mode (also called isochronous service). The basic characteristic of the AAL-1 is the provisioning of a time-synchronization functionality to offer a CBR service. Additionally some error control mechanisms are performed, like sequence numbering in order to detect lost or misinserted ATM cells and Forward Error Correction (FEC), and a specific mechanism allows the structuring of the information which is transported by using the AAL-1 service. Currently, there are 3 services offered by the AAL-1:

1. The CBR Circuit Emulation (CE) service supports the transmission of isochronous digital information (SDU size is one bit). The CBR CE service encompasses two different options: (i) asynchronous circuit transport of unstructured G.703 signals (2 Mbit/s or 34 Mbit/s); and (ii) synchronous circuit transport of 64 kbit/s time slots to support the transport of G.704 signals. This service supports the interworking between B-ISDN and 64 kbit/s ISDN.
2. Video and voice-band signal transport services: The main differences to the CBR CE service to these two services, are the SDU size (one bit for CBR CE and one octet for the video and voice transport), and the provision of a specific FEC mechanism for the video signal transport. The video service is applicable for the transport of MPEG-2 CBR information streams.

   It has to be noted, that there are still some open issues like handling of voice signals in an ATM cell stream (problem of partly filled cells), transmission of MPEG-2 VBR information stream, and furthermore the envisaged FEC mechanisms are still under discussion.

The *AAL-2* protocol type should support CO VBR traffic and is intended to support transmission of VBR video codec signals for example. However the specification of this service is very premature in standardization, and will not be considered within this paper.

The *AAL-3/4* protocol [16] offers a CO service, not including any timing aspects, and is intended to support applications for data transmission. The basic functionality of the AAL-3/4 is SDU delimiting and multiplexing of higher layer SDUs on the ATM connection. From the very beginning on the AAL-3/4 was foreseen to offer a CL data service over the CO ATM. Since it is an architectural requirement that the AAL should not deal with network layer issues, an additional layer is foreseen above the AAL-3/4 to offer a CL service. This CL protocol layer encompasses mainly the addressing functionalities in order to offer a BCDS (Broadband Connectionless Data Service) ITU-T terminology or CBDS (Connectionless Broadband Data Service) in ETSI terminology. Thus it makes no sense to consider the AAL type 3/4 service on its own.

The prime objective of the *AAL-5* [11] is high speed data transmission with reduced overhead. It is a compromise between overhead and functionality. The strong wish to develop a high speed AAL protocol type, which should support existing protocols, was the reason to specify a reduced function AAL protocol type[3]. Currently AAL-5 offers a service for the transport of B-ISDN signalling information[4], and a Frame Relay (FR) service (see below). It is important to note that there are activities, as in the ATM Forum for example, where the possibilities of AAL-5 for the transport of real-time sensitive multimedia information is under investigation, although originally the AAL-1 is foreseen for this kind of communication support.

Since the ATM network itself is not considering the specific characteristics of the AAL services (time and semantic transparency of ATM networks), the service specific AAL protocols are implemented in the subscriber premises equipment, within Interworking Units (IWUs) or within specific "servers". Furthermore it has to be recognized, that the services offered by the AAL are still very basic and generic services. However, they are the state-of-the-art in the development of the broadband technology and have to be seen as the first step in the development of future enhanced B-ISDN tele-services to support the needs of the distributed multimedia applications of tomorrow.

The following B-ISDN bearer services are currently under standardization and are described below: (i) Virtual Path (VP) ATM bearer service; (ii) Broadband Connection Oriented Bearer Service (BCOBS); (iii) Connectionless Broadband Data Service (CBDS)[5]; (iv) Frame Relay (FR) Service; and (v) further service specifications.

3.2. VP/ATM bearer service

Basically we can say, that the Virtual Path (VP) ATM bearer service is the very basic service of an ATM network and is based on the offered service of the ATM layer. The specification of this "Virtual Path Service for Reserved and Permanent Communications (VPRPC service)" is standardized in F.813 and [15] respectively.

The service allows communication in both directions between two users in a point-to-point configuration and is based on VP connections only. The users have the

---

[3]However, from a service point of view, AAL-3/4 and AAL-5 are offering the same layer service. The main differences between these two protocol types are: the AAL-5 performs minimum error control mechanisms in comparison to the AAL-3/4; they perform different mechanisms for SDU delimiting; and the AAL-5 does not offer a higher layer SDU multiplexing capability.

[4]This service is used by the so called Q.2931 (UNI) and Q.2761 (NNI) respectively signalling protocols (Q.2100).

[5]It is important to note, that the CBDS service offers more then just bearer capabilities; e.g. address screening.

possibility to transfer information with CBR or VBR characteristics. However, the resource allocation within the network is done on peak bandwidth allocation per VP only. There is currently no statistical multiplexing. Several VPs can be multiplexed at a specific interface (up to 356 VPs at an UNI according to I.361) and the establishment of a VP connection is done on a semi-permanent basis. An additional important characteristic of this service is the possibility of time-controlled set-up and tear-down of the semi-permanent connections. This is performed via network management functions.

This service is based on the ATM layer service and the user provides the information structured in blocks of 48 octets which are than transferred transparently by the ATM network. Thus it is completely the responsibility of the user how to structure the information which is transported by the ATM layer, and to perform specific mechanisms for error control, congestion, etc.

3.3. Broadband Connection Oriented Bearer Service

The Broadband Connection Oriented Bearer Service (BCOBS) should support unrestricted CO information transfer between UNIs at any bit rate over VP and Virtual Channel (VC) connections with 155 or 622 Mbit/s, and includes signalling capabilities (on a separate VC). The communication configuration is not limited to point-to-point only, but should also support multipoint-to-multipoint configurations[6]. This service is standardized in [8] and F.811 respectively.

It is important to note, that the current BCOBS specification within ETSI and ITU-T has a much wider scope then originally intended. Currently, a lot of other services are included like the real-time services to support the transmission of isochronous G.703 and G.704 signals. This problem is now identified within the standardization, and they are currently in the process to derive a revised service specification of the BCOBS (VC and VP connections with signalling capabilities). All other service specifications will be specified in separate documents. These services are briefly summarized in Sect. 3.6.

3.4. Connectionless Broadband Data Service

The realization of the Connectionless Broadband Data Service (CBDS)[7] service within an ATM network is done by a dedicated Connectionless (CL) protocol[8] which uses the AAL-3/4 layer service [17,18,19]. The basic service, which is realized by a CBDS service as defined by I.364 is the CL routing of user information (with a maximum length of 9188 octets). Since there is no exchange of signalling information to set-up a specific route through the network, every data unit has to carry the full address information to reach the appropriate destination. The coding of the addressing information is following the ITU-T E.164 numbering scheme.

---

[6]However, the technical solution for the support of multipoint-to-multipoint is still not clear.
[7]The CBDS service is equivalent to the SMDS service.
[8]Which is called CLNAP (UNI) and CLNIP (NNI) respectively.

Additionally, this service includes further functions, i.e. supplementary services, in order to fulfill essential requirements of commercial applications. These functions are address validation, address screening, access class enforcement and group addressing. For this reason, the CBDS is more then a bearer service.

Since the same service is offered by both technologies, DQDB and ATM, a simple interworking could be reached. One of the very first commercial usages of CBDS is the LAN-to-LAN interconnection.

3.5. Frame Relay Service

The Frame Relay (FR) / Frame Mode Bearer Service (FMBS) is a CO service, analog to the ATM, but with the main difference that the information unit which is transported by a FR service is of variable length (and of course the much lower bandwidth of 2 Mbit/s). The term "Frame Mode Bearer Service (FMBS)" (as defined in ITU-T I.233) should be seen as a general term which encompasses Frame Relaying Bearer Service (FRBS) (I.233.1) and Frame Switching Bearer Service (FSBS) (I.233.2). In relation with broadband networking, only the FMBS is relevant [13,9].

The FR service is offered either by pure FR networks or within ATM networks respectively. [13] describes the realization of the FR service by using the ATM technology (i.e. a specific AAL-5 service). The AAL-5/FR-service is equivalent to the offered services of the Q.922[9] core functions; this means, that only the so called core services of the generic FR service are realized, as described in the annex of I.233.1; which means basically no error correction mechanisms.

3.6. Further Service Specifications

There are several services foreseen to support various communication requirements as CBR, VBR, with and without timing information transfer, etc. Currently, these services are tackled within the BCOBS service as described above. It has to be noted, that these services are offering more than just a bearer capability. We summarize these services in two sub-categories here:
1. Real-time services to support CBR information streams: These services are based on the AAL-1 layer services (CBR CE, voice band signal transport, and video signal transport) in order to support the transmission of isochronous signal support (G.703, G.704, ISDN, etc.), and audio and video signal transport.
It has to be noted, that of the three AAL-1 services, only the specification of the CBR CE service has up to now reached appropriate stability, and thus this is one of the offered broadband services of todays broadband products.
2. Services to support VBR, non real-time information transfer, like the FR service based on the AAL-5 layer service as described before. Other service specifications like X.25 support are not yet worked out in much detail.

---

[9]Q.922: ISDN Data Link Layer Specification for FMBS.

3.7. Will B-ISDN Network Services of today fulfill the needs of Distributed Multimedia Requirements ?

In [27] it was clearly presented, that the current service specifications of the B-ISDN are not satisfying all the communication requirements as requested by the future distributed multimedia applications. Several issues are still under consideration like appropriate error control mechanisms; appropriate support for real-time information transfer; group communication support; orchestration (i.e. synchronization) of multiply related information streams; enhanced protocol support such as end-to-end QoS negotiation, re-negotiation and indication of QoS degradation, etc.

Additionally, little attention has so far been paid to the definition of a coherent framework that incorporates QoS interfaces, management and mechanisms across all the layers of the future broadband protocol modell. A first stage in establishing an integrated Quality of Service Architecture (QoS-A) for distributed multimedia computing, which encompasses ATM, AAL and the higher layers, was done in [26].

It is a necessary to provide a complete set of B-ISDN services which will really meet the needs of distributed multimedia communications. The current status of activities within standardization in order to specify an appropriate set of future broadband tele-services is described below.

4. Tomorrows Tele-Services

In the following we will summarize the status of the specifications of advanced broadband services in the three most important broadband standardization organizations: ITU-T, ETSI, and ATM Forum.

4.1. ITU-T and ETSI Services

Table 1. provides an overview of the tele-service specifications currently under work within ITU-T and ETSI.

Table 1. Teleservices under work within ITU-T and ETSI

Tele-service	ITU-T Rec	ETSI
Broadband Video Telephony Services	F.722	DE/NA-010029
Broadband Videotex Services	F.310	DE/NA-010030
Broadband Video Conference Services	F.732	-
Broadband TV Distribution Services	F.821	-
Broadband HDTV Distribution Services	F.822	-
Multimedia Distribution Services	F.MDS	-
Multimedia Delivery Services	F.MDV	-

The basic characteristics of these services, as currently seen by ITU-T, are summarized below. Since "multimedia" is a very often used term, when discussing B-ISDN tele-services, [14] encompasses the basic issues to define the term "multimedia" and how multimedia services can be described.

4.1.1. Broadband Video Telephony Service.

The broadband video telephony service (F.722) is likely to be used as an enhancement to the telephone service to allow multimedia person-to-person communication. This service provides real-time bidirectional end-to-end communication of moving colour pictures with high spatial and temporal resolution and video quality equivalent to conventional TV-standards (PAL, SECAM, NTSC) or lower, a voice-sound quality (up to 15 kHz bandwidth, stereo) and means for the transfer of data, e.g. files, still images, graphics, text and end-to-end control messages.

Because of the integrated multimedia communication this service provides, besides the means for "face-to-face" dialogue, a basis for cooperative work. This service may be used for the following applications: "face-to-face" dialogue involving at least head-and-shoulder images; dialogue including interactive viewing of documents such as sketches, diagrams or charts; audio-visual tele-education; health "vising"; teleshopping; tele-advertising; video-surveillance of sick persons, babies, buildings, animals etc.

4.1.2. Broadband Videotex Service.

A broadband videotex service (F.310) [21] is an interactive service which provides for users of appropriate broadband terminals the capabilities to communicate with data bases via telecommunication networks, to retrieve multimedia information (text, data, graphics, still images, audio, moving pictures, etc.). The service should be very simple (i.e. user friendly) and should provide facilities for the users to create and modify information in the data bases.

In relation to narrow-band videotex this service should bring a lot of enhancements in spatial resolution, colour, speed of picture composition, etc. Potential applications for this service could be: remote education and training, tele-shopping, news retrieval, etc.

4.1.3. Broadband Video Conference Service.

The broadband video conference service (F.732) provides person-to-person or group-to-group conferencing capabilities for the transfer of different multimedia information types primarily including voice (sound), full motion video, animated graphics, still images, text and other video information, to support conferencing between two or more locations. Additional to the transfer of multimedia information, the exchange of control information (speaker identification, floor request, etc.) is an essential part of this service.

For the connectivity of terminal equipment at three or more locations, a specific interconnection facility is required, which is called Multipoint Control Unit (MCU), to which all locations are connected individually. The MCU provides proper distribution of the information among the connected locations and takes part in maintaining the proper procedure among the connected terminals. Example applications for this service are: video conference providing head-and-shoulder images, studio-studio conferences, remote education and training, tele-advertising; etc.

4.1.4. Broadband TV and HDTV Distribution Services.

The broadcast TV/HDTV service (F.821, F.822) provides the capability of distributing TV programmes with the quality equivalent to 525 lines or 625 lines NTSC, PAL or SECAM, or HDTV standards to the customer's premises. The selection of the TV channel is controlled by the customer. Several different channels are presented simultaneously, and viewed independently by a number of persons, or recorded on video tape recorders, or selections can be made by the customer's terminal equipment.

In normal operation, calls are only initiated by the customer or by customer equipment (e.g. programmed video tape recorder). Although, the customer may program the network to set-up calls automatically at some later time to an unattended video tape recorder at his premises. Authorization may or may not be required depending on charging mechanisms and package of services. The application of this service is obviously entertainment.

4.1.5. Multimedia Distribution Services.

In addition to the TV program distribution in F.821 and F.822, it seems desirable to include ordinary video information distribution. This service (F.MDS) [23] is intended to support the distribution of multimedia information from one user to several specified receiving users. These receiving users are specified through appropriate signalling procedures. The local information management, i.e. selection of a specific information channel, recording, printing, etc., is under the responsibility of the receiving user.

This service is the basis for the provisioning of new services by utilizing wide-area connectivity and end-to-end switching function of the network. Typical applications of such a service are: closed user group video distribution for entertainment, advertising, education, business information, and multimedia surveillance from multiple sites.

4.1.6. Multimedia Delivery Services.

This service (F.MDV) [22] should support the transfer of multimedia information faster then needed by the video signal itself; i.e. memory-to-memory transfer capabilities from one information sending user to one or more information receiving users. Local management (monitoring, printing, etc.) is up to the receiving user.

Possible applications of such a service are: high speed VTR transfer, multimedia delivery from on-line libraries, multimedia news delivery, multimedia e-mail, etc.

4.1.7. Further Service Specifications.

It has to be noted that there are two new activities initiated within ETSI in order to derive the specifications of a so called "Available Bit Rate" (ABR) concept and a "Video on Demand" (VOD) service.

The objective of the "Available Bit Rate" concept[10] (DE/NA-010031) is to optimize the use of the available bandwidth by the transmission of VBR data in an ATM network (e.g. for LAN interconnection). A definition of this concept is currently not stable. However, discussed issues are: no real-time constraints (cell delay and cell delay variation), minimum guaranteed bandwidth, no declaration of the traffic characteristics, use of preventive congestion control schemes, and provision of a low cell loss rate.

4.2. ATM Forum Services

Also for the ATM Forum, it has to be noted, that the tele-service specification is just at the beginning. Currently, the following Audio Visual Multimedia Services (AMS) have been identified by the ATM Forum[11], and are under investigation to achieve detailed specifications [4]:

1. Real-time transport service: The objective is to identify the set of requirements for a generic transport service which would support real-time delivery of digitally encoded video or audio streams. The transport service should be independent of the particular video or audio encoding used [31].
2. Audio Visual Service (AVS) & Multimedia (MM) on desktop: Conversational service, with information types moving picture, sound and data. This service is equivalent to the ITU-T "Broadband Video Telephony Service" (F.722).
3. Video conferencing: Conversational service, with information types moving picture, sound and data. This service is equivalent to the ITU-T "Broadband Video Conference Service" (F.732).
4. Audio/Data service: Conversational service, with information types sound and data.
5. Broadcast video service: Distribution services without user presentation control, with information types moving picture and sound. This service is equivalent to the ITU-T "Broadband TV/HDTV Distribution Services" (F.821, F.822).
6. Video on Demand (VOD): Retrieval service, with information types moving picture and sound.
7. Available Bit Rate (ABR) concept in line with the same study item in ITU-T and ETSI as already mentioned above.

For the description of the services above, the ITU-T I.211 service description taxonomy is used. As we can easily see, the classification is very generic and at this level does not show for instance the difference between the services "Video Conferencing" and "AVS&MM on desktop". This will only show up with a further definition of the services. Thus, it has to be noted that the existing ITU-T service description taxonomy is not satisfying the needs of broadband communications.

---

[10]Terms also in use: best effort, class-y, VBR+.
[11]Responsibility of the Service Aspects and Applications (SAA) working group.

4.3. Related Work

Although there are many activities dealing with service specification and realization, we present only two important European projects in this area: BERKOM and RACE/CIO. Both of these projects aim to develop generic communications architectures in addition to service specifications and realizations.

4.3.1. BERKOM.

In 1985 a project, called BERKOM [3,7], was started in Berlin, in order to identify and prototype new broadband services and applications for the future B-ISDN. Now BERKOM is changed to an own organization, with strong links to the German PTT. The main results of this project are the definition of a communication architecture and the realization of various broadband services. The two main services which are specified, implemented and demonstrated in a multivendor and multinetwork environment are a Multimedia Collaboration (MMC) service, and a Multimedia Mail (MMM) service.

The BERKOM MMC service [1] allows users to engage in multimedia conferences. Within a conference, users can share applications and engage in audiovisual communications. This service encompasses the following components: administration of a conference directory; management of conferences; sharing of applications; and audiovisual conferencing. The BERKOM MMM service [2] supports multimedia message transfer in an open heterogeneous environment. The service support the composing, sending, receiving and provides a framework for presenting multimedia messages.

4.3.2. RACE/CIO.

CIO is a RACE project (R2060) with the main objective to specify and implement two advanced multimedia tele-services on various end systems. These two tele-services are: multimedia mail, and a Joint Viewing and Tele-Operation Service (JVTOS).

Beside multimedia mail, JVTOS [5] was conceived to support the computer-based equivalent of a personal meeting, where people convene to jointly work on a specific topic, for example by exchanging multimedia information types.

4.4. Summary

The previous sections have shown, that the specifications of the tele-services within ITU-T, ETSI and ATM Forum are very premature. So far the main activities were aimed at completing the standardization process for the basic broadband technology and the B-ISDN bearer services. The realization of appropriate broadband tele-services is starting but much remains to be done.

The disparity in the progress of the two is to be wondered at - although for technical reasons this has been unavoidable - because the real business will be the future tele-services. In broadband networks, the performance of the network itself is ever increasing. To achieve this performance, it is quite clear, that the complexity of the network itself must be limited. This is exactly what is demonstrated by the new ATM technology. The network itself offers very basic services (bearer services) with high performance and high reliability, and more complex services (tele-services) are realized outside the pure network or within dedicated "servers" within the network.

It must be clear, that between todays bearer services of the B-ISDN and the provisioning of enhanced tele-services as video-conferencing for example, there is a big gap of functionality. Two important issues need to be tackled: (i) to study this technical area in order to identify technical solutions for the provisioning of tomorrow's broadband telecommunication services and (ii) to evaluate at the same time the usefulness of such services for real customers.

For the second objective it is a prerequisite, that customers are involved in early design phases of new broadband tele-services and are willing to evaluate such services in their real environment, in order to identify the market profit of new services. For both objectives, it is quite obvious that an early prototyping has to take place, in order to have the possibilities for critical evaluations by customers.

Since it has to be expected, that the whole service development will be very dynamic, the Support Environment (SE), which allow the realization of new services, must offer great flexibility. Neither the hardware platform, nor various communication technologies, should limit the service specification and realization. Even more, several different people, in the public and in the private area, should have the possibility to contribute to the development of a dedicated service. This development of services and applications should be done in parallel to the development of the underlying communication technology. The difference in various communication technologies, should be transparent to the services; i.e. just be seen in terms of degraded or improved QoS.

5. From Service Specifications to an Application Support Environment

5.1. Support Environment Candidate Systems

Services are usually constructed using software components developed by different software vendors. The application interoperability is achieved by an environment enabling different software components to interact across different network domains and supporting distributed processing and multimedia communications. We call such an environment Multimedia Support Environment (MM-SE). An appropriate MM-SE is aiming to be independently evolvable from the underlying switching and transport infrastructure. This allows for the construction and deployment of services and applications independently of specific technologies.

There are several potential candidate systems which could serve as MM-SE: OSF DCE/DME, OMG/CORBA, APM/ANSAware, HP/OpenView, IBM/Netview and of course various proprietary MM systems like the BERKOM and CIO systems. Also important is the number of activities going on for integrating architectures mainly focusing on management issues with those basically related to the control of services; e.g. the integration of IN and TMN [29]. In this area, the Telecommunications Information Networking Architecture Consortium (TINA-C) [6] effort seems to be one of the relevant attempts to fully merge the recent achievements of management architectures, service architectures and distributed processing.

5.2. Requirements on an Architecture and its MM-SE

In order to select the most appropriate architecture and application support platform (MM-SE) a number of requirements on an architecture in general and on the MM-SE in particular should be considerd (the first 4 requirements represent strategic requirements the last three are technical requirements): An appropriate MM-SE should
1. be a comprehensive architecture geared for telecomms, and which is capable and willing to use the results of other related architectures.
2. be a result of open discussion of representatives from all possible interested parties: computer manufacturers, software companies, telecom operators, equipment and service providers and users. The results must have credibility and there should be no chance of the work being unsupported because of competition between players (i.e. open system).
3. be applicable for service creation and execution, service management and network management. That means it should respect both TMN and IN models (i.e. universal environment).
4. not alienate or be independent of the greater non-telecomms information processing world. This would allow the migration and interaction of applications and components between the telecoms operators' domains and the private information processing domains.
5. be based on the current information processing state-of-the-art; e.g. object-oriented, use of well defined interfaces, etc.
6. allow the construction of large systems and the federation or interoperation of semi-independent systems.
7. provide a comprehensive set of system services, which are usable in service construction and management.

These requirements clearly demonstrate that it is essential that an appropriate MM-SE should conform to international and industry standards and should be built on modern concepts. The major influences are therefore ISO RM-ODP [20] and OMG/CORBA [30]. Associated with these are the analysis of concerns according to viewpoints and the definition of appropriate Interface Definition Languages (IDL) in relation with the object-oriented modelling concepts [25].

5.3. The future MM-SE

After considering the requirements stated above, and comparing it with the aim and scope of the various candidate systems, we find that the TINA-C architecture defines the architecture and the environment which is most promising for appropriate system support. At this point, the reader is justified in asking how the TINA-C concepts help in defining the layer required between simple bearer services and advanced tele-services. To demonstrate this, we shall now focus on a portion of the TINA-C architecture and how this is used to provide appropriate tele-service support. This portion of the architecture will be refered to as the distributed Multimedia

Support Environment (MM-SE) and is called Distributed Processing Environment (DPE) in TINA-C.

This section presents a justification for proposing the TINA-C Framework Architecture and its DPE as the basis for the broadband tele-services of the future. The justification in terms of a comparison of the list of requirements (as stated above) and how well they are fulfilled by potential alternatives to the TINA-C DPE is done elsewhere. In the following only the main findings are presented.

In this comparison we come to the conclusion that the DCE/DME, CORBA or ANSAware would be good candidates. HP/Open View and IBM/Netview are targeted for network management issues mainly and not for general system support. DCE/DME, CORBA and ANSAware are more oriented for general system support (management and end-user services). CORBA and ANSAware are related to each other and the TINA-C DPE is likely to be a combination of these.

The TINA-C DPE Interface Definition Language (IDL) is based on OMG/CORBA IDL. CORBA services include DCE/DME and ANSAware services, which is not surprising since the companies behind these two are also in OMG. ANSAware includes an option to use DCE Remote Procedure Call (RPC) and there is an intension to support the CORBA IDL. Various combinations of ANSAware, DCE and CORBA are being considered as a basis for the DPE. The future de-facto MM-SE standard will therefore be an amalgamation of the three, whatever it will be called and that is exactly what the TINA-C DPE will be.

A key point of this paper is that the TINA-C architecture fits the bill for supporting the construction of distributed multimedia applications and telecommunication services. To provide more detailed understanding to TINA-C, a brief overview is provided in the following.

6. TINA-C Overview

The Telecommunications Information Networking Architecture Consortium (TINA-C) [6], which consists of telecommunication operators, telecommunication and computer vendors, was assembled in the end of 1992[12]. The objective of TINA-C is to provide an architecture, based on distributed computing technologies [20], enabling telecommunication networks to support the rapid and flexible introduction of new services and the ability to manage services and the network infrastructure in an integrated way. Services are constructed using software developed by different software vendors. The application interoperability is supported by the DPE enabling different software components to interact across different network domains. The TINA-C architecture aims to be independently evolvable from the underlying switching and transport infrastructure. This allows for the construction and deployment of services and applications independent of specific technologies.

The TINA-C architecture is composed of a Logical Framework Architecture providing basic concepts mainly for distribution and interoperability, a Service

---

[12]Members include: Alcatel, AT&T, Bellcore, BNR Europe, BT, CSELT, DBP Telekom, DEC, Ericsson, FT/CNET, HP, KDT, NEC, Royal PTT Netherlands, Siemens, SIP, Stratus, Telefonica, Telekom Australia, etc.

Architecture providing concepts for service specification, design and management, and a Management Architecture essentially dealing with network resource management. These three are described in more detail below.

6.1. Logical Framework Architecture

The TINA-C Logical Framework Architecture is based on the ODP viewpoints [20][13]. The information viewpoint focuses on the semantics of information and information processing activities in a system. The TINA-C information modelling concepts define information bearing entities (objects), how the relationships between the entities can be described and define the constraints and rules that govern their behavior. The computational viewpoint focuses on the functional decomposition of a system into software components (or computational objects) which are candidates for distribution.

The engineering viewpoint focuses on the infrastructure required to support distribution transparencies. The TINA-C engineering modelling concepts define how software components are bundled in placement and activation units and how these units communicate.

It is important to separate the Logical Framework Architecture from the DPE. Some of the requirements may only be compared to the DPE and this is what is done in Sect. 5.3. The TINA-C frameworks architecture is more than just the DPE. Even if the TINA-C DPE is replaced by one of the candidate systems, the rest of the architecture is still valid because it was established from discussion by representatives of all interested parties and is specifically aimed at supporting telecomms services and management.

6.2. Service Architecture

The TINA-C Service Architecture is built upon the Logical Framework Architecture. It provides a set of concepts, principles, and guidelines for constructing, deploying, and operating telecommunications services which may be based on a number of reusable components. This architecture is applicable to a wide range of service types, including: point-to-point and multipoint bearer services (called transport services in the TINA context), supplementary services, and tele-services like management services (network & service management), access services (policies and mechanisms to use other services) and information services.

In TINA-C, a service consists of a number of interacting objects that reside in a distributed processing environment (MM-SE). The architecture addresses the objects that are required to build a service, how they should be combined, and how they should interact. The architecture also addresses what objects are needed in a MM-SE to deploy, configure, instantiate, manage and use services. At present the architecture

---

[13]It has to be noted, that although ODP encompasses 5 viewpoints, TINA-C decided, that the ODP enterprise and technology viewpoints are not relevant for their considerations.

addresses two major concerns: (i) a service session model[14], which provide concepts for establishing, using and releasing sessions; and (ii) a service management model, which addresses the management of sessions (subscriptions, fault, performance, accounting, security and configuration).

6.2.1. Service Session Model.

Although services by their nature are different from each other, they all have a fundamental property in that they provide a context for relating activities. Such a context is termed a "session". It is possible to identify a number of different contexts, or session types, that a service will have to maintain. Three types of sessions have been identified:

1. A *service session* (or service instance) is the single activation of a service, and relate the users of that service together so as to share entities, such as a blackboard in groupware services, and communication media in transport services.
2. A *user session* maintains state about a user's activity with a service session. For example, whether a user is currently suspended or not.
3. A *communication session* relates the connections in a transport service to the parties involved.

These three sessions describe the relationship of a user with dedicated services. A user can be involved in multiple independent service sessions, such as using a database service and a conference service simultaneously. A service session has one or more users associated with the service, and for each associated user there will be a user session unique to the service session.

Usually service sessions have an association with communication sessions. A user session may have a direct association with the communication session if the service involves special real-time communication (i.e. streams). However, users need not be associated with the communication session; i.e. not all the users involved in a service need to have an active communication channel. For example, when a conference service is left on overnight, the association between a user session and a service session should be maintained, but an expensive transport connection need not be maintained. Hence an association with a communication session need not exist. A communication session is associated with exactly one service session, and must be associated with at least one user.

6.2.2. Service Management Model (Access Concepts and Principles).

Within the context of sessions, it is useful to define the types of users of sessions. These users need to have flexible access to services, in terms of what locations they access from and the types of terminal they use. To support flexible access to services the following concepts are identified:

1. *User Agent (UA)*: A UA represents a user within a network. It acts on behalf of the user when outgoing and incoming session setup requests are processed. A UA

---

[14]The concept of "session" is similar to today's concept of a "call" within the telecommunication world.

can be dynamically associated with any terminal. Thus independent of the role (e.g. end-user, network manager), all people interact with services through a UA.
2. *User Groups (UG)*: A UG is an addressable collection of user agents, and as such calls can be made to and from groups, or more correctly, to/from a user agent within the group. A UG consists of several UAs, and one UA could be member of several user groups.
3. *Terminal Agents (TA)*: A TA represents a terminal attached to a network, or is accessible from a network (e.g. a mobile phone). Terminals, which are the physical devices which are used to access services, may or may not be TINA-C compatible. An example incompatible terminal is a traditional telephone. The main responsibility of a terminal agent is to manage all of the details and idiosyncracies of the device it represents. For incompatible terminals it is responsible for interpreting the terminals proprietary signals. This may require converting the signal into an invocation on an object in the TINA-C system.
A terminal may be associated with many user agents, possibly at the same time. For example, during a meeting all attendees associate their user agents with the telephone in the meeting room.

6.3. Management Architecture

The TINA-C Management Architecture defines a framework for designing reusable modeling concepts of the TINA-C Logical Framework Architecture, with applicable concepts from TMN, OSI Systems Management, and other relevant inputs such as IN, NMF and OSF.

7. Conclusion

In this paper we have shown, that the term "service" is used in many different fields and by many different people with correspondingly differing comprehension. We have tried to highlight the understanding of "services" in the telecommunication area, as seen by the standardization bodies. Then we have provided an overview of todays broadband network services, and the scope for the next activities for realizing modern enhanced communication services. We have shown, that the state of advancement of broadband bearer service development is good, but the development of the enhanced tele-services is just initiating.

To close the gap between todays bearer services of the B-ISDN and the provisioning of enhanced tele-services a comprehensive architecture and a flexible distributed Multimedia Support Environment (MM-SE) is required. Although there are several possible candidate systems to realize a MM-SE for the development of new services and applications, we think that TINA-C's Architecture and DPE is fulfilling the requirements, which we have stated in this paper. We propose to see TINA-C as an appropriate platform for the development and realization of tomorrows broadband telecommunication services and the management of the modern broadband networks.

The next important step, is to evaluate the envisaged tele-services from the standardization bodies by implementing them as soon as possible by using the available broadband bearer services. This is the basis for the identification of the valuability of various tele-services in the new communication environment and to influence the standardization process accordingly. With such an approach, two goals can be reached: (i) contribute to technical solutions for the realization of new broadband tele-services, and (ii) the future customers became familiar with a new technology.

References

[1] M. Altenhofen et al. The BERKOM Multimedia Collaboration Service. Working Document Release 2.0, BERKOM, June 1993.

[2] C. Blum et al. The BERKOM Multimedia-Mail Teleservice. Working Document Release 2.2, BERKOM, May 1993.

[3] B. Butscher and R. Popescu-Zeletin. BERKOM A B-ISDN Pilot Project. Collection of slides (presented at the 3rd IFIP conference held in Berlin), DETECON Technisches Zentrum Berlin, 1991.

[4] P. Coppo and M. Ullio. Service Identification in the SAA Group. Draft Contribution ATM Forum/94-0159, March 1994. Contact: paolo.coppo@cselt.stet.it.

[5] G. Dermler and K. Froitzheim. JVTOS - A Reference Model for a New Multimedia Service. In *4th IFIP Conference on High Performance Networking*, Liege, Belgium, December 1992.

[6] L. A. de la Fuente, J. Pavon, and N. Singer. Application of TINA-C Architecture to Management Services. In *Conference on Intelligence in Broadband Services and Networks*, Aachen, Germany, August 1994.

[7] G. Domann. BERKOM Test Network and BISDN/CATV Concept. Electronical Communication, 62 (3/4), 1988.

[8] ETSI/NA1. Broadband Conenction Oriented Bearer Service (BCOBS) description. Draft ETS DE/NA-10019 (V3.1), Helsinki, October 1993.

[9] ETSI/NA2. Frame Relay Service - General Description. Draft ETS 300 399, September 1993.

[10] ETSI/NA5. B-ISDN ATM Adaptation Layer (AAL) specification, type 1. Draft ETS 300 353 (DE/NA-52617), March 1993.

[11] ETSI/NA5. B-ISDN ATM Adaptation Layer (AAL) specification, type 5. Draft ETS DE/NA-52619, September 1993.

[12] ETSI/NA5. Optionality Aspects of ATM Adaptation Layer type 1. Technical Report, November 1993.

[13] ETSI/NA5. Support of Frame Relaying Bearer Service in B-ISDN and Frame Relay Interworking between B-ISDN and other Networks. Draft ETS DE/NA-53204, September 1993.

[14] ETSI/NA1. Base Document on Multimedia Services. Technical Report ETR/NA-12409 (V4.2), February 1994.

[15] ETSI/NA1. Virtual Path Service for Reserved and Permanent Communications (VPRPC service). Draft ETS DE/NA-10020 (V4), April 1994.

[16] ETSI/NA5. B-ISDN ATM Adaptation Layer (AAL) specification, type 3/4. Draft ETS 300 349 (DE/NA-52618), April 1994.

[17] ETSI/NA5. CBDS over ATM. Technical Report DTR/NA-53203, Paris, September 1994.
[18] ETSI/NA5. CBDS over ATM: Framework and protocol specification at UNI. Draft ETS (Version 1), Helsinki, March 1994.
[19] ETSI/NA5. CBDS over ATM: NNI protocol specification. Draft ETS (Version 1), Helsinki, March 1994.
[20] ISO. Basic Reference Model of Open Distributed Processing - Part 1: Overview and Guide to Use. Draft Recommendation X.901; ISO 10746-2.2, ISO/IEC JTC1/SC21/WG7, June 1993.
[21] ITU-T. Broadband Videotex Services. Draft Recommendation F.310, ITU-T/SG1, ITU-T Report COM I-R 49-E, 1993.
[22] ITU-T. Multimedia Delivery Services. Draft Recommendation F.MDV, ITU-T/SG1, ITU-T Report COM I-R 54-E, Annex 2, 1993.
[23] ITU-T. Multimedia Distribution Services. Draft Recommendation F.MDV, ITU-T/SG1, ITU-T Report COM I-R 54-E, Annex 1, 1993.
[24] H. Leopold, G. Blair, A. Campbell, G. Coulson, P. Dark, F. Garcia, D. Hutchison, N. Singer, and N.Williams. Distributed Multimedia Communication System Requirements. OSI95/Deliverable ELIN-1/P/V4, Alcatel Austria Forschungszentrum and Lancaster University, Ruthnergasse 1-7, A-1210 Vienna, Austria, April 1992.
[25] H. Leopold, G. Coulson, K. Frimpong-Ansah, D. Hutchison, and N. Singer. The evolving relationship between OSI and ODP in the new communications environment. In *2nd int. conf. on Broadband Islands*, Athen, Greece, 14-16 June 1993.
[26] H. Leopold, A. Campbell, D. Hutchison, and N. Singer, Towards an Integrated Quality of Service Architecture (QOS-A) for Distributed Multimedia Communications. In *4th IFIP Conference on High Performance Networking*, Liege, Belgium, 14-18 December 1992.
[27] H. Leopold, A. Campbell, and N. Singer, Will B-ISDN Services meet the needs of Distributed Multimedia Communications ?. In *4th IEE Conference on Telecommunications*, Manchester, UK, 18-21 April 1993.
[28] H. Leopold, A. Campbell, N. Singer, and D. Hutchison, Distributed Multimedia Communication System Requirements. In *The OSI95 Transport Service with Multimedia Support*, pages 64-81, Springer, 1994. ISBN 3-540-58316-5.
[29] Thomas Magedanz.: IN and TMN: The basis for future information networking architectures, computer communications, 16 (5):267-276, May 1993.
[30] OMG, The Common Object Request Broker: Architecture and Specification, Technical Report 91.12.1, December 1992.
[31] A. Patel: Outline for the SAA AMS Baseline Text Document, Draft Contribution ATM Forum/94-0279R1, Munich, Germany, May 1994. Contact: ashraf@sabre.com.
[32] M. De Prycker, R. Peschi, and T. Van Landegem: B-ISDN and the OSI Protocol Reference Model, In *3rd IFIP Conference on High Speed Networking*, Berlin, March 1991.

[33] N. Williams and G.S. Blair: Distributed Multimedia Applications: A Review, computer communications, 17(2), February 1994.

Managing Shared Ephemeral Teleconferencing State: Policy and Mechanism

Scott Shenker[1], Abel Weinrib[2] * and Eve Schooler[3] **

[1] XEROX PARC, 3333 Coyote Hill Rd., Palo Alto, CA 94304
[2] Bellcore, 445 South St., Morristown, NJ 07960-6438
[3] USC/Information Sciences Inst., 4676 Admiralty Way, Marina del Rey, CA 90292

Abstract. In recent years there has been dramatic progress on the enabling technologies for workstation-based multimedia teleconferencing applications. We expect that such applications will soon become an important component of many future social and business interactions. Teleconferencing applications have aspects of their state, such as membership, types of media being used, and encryption, that are under joint control. Much of this state is *ephemeral*, in that it is of importance only for the duration of a session, and does not have importance outside of the session itself. In this paper we focus on the specification and realization of policies for managing this shared ephemeral teleconferencing state. We first define a broad family of policies which has three dimensions: initiation, voting, and consistency. We then present a mechanism that implements this family of policies for two different communication models.

1 Introduction

There has been much work in recent years on workstation-based multimedia teleconferencing applications. Significant progress has been made on enabling technologies such as packet transport for audio and video, resource management, connection establishment, scalability and privacy. These applications are now highly usable and, after a long gestation as research prototypes, are now vigorously entering the mainstream commercial market. With the increasing availability of audio and video equipment on workstations, and with much faster networks being installed, we expect teleconferencing applications to become an important component of many future social and business interactions. However, while there has been considerable work on underlying technologies for workstation-based teleconferencing, there has been less attention given to higher level aspects of teleconferencing. In this paper we focus on one of these higher-level components: the specification and realization of policies for managing shared teleconferencing state.

Multimedia teleconferencing applications have aspects of their state that are, in general, under joint control. The membership of the conference, the types of

---

* Present address: Intel Corporation, 2111 N.E. 25th Avenue, Hillsboro, OR 97124.
** Present address: California Institute of Technology, 256-80, Pasadena, CA 91125.

media being used, the detailed specification of the media streams (for example, the video encoding), which member has the floor, and how the bill is to be shared are all aspects of this shared *session* state about which the participants must often decide jointly. Much of this state is *ephemeral*, in that it is of importance only for the duration of a session, and does not have importance outside of the session itself; thus, the required level of correctness and consistency can be more relaxed than that required for many applications built using standard database technology. Joint control of ephemeral state is a largely unexplored issue, since the database literature typically ignores questions of policy and focuses instead on satisfying stringent correctness conditions. In this paper we define a family of policies for managing this shared ephemeral teleconferencing state, and then discuss mechanisms for implementing these policies. While our primary motivation comes from teleconferencing, our results are relevant to other arenas that have joint ephemeral state, such as key management for privacy and temporary access control for security.

Even though a teleconference is mediated by technology, it is important to remember that teleconferencing is fundamentally a human endeavor. As with all human endeavors, the exercise of joint control is primarily a social issue, not a technical one; that is, social conventions rather than objective standards dictate how control should be exercised. Given the incredible diversity of such social conventions, we expect teleconferences to utilize a wide variety of decision making criteria, or *policies*, to control this ephemeral shared state; some common examples of such policies are unilateral control, majority rule, and anarchy, but a quick inspection of our election laws, country club membership bylaws, and academic tenure decision processes reveals a much richer set of possibilities. To avoid separate implementations for each social convention in this diverse collection, the mechanism by which such shared state is controlled should support a wide variety of these policies.

Current teleconferencing applications do not meet this standard. To take the most ubiquitous example, the phone system has one particular model for such joint control of membership[4]; there is very tight control on membership since the two ends of the phone call are fixed. On the other hand, the conferencing currently enabled by unencrypted multicasting in the Internet[5] allows a very fluid membership, with participants free to come and go as they please. These policies are embedded in the mechanism itself; one cannot have phone calls with loose membership styles and one cannot have tight membership control in unencrypted multicasted conferences[6]. In the teleconferencing literature, there

---

[4] We are ignoring, for the moment, operator assisted conference calls and "gab-line" services in which individuals can call into a bridge within the network.

[5] The most common example of this is conferencing using the *sd* session directory tool and the *nv* and *vat* video and audio tools.

[6] Here, we are considering open unencrypted multicast. Using multicast to transport the media streams that make up the conference does not preclude other forms of joint control of membership. In particular, using encryption and key management provided by a session layer allows one to separate distribution from membership, and thus implement any desired membership control.

has been little discussion about the spectrum of such policies of joint control; instead, there are merely instantiations of particular policies implicitly defined by the various teleconferencing mechanisms. However, we think it crucial that such policy choices be explicitly expressed, rather than implicitly defined, and that the teleconferencing infrastructure support a wide variety of policy choices.

In this paper we carefully separate policy from mechanism. In Section 2, we introduce a framework for expressing a broad family of policies for joint control of ephemeral state. These policies describe who can propose changes to state, and the degree of consensus needed to enact these changes. The policies also describe to what extent the views of state must be consistent when voting and when all changes to state have been executed. This family of policies is not completely general, but we expect that its scope is sufficient to adequately serve most applications. In particular, it expresses a spectrum of policies which span the traditional "tight" control of phone calls to the "loose" control exhibited in today's Internet multicast teleconferences.

In Section 3 we describe two communication models upon which mechanism can be built; these models differ in how the transport is addressed. We propose a mechanism that implements our policy family for both communication models; the mechanism specifies how members[7] of the conference send control messages among themselves using the available communication service. The mechanism combines a two-phase commit algorithm with a timestamp ordering algorithm. Every proposed state change is initiated by a member, the *initiator* of the change. Depending on the policy, the initiator may need to contact some or all members to collect votes as to whether the change should take effect and/or to set locks that exclude other changes for guaranteeing various consistency policies. Finally, the initiator announces the change to all the members of the session. These changes are applied by the members; the order in which these changes are applied can depend on the timestamps. At this point, the change has become *executed*. Care has been taken to ensure that the mechanism scales reasonably well in the limit of many members and imposes relatively little overhead on the teleconference.

Implementing the family of policies is straightforward when the session state is centralized. However, the teleconferencing applications we consider are inherently decentralized, with each member of the teleconference having its own version of the session state. The distributed nature of teleconferences lends particular significance to the policies regarding state consistency. The correctness of the mechanism discussed in Section 3 depends upon certain assumptions about the computing and communication context in which they will operate. These assumptions are not universally valid. However, our foremost goal is to provide precise expressions of the various policies; we are willing to concede that the actual implementing mechanism, when operating in a distributed world with unreliable and unpredictable communications and with finite buffers, may some-

---

[7] We use the term *participant* to refer to a person who is taking part in a teleconference, and the term *member* to refer to the computer program that is executing the agreement algorithm on behalf of the participant.

times fail to faithfully implement those policies. In Section 3.4 we discuss how the possibility of failure can be accommodated to some extent by modifying some of the implementing mechanism and some of the policy requirements to enable sessions to be aborted easily. Since the teleconferencing state is ephemeral, we believe that optimistically relying on the validity of our assumptions, and then facilitating termination of sessions when there is a problem, is an acceptable risk. The alternative would be to use heavyweight fault-tolerant protocols which make very few assumptions about the computing and communication context in which they operate, but whose overhead could well be prohibitive. In short, we choose to approximately implement explicitly and precisely stated policies; we do not merely implicitly define policies by an implementation.

We turn to related work in Section 4. We are not aware of any other formal discussions of the policy issues inherent in controlling shared and ephemeral distributed state. However, there are many existing workstation-based multimedia teleconferencing applications and each of these implicitly define policies. Furthermore, there have been informal discussions of the various styles of teleconferences: open and closed, public and private, and tight and loose. We conclude in Section 5 with a brief summary of our work.

2 Expressing Policy

This section first introduces our notation. We then describe how policy is specified, identifying three aspects to policy: initiation, voting and consistency.

2.1 Notation

Let S denote the shared state that is under joint control of the members of the session. The state S is decomposed into state variables $\{S_i\}$. These variables describe such aspects as the set of members of the session M, a description of the session policies, and other state variables that define domain-specific parameters such as media encodings, encryption keys, and the like. We let T_i be the range of possible values for S_i and define the session state space $T = T_1 \times T_2 \times \ldots$. Each member maintains its own local values for the state S, providing to each participant its own view of this state.

During the course of a teleconference, members will occasionally choose to change this shared state. We denote a general change operation by *op*. Changes to S are specified as a composition of separate elemental operations to the state variables S_i that comprise the state. We define the set of operations that change a state variable S_i to be $\mathcal{O}_i$, and denote a particular operation in that set by O_i. Each elemental operation O_i changes, and depends on, only the state variable S_i, so $O_i : T_i \mapsto T_i$. Although not the most general formulation of a change operation,[8] it appears adequate for a significant range of interesting cases. Thus, in our formulation, a general change operation *op* can be uniquely expressed as

---

[8] In particular, one could imagine conditional change operations that depend on other variables. For instance, the following change command is not expressible in our formalism: "if you are using video codec A then decrease your contrast but if you are using video codec B then leave your contrast unchanged".

$op = O_i \circ \ldots \circ O_j$ where no subscript is repeated. Furthermore, note that for all O_i, O_j, with $i \neq j$, the operations will commute ($O_i \circ O_j = O_j \circ O_i$) because they act on different state variables. We use the notation that $STATE(op) \subseteq S$ refers to the set of state variables acted upon by op.

2.2 Policies

Policies are specified along three dimensions: initiator policies, voting policies, and consistency policies. In this subsection we describe these in turn.

Initiator. The first dimension of policy is which members may initiate certain change operations. For each elemental operation O_i we define the initiating set of members $I(O_i)$; only those members in $I(O_i)$ can initiate the operation. For a general change operation $op = O_i \ldots \circ O_j$ we then define $I(op) = I(O_i) \cap \ldots \cap I(O_j)$; a member can initiate a general operation if and only if it can initiate each of the elemental operations. For instance, one such policy is that only a member can initiate deleting itself. Our mechanism, discussed in Section 3, treats in a special way those variables whose changes can be initiated only by a single specified member. For each member $m \in M$ there is a unique largest (but possibly empty) set U_m of state variables such that $STATE(op) \subseteq U_m \Rightarrow I(op) = m$. The sets do not intersect: $U_m \cap U_n = \phi$ when $m \neq n$. While in this case there is a single member who may initiate the change, the change is not necessarily unilateral—the voting policy (discussed below) may well still require that a vote be taken. We define the union of all such sets to be $U = \cup_m U_m$.

Voting. The second dimension of policy is voting. For every elemental change operation O_i, there is an associated voting rule $V(O_i)$ which takes a vector of votes, with each element taking a value from {YES, NO, ABSTAIN, NO-REPLY}, and returns 0 if the vote fails and returns 1 if the vote carries. Voting rules can span the gamut from requiring unanimous consent to requiring no consent at all. The first three responses (YES, NO, ABSTAIN) are explicit responses from other members; the NO-REPLY value indicates that no explicit response was received. In general, as a function of the vector of responses, a voting rule may depend on how specific members respond, not just on the total number of votes of each type. A NO-REPLY response could be caused by the failure of a member process, by communication failure or delay, or because the member wasn't asked. For instance, if a voting rule only requires two YES votes to carry, then the initiator might choose to ask for votes from only two, or a few, members. A voting rule may explicitly take into account NO-REPLY responses (for instance in requiring a quorum), may pessimistically map them to NO replies, or may simply ignore them.

To be more precise, votes are collected for a fixed amount of time. Once the voting is over, NO-REPLY is assigned as the votes for any members not heard from and the voting rule is applied to determine whether the vote carries or fails. Thus, the result of the vote does not depend on the order in which votes are received. The outcome of the vote may be decided before the voting period expires: if at any point during the time votes are being collected the vector of {YES, NO, ABSTAIN} votes received combined with all possible values[9] for the votes not yet received (assuming that the initiator knows a priori who might vote) yields a passing vote, then the vote passes. An operation has a *null* voting rule if no other members need to cast votes in order for the measure to pass (i.e., requires no approval from anyone other than perhaps from the initiator[10], and so V returns 1 no matter what votes were cast).

For a general change operation $op = O_i \circ \ldots \circ O_j$ the resulting voting rule is written as $V(op) = V(O_i) \circ \ldots \circ V(O_j)$; the composed voting rule is satisfied only if the responses satisfy *all* of the subsidiary voting rules in turn. Since any op has a unique decomposition into $op = O_i \circ \ldots \circ O_j$, this voting rule is well defined. An operation involves no voting if the voting rule for each elemental change operation is null.

Members are polled (if required by the voting policy); if the set of responses satisfy the voting criterion then the change is executed. A voting rule is specified for a particular operation on a particular state variable. For instance, the voting policy for adding a member might require that a majority of the current members vote YES, while the voting policy for deleting a member might require that either the member agrees or all but one member vote YES. Moreover, a voting rule may be asymmetric; for instance, the rule $V(O_i)$ could return 1 when either a particular member m votes YES or when a majority of members vote YES. However, we have made the simplifying decision that the voting rules do not depend on which member initiates the change.

Consistency. The third dimension of policy is consistency; this dimension is of particular importance when, as assumed in this paper, the teleconference is to be controlled in a distributed manner. Recall that each member has its own view of the session state S. There is no other definition of the session state besides the state belonging to each member; in particular, there is no "truth" against which to compare these individual versions. We can only compare these individual versions to each other and require various degrees of consistency.

In general, even when messages are reliably delivered, these individual views will become different when messages related to two or more changes are received in different orders by the members. One policy issue is the extent to which participants want their views to be *eventually* consistent; that is, consistent

---

[9] One would expect that well-formed voting rules will, in general, be monotonic in their arguments {YES$\geq$ABSTAIN,NO-REPLY$\geq$NO}, thus allowing the test to be simply assigning NO to the votes not yet received, rather than all possible values.

[10] Some voting rules are null if the initiator approves but are not null if the initiator cast a NO vote; for example, consider the case where there must be at least one YES vote. In any case, the initiator, knowing its own vote, can decide if the voting rule is null.

after all change operations have taken effect. We expect that there will be, in general, a subset of members for which this condition is to hold; we will call these *eventual* members and denote the set of them by M_e. In addition, there will be a subset of state for which this condition should hold; we call this subset E. The final issue is how to constrain the state that all members in M_e eventually agree upon. To this end, we define *causal* ordering: an ordering of change operations for a member is causal if and only if for each change operation that the member sends out (i.e., executes on the session state), say at time t, all executed change operations that have arrived at the member before time t are ordered before that change operation. Equivalently, if a member has seen the message executing change operation op^α before it executes the change operation op^β, then a causal ordering of the change operations will have op^α before op^β.

Eventual consistency requires that all eventual members are in agreement on the state in E a given time T after the last change operation was initiated (at time t_0). Furthermore, the state upon which the eventual members all agree is given by a causal ordering of some sequence of the change operations $op^a \circ op^b \circ \ldots$ applied to the initial state, and this sequence includes all of the change operations there were executed.[11] More formally, we define eventual consistency as:

Eventual Consistency (EC) There is some time T such that all eventual members agree on all state variables in E at all times $t > T + t_0$. Furthermore, this state is the result of a causal ordering of the executed change operations.

Note that while all eventual members will eventually agree on the value of the state variables in E, it is possible that while casting votes on the various change proposals the members have different views of E. Thus, another form of consistency involves the views participants have when votes are cast. More specifically, we identify two subsets of state variables, C and D, and a subset of members M_c. C is the *critical state* (the state variables that are viewed consistently), D is the *delta state* (the state variables that are acted upon), and M_c lists the *critical members*. The consistent voting condition is then:

Consistent Voting (CV) If $STATE(op) \cap D \neq \phi$ then all votes regarding operation op cast by critical members (i.e., members in M_c) are cast with an identical view of the state variables in set C.

Informally, the critical state C is the state upon which decisions are based, and the delta state D is the state being changed. The CV condition requires that whenever a change is made to D the members in M_c agree on C. For this condition to make sense, when critical members are voting they must share a common view of the critical membership, M_c. Thus, M_c must be part of the critical state C. In addition, we introduce the restriction that only critical

---

[11] As we shall see below, the mechanism we define in Section 3 actually produces an ordering that only reorders non-causally related change operations that are executed at times closer than the clock skew between the processors.

members may initiate changes to the state in C or D, i.e., $STATE(op) \cap (C \cup D) \neq \phi \Rightarrow I(op) \subseteq M_c$. This restriction is not unreasonable, since critical members in a sense "control" the critical state, and allows for considerable simplification of the mechanism introduced in Section 3. Finally, notice that we have defined the consistency conditions independent of the operation being performed; at the cost of additional complexity, one could define a more general framework in which the consistency conditions are applied depending on the operation. Naturally, the added complexity would be a function of the number of operations allowed on the session state, along with the overhead of tracking potentially different subsets of state variables associated with each operation, e.g., C, D, M_c.

Our mechanism does not require further restrictions on the various sets that define the consistency policies. However, useful policies may well further restrict the space. For instance, a common choice may be to have $M_c \subseteq M_e$, $C \subseteq E$, and $D \subseteq E$ because it seems likely that critical members would also care about eventual consistency and that the carefully controlled sets C and D should appear consistent to the eventual members in addition to the critical members.

In summary, the consistency policies define three interesting subsets of state: E, the eventually consistent state, C, the critical state, and D, the delta state that requires consistent critical state when voting about changes. Similarly, we have two interesting subsets of members: M_e, the eventually consistent members, and M_c the critical members. The policy statements are that (1) all critical members must have a consistent view of C when voting on changes that affect D, and (2) all eventually consistent members eventually agree on E. Of course, there can be other gradations of membership (e.g., listening only members, nonvoting members, etc.); these are specified by the particular policies adopted by a teleconference. The membership distinctions we are making here are special in that they refer to the degree of consistency of the views of the state, not the content of that state.

Example. Before turning to how to implement this family of policies, we briefly discuss an example in which a core of critical members controls the membership of a multimedia teleconference. The shared state S consists of seven variables: critical members, eventual members, all members, media stream descriptor (describing the media streams that comprise the conference), encryption key (the key currently being used to encrypt/decrypt media streams), common-names of members (the names used by the applications to identify the various members), and next-to-speak (a state variable identifying the member who gets the floor following the current speaker). All members maintain an eventually consistent view of the eventually-consistent state, so $M = M_e$ and we shall not distinguish between them from now on.

Critical membership M_c is in C and D, and M_e is in D. Thus, only critical members may initiate changes to either type of membership. Adding a member requires a unanimous vote of the critical members; the vote is taken with a consistent view of who is voting (the critical members). Deleting a member is initiated by a critical member, but can occur if and only if the member assents.

The media stream descriptor and encryption key is in $E-(C\cup D)$ (i.e., needs to be eventually consistent, but critical members do not need a consistent view of these state variables to vote on changes to D and critical members do not need a consistent view of C to vote on changes to these state variables). The voting rules on the encryption key are null (no vote need be taken) and any member can initiate a change. The common name of member m is in $E-(C\cup D)$ and also in U_m with a null voting rule (member m, and only member m, may unilaterally make a change to its common name). The next-to-speak state variable is outside of E (since this changes so rapidly, the eventual consistency condition is not considered relevant), and requires a majority rule vote that can be initiated by any member.

This example illustrates how, in a single teleconference, many different aspects of policies can be used, thus motivating the development of a mechanism that can implement the entire family of policies we have defined. We now turn to a discussion of such a mechanism.

3 Implementing Policy

This section describes a mechanism for implementing the family of policies discussed above. We are assuming *cooperative* and functioning members. We have made no attempt to design a mechanism that is invulnerable to attack from one of the members, focusing on *implementation* of policy not *enforcement* of policy. While we do not assume cooperation from nonmembers, we do assume either the absence of or the protection from malicious attacks from nonmembers—a session will be safe only if there is underlying protection against malicious attacks such as nonmembers spoofing as members or flooding the members with messages[12]. We provide a logical description of the mechanism, leaving implementation specific details for the future. For instance, one can achieve the logical equivalent of a limited distribution by broadcasting encrypted versions of the messages and then carefully controlling the key distribution.

The appropriate mechanism necessarily depends on the context in which it is implemented. We first describe our assumptions about the computing and communication infrastructure. In particular, we identify two different communication models to consider. We then present an implementing mechanism that works for both models.

3.1 Communication and Computing Infrastructure

Associated with each member of a teleconference is a computing infrastructure that has a *local clock* and an *operation buffer*. The local clock is used to timestamp certain messages, and should be thought of as being roughly the member's local notion of time-of-day; however, in order to preserve causality, we will insist that a member's clock is never less than the greatest timestamp it has received (this is

---

[12] We do not explicitly discuss ways to provide security and privacy in this paper.

accomplished by always inserting into the local clock the greater of current time and an arriving timestamp). We further assume that the clock skew between the local clocks of members is bounded by a constant time τ_{skew}; that is, if at any particular instant the clock readings are denoted by $\{t_i\}$, then $\|t_i - t_j\| \leq \tau_{skew}$[13]. The operation buffer will be used to store the sequence of operations applied to the shared state, and in some cases the timestamps will be used to order the operations. We assume that the operator buffer is unlimited[14]. We assume further that all initiated changes are either cancelled or executed (to be mechanistically described below) within a time τ_{decide} of being initiated.

There is also an underlying transport service[15] that is used by the teleconferencing application to send a variety of control messages. Here, we are concerned only with the transport *abstractions* made available to the teleconference application, not with the details of how these abstractions are realized on the underlying network. In this paper we assume a reliable communication model in which all messages arrive at their intended destination on time (within $\tau_{deliver}$), and all messages from a particular sender to a particular receiver arrive in the order in which they are sent; we do not require that messages from different senders arrive in the same order as sent. (In extensions to the work reported on in this paper we are also studying unreliable communications models; these models do not naturally support mechanisms that provide the full range of policies introduced in Section 2, but often exhibit attractive scaling for large session sizes.)

We consider two different models for how the underlying transport is addressed. In the *Explicitly Named List* (ENL) model, messages are delivered only to those members that are known by the sender to be members. One way to implement this would be as a sequence of unicast messages; another would be using a sender-defined multicast tree. In the *Shared Bus* (SB) model, messages are delivered to all those who choose to receive the messages. This could be implemented using receiver-initiated multicasting, where the sender just sends to a multicast group identifier without knowing explicitly who is a member of the group; it is up to the receivers to join the multicast group if they want to receive messages. Joining a multicast group is entirely different from joining the session—there is no policy on joining a multicast group, one merely informs the routing protocol that one wishes to receive traffic addressed to that multicast group. Note that the key difference is that in the ENL model the sender controls the delivery, whereas in the SB model the receivers control the delivery. We assume that joining a group takes no more than time τ_{join} to take effect; all messages sent more than time τ_{join} after a member has indicated a desire to join the group will be directed to the member.

---

[13] This bound is preserved by the modification to the local clocks described above.

[14] While formally unbounded, the operator buffer will be unlikely to grow large given reasonable values for the various time constants τ defined in this section.

[15] Note that this paper focuses solely on the control of a multimedia conference; thus, this discussion concerns only assumptions about the transport for control messages associated with our mechanism, *not* the transport of the media streams that make up the conference.

In the following subsection we first describe the mechanism for the ENL case; we then show that the same mechanism works for the SB model. We are making certain assumptions about the context in which our mechanism operates. If any of these conditions are violated, then the mechanism may fail to properly implement the policy. Such failures might make continuation of the session impossible or undesirable. We state now, but will discuss in more detail later in Section 3.4, that to recover from such a situation, members are always able to unilaterally leave a session, thus enabling recovery to a safe state.

3.2 Explicitly Named List

There are three dimensions to policy: initiation, voting, and consistency. The implementation of the initiation rules is essentially a local matter—we assume that each member knows the policy and abides by it (we are discussing implementation of policy, not enforcement of policy). The implementation of the voting rules is similarly straightforward; the initiator solicits votes on proposed change operation from the other members, and then based on the responses decides whether or not to execute the operation as dictated by the voting rule. However, implementation of the consistency conditions is not so straightforward. The traditional way of ensuring that distributed databases retain consistent views is the two-phase commit paradigm. We will use a variant of that approach to meet our consistent voting condition. The eventual consistency condition is less standard, but timestamp approaches can be employed (see [1]); we adopt that approach here to meet our eventual consistency condition.

As a preview, we provide a brief overview of the mechanism. Each change in state is initiated by a single member, and this change is assigned a unique identifier for tying together the various messages associated with a particular operation. If the change requires a vote, then the initiator is responsible for soliciting and then counting the votes. If, for reasons of consistency, locking must occur during the change process, then the initiator is responsible for installing and then releasing[16] the lock. Change operations are associated with a timestamp, and members keep a buffer of the change operations so that out-of-order receipts of change operations can be detected and corrected by reapplying the ordered list of operations to the state.

Changes in membership require a somewhat more complex mechanism than other state changes. For clarity, we first describe the basic mechanism, where membership is constant, and then discuss changes in membership. We then discuss how this mechanism applies to the SB model. This section ends with a discussion of when the reliable model fails and how the consistency condition can be modified slightly to allow unilateral departures of critical members.

---

[16] Because the initiator may fail while holding a lock, a complete protocol will include specification of a time out for locks. We defer this level of detail to future work.

Basic Mechanism: Static Membership. The mechanism utilizes several basic message types:

Poll(Id: id, Operation: op) Asks for a vote on the proposed operation; id is the identifier that ties together messages associated with a particular operation. The response is {YES, NO, ABSTAIN}.

Lock(Id: id) Asks for exclusive lock from all critical members. Only critical members respond, and the response is {OK, BUSY}.

Poll_Lock(Id: id, Operation: op) Asks for exclusive lock from all critical members and for a vote on the proposed operation from all members. The response from critical members is {YES, NO, ABSTAIN, BUSY}. The response from noncritical members is {YES, NO, ABSTAIN}.

Announce(Id: id, Operation: op, Time: timestamp) The operation is to be applied to the local state of each member. No response. The timestamp refers to the time the message is sent (according to the clock of the initiator), and will be used to order operations. Upon receiving such a message, each receiver in M_e must set its local time to be the maximum of its current value and the timestamp of the message[17].

Commit(Id: id, Operation: op, Time: timestamp) The operation is to be applied to the local state of each member, and the associated lock, if any, is to be released. No response. The timestamp refers to the time the message is sent, and will be used to order operations. Upon receiving such a message, each receiver in M_e must set its local time to be the maximum of its current value and the timestamp of the message.

Release(Id: id, Operation: op) The associated lock, if any, is to be released. No response.

Response(Id: id, Response: response) Response to Poll, Lock, or Poll_Lock message.

A member starts with some initial state $S(t^0)$ and then receives a sequence of change operations $op^\alpha = O_1^\alpha \circ O_2^\alpha \circ \ldots$ with associated timestamps t^α. Each member in M_e has an operation buffer that stores the various operations that are to be applied to its state. This buffer keeps track of each state variable in E separately (members in M_c need only keep a buffer for state variables in $E - (C \cup D)$); we denote the buffer for state variable S_i as B_i, and the sequence of stored operations is represented by $< O_i^\alpha, O_i^\beta, \ldots >$, where the operations are sorted in order of increasing t^α. The current state of S_i can be computed by composing the sequence of operations and applying them to $S_i(t^0)$. When an Announce or Commit message is received containing an operation $op^\alpha = O_1^\alpha \circ O_2^\alpha \circ \ldots$, the corresponding O_i^α is placed in order in the sequence stored in B_i. The states are then recomputed. Because message delivery times are bounded, and clock skews are bounded, these buffers need not store change operations for more than time[18] $\tau_{buffer} = \tau_{deliver} + \tau_{skew}$. The buffer stores all operations with timestamps

---

[17] This is necessary for causality, and preserves the bound on clock skew.

[18] The time that change operations must be stored is measured in real time, not the timer being used to compute time stamps which may jump by up to τ_{skew} upon receipt of a message.

more recent than $t - \tau_{buffer}$ and the state resulting from all operations with timestamps older than $t - \tau_{buffer}$ is computed to form $S_i(t - \tau_{buffer})$, at which point these operations can be discarded.

There are some special cases where such a buffer is not required because change operations arriving in **Announce** or **Commit** messages can be immediately applied to the current state and then discarded. Changes to state variables outside of E need not be buffered. Furthermore, if a state variable is in U, then no buffer is needed because all messages changing that variable will arrive in order (since they are sent by a single member and we assume that messages from the same sender arrive in order). Furthermore, if for a particular state variable all operations commute, then buffering of changes to that state variable is unnecessary.

Each critical member is either in the *busy* (i.e., locked) state or the *nonbusy* (i.e., unlocked) state. Upon receiving a lock request (in either a **Poll_Lock** or **Lock** message), a nonbusy critical member enters the busy state. Upon receiving a **Release** or **Commit** message for the associated lock, the critical member enters the nonbusy state. Here, we have chosen to define a single lock for the entire critical state, including both C (the "readset") and D (the "writeset"). Other options exist, such as having separate locks for these two sets or even separate locks for the different elements of the sets. These options would allow for possible additional concurrency, at the cost of significantly more complexity. The advantage to be gained from this additional concurrency is dependent on the frequency of operations that compete for elements of sets C and D.

Any proposed operation *op* can be categorized according to (1) does it require a vote (do any of the elemental operations require a vote), and (2) does it change any state variables in $C \cup D$? Consider the case where the operation *op* changes some state variable in $C \cup D$, and requires voting. Then, the initiator (which must be a critical member in M_c) sends out a **Poll_Lock** message to all members. The members respond with one of {YES, NO, ABSTAIN, BUSY} in their **Response** message (noncritical members may not respond BUSY because they do not maintain locks on the critical state). If any critical member responds with BUSY, or if the vote fails, then the initiator sends out a **Release** message and the proposed operation does not take place. If the vote passes, and no critical BUSY responses are received, then the initiator sends out a **Commit** message and all members execute the operation *op* on their view of the state.

If the operation changes some state variable in $C \cup D$, but does not require voting, then the change is initiated with a **Lock** message, and the responses from the critical members are either OK or BUSY. If all these responses are OK, then the initiator sends a **Commit** message; otherwise it sends a **Release** message. Note that for each of these two cases, the **Commit** messages can only be sent after all responses are collected, but the **Release** messages can be sent after the first BUSY response is received.

If the operation does not change any state variable in $C \cup D$, but does require voting, then the initiator sends a **Poll** message and, after collecting enough votes

	VOTE	NO_VOTE
$STATE(op) \cap (C \cup D) \neq \phi$	PL—Rs—{C,Rl}	L—Rs—{C,Rl}
$STATE(op) \cap (C \cup D) = \phi$	P—Rs—A	A

Table 1. Message Exchanges: Poll *messages are indicated by* P, Poll_Lock *messages are indicated by* PL, Lock *messages by* L, Announce *messages by* A, Commit *messages by* C, Release *messages by* Rl, *and* Response *messages by* Rs. *Brackets are used to indicate alternative messages, depending on whether the operation is successful or not.*

to know if the vote passes, the initiator sends an Announce message. If the vote fails, no message is sent.

If the operation does not change any state variable in $C \cup D$, and does not require a vote, then the initiator merely sends out an Announce message.

These message exchanges are summarized in Table 1, where we use the notation that $STATE(op)$ refers to all possible state variables changed by op.

In describing this algorithm, we implied that all members receive the same message. It would be algorithmically equivalent to send Poll_Lock messages only to critical members and send corresponding Poll messages to noncritical members. Similarly, Lock messages need only be sent to critical members, and noncritical members need only be sent the Announce message if the lock request succeeds. This is a question of implementation.

The mechanism described above satisfies the Eventual Consistency condition and the Consistent Voting condition. The EC condition holds for the following reason. Because messages are reliably delivered, all final Commit or Announce messages will be received by time $T = \tau_{deliver} + \tau_{decide}$ after the last message is sent. Whenever there are no messages outstanding, all members of M_e have the same view of E because their buffers have the same contents and the operations are applied in the same order. The causality condition holds on the ordering because if a member has seen the Announce or Commit message for one change operation before sending an Announce or Commit message of a change for which it is initiator, then the timestamp on its message must be later since the local clock is made equal to the timestamp on an arriving message if that timestamp is later than the local clock.

The CV condition holds because Lock messages are used to ensure that the critical state C is not being changed while critical members are voting on changes to D. In addition, the lock ensures that there is at most one proposal to change D or C in progress at any time. Therefore, all critical members process all operations which change $C \cup D$ in exactly the same order, so eventual consistency is achieved for the critical members for state in $C \cup D$ without any further work (i.e., without requiring that critical members maintain buffers for operations on state in $C \cup D$). If some of the state in C or D is also in the eventually-consistent set E, then eventual members who are not in M_c (i.e., members in $E - M_c$) must buffer changes to that state and order them by timestamps just as for any other state in E.

Notice that the sets C and D appear symmetrically in the algorithm we have defined because we use a single lock to provide mutual exclusion on both of the sets. As discussed above, we could define a more complicated algorithm that makes use of additional types of locks or that makes use of knowledge that certain operations commute; such optimizations would have the potential to allow for more concurrency, but at the cost of significant additional complexity, especially if performance bottlenecks cannot easily be generalized.

This concludes our description of the mechanism when membership is constant. We now describe the additional mechanism needed to accommodate membership changes.

Membership Changes. There are three kinds of membership changes: members leaving an existing teleconference, members joining an existing teleconference, and the establishment of a new teleconference.

No additional mechanism is required for a member to leave a session. However, we need to specify that no member initiates a change to membership that is not meaningful; in particular, no member initiates the leaving of someone that the initiator does not consider to be a member, and no member initiates the joining of someone it already considers a member. This prevents inconsistencies if a member leaves and then immediately rejoins the group using two different initiators[19].

The process of joining a group is somewhat more involved. The prospective member (called a candidate) is assumed to be in contact with the initiator of its membership. No such initiation will occur unless the candidate wants to join (we assume that it makes no sense for a member to be forced to join). When the **Commit** message announcing the addition of a new member is sent, the initiator also sends out a state status message to the candidate supplying it with the initiator's view of the current state (contents of operation buffers and initial states as well as the current values of all state variables S_i). The initiator will forward to the new member all **Announce** or **Commit** messages that arrive after it sends the state status message for a time period $\tau_{forward}$.[20] This time constant must be at least $2\tau_{deliver}$, thus ensuring that the new member will receive all **Announce** or **Commit** messages sent to the group that arrive at the initiator after the new member has joined. Forwarding of **Poll** messages is optional, but would allow the new participant to participate in voting more quickly. Forwarding of **Lock** messages is not necessary since they are relevant only to critical members; since the critical membership M_c is included in C, the

---

[19] Without this rule, the rejoining might have an earlier timestamp than the leaving (because of clock skew), resulting in an eventual state with the member in question thinking it was a member but the rest of the group thinking it was not.

[20] This algorithm requires that the initiator of a new member remain a member for the time period $\tau_{forward}$. This requirement does not appear too onerous given the expected small amount of time involved.

joining of a critical member has its own **Lock** message so either the locks will conflict, or the later lock messages will be sent directly to the new member.

Initiation of a session may be accomplished by sending **Poll** messages to all of the proposed members for an operation that builds the proposed session state. This state includes a voting rule that determines whether the session will be established that is a function of how the proposed members respond (whether YES, NO, ABSTAIN, or NO-REPLY). For a successful proposal, only those proposed members that respond with YES are sent **Announce** messages and become members of the session. Alternatively, a session can be initiated by starting with a session containing a single member and adding members one-by-one.

3.3 Shared Bus

In the mechanism described above, there was no explicit mention of the addressing scheme except when various time constants were invoked to motivate why the mechanism indeed satisfied the EC and CV conditions. We claim that the same mechanism can be applied to the SB case; of course, we continue to use the reliable point-to-point communication for the forwarding of messages and for membership joining exchanges. The only difference in the time constants is now that the time for which messages must be forwarded is $\tau_{forward} = \tau_{join} + \tau_{deliver}$ (rather than $2\tau_{deliver}$); this ensures that all messages not forwarded were sent after the joining has taken effect in terms of routing. The time constant T in the EC condition (the time after which all members agree on state) is still given by $T = \tau_{deliver} + \tau_{decide}$.

3.4 Failures

Our arguments concerning the correct operation of the algorithm described above are based on the assumption that the transport of messages is reliable with a bounded delay. Transport algorithms exist that provide reliability, but at the cost of potentially unbounded message delay; on the other hand, a fixed upper bound on message delay can only be provided in real networks by dropping late messages. Thus, our reliable communications assumption may occasionally fail in any real implementation. We have made the design choice that it is better to tolerate occasional violations in the consistency conditions when communication is unreliable than to use a more complicated algorithm that maintains correctness in the face of general network or host failures; this latter choice would require that applications tolerate potentially unbounded delays, which is not reasonable. Shared ephemeral state exists solely for the purposes of the session, and so the worst that can happen is that the session becomes useless due to some inconsistency. While it may be tolerable for the algorithm to occasionally enter a useless state when communication becomes unreliable, it is important, however, to allow participants to extricate themselves from such a wedged state.

In the algorithm as described up to now, critical membership is in the critical state ($M_c \in C$), and membership in general may be in the critical state as well.

Thus, if some critical member responds with a BUSY to a **Lock** request associated with the operation to remove a critical member m, or the critical member does not respond at all, then m can not leave. If this is merely a temporary condition, and a subsequent attempt to leave succeeds, then the slight delay to leaving is probably tolerable. However, due to some software error or communication or host failure, the critical member may continue to respond with BUSY (or not respond at all) for an indefinite period, resulting in m being trapped in the session. To rectify this, we modify the policy associated with membership changes and change the description of voting correctness. A member may leave unilaterally if need be (although it may still be advantageous for a member to leave using the standard **Lock** protocol) by sending out an **Announce** message to inform the rest of the session members of its unilateral departure from the session. If such a message is received while an initiator is collecting votes, the departing member's vote is recorded as an NO-REPLY (unless a vote for the proposal was received earlier) and the departure is processed after the lock is released.

Thus, the consistent voting condition is changed: all critical members cast their votes with identical views of C, except for critical (and possibly other) membership, and all votes are *plausible*. A vote is *plausible* if it is cast with some view of membership that may or may not include those members that have unilaterally departed. While each vote is cast with some individual view of membership that lies between the full one (the one before all the unilateral departures occurred) and the reduced one (the one after all the unilateral departures occurred), there is not necessarily a single global view of the membership that would explain all the votes.

4 Related Work

There are several multimedia teleconferencing applications that are relevant to our discussion here. The Touring Machine^{TM}[21] system project at Bellcore [2][3] was one of the first projects to introduce the concept of a session abstraction with specification of associated session policies. The session is implemented in a centralized fashion; thus, policies related to consistency of the shared state were not required.

The Rapport system [4], the Etherphone system [5], the BERKOM system [6], and similar multimedia conferencing system have not focused on issues related to specification of session policy. They tend to use a fixed "call model" to realize the communications abstraction they offer to a participant.

MMCC [7] and IVS [8] are both designed to run over the Internet. MMCC provides a mechanism for an initiator to rendezvous with other potential members to set up a conference using the packet video and audio conferencing tools *nv* and *vat*. IVS also supports rendezvous, but uses its own H.261 compliant coding for packet video and audio transport. Both of these tools have relatively static session policies.

---
[21] Touring Machine is a trademark of Bellcore

Finally, the IP-multicast based packet conferencing tools *nevot*, *nv*, and *vat* (and probably others that are appearing as we write), fit into our shared bus model, but with unreliable transport. They lend themselves to relatively open conferences, and do not have ways to support the range of policies introduced in this paper. Each member uses unreliable multicast to send out its state—primarily to inform others that it is a member and what is the participant's real name. The session directory tool *sd* is used to inform interested participants of the existence of ongoing conferences, and automates starting up the required tools.

It is important to distinguish mechanisms that coordinate the actions of participants in such conferencing tools from the policy issues we have discussed here. We addressed the issue of how to jointly reach a decision about the shared state, *not* how to turn that state into appropriate actions by the members. For instance, we consider how to decide who can speak in a teleconference but we do not discuss the mechanisms by which those members who are not allowed to speak are prevented from doing so.

We are not aware of other work that formally addresses the spectrum of policies for managing shared ephemeral state. However, there has been much informal discussion of these policy considerations. These informal discussions refer to several dimensions of policy, such as loose/tight, open/closed, and public/private. Loose control versus tight control refers to both the degree of consensus needed (the voting policies) and how consistent the views of the state needs to be. Open versus closed control typically refers to the membership policies; at the extremes, closed sessions do not allow new members, and open sessions allow anyone to join. Public versus private, to the extent that it is different than the open versus closed distinction, refers to who can listen to the session rather than who can join. In this paper we have defined a family of policies which addresses all of these dimensions, and allow us to vary the policies along these dimensions independently.

Often the policy attributes of being loose, open, and public are associated with large groups, whereas the policy attributes of being tight, closed, and private are associated with small groups. In principle, there is no inherent connection between the size of the group and the policies being used; in fact, none of our policy definitions refer to the size of the group. However, practical issues do force at least some connection between size and policy. The locking technique used to achieve consistent voting is likely to incur significant problems when applied to very large groups. However, the number of (controlling) critical members that must execute the locking protocol may be small in certain situations even when the total membership grows large, thus removing this problem. In addition, the voting mechanism will result in an implosion of **Response** messages at the initiator if the group is large and the voting rules require that all members vote. Even the assumption of reliability will become more problematical as the size of the group increases (since the chance that a message reaches all participants will decrease). We have yet to explore the extent to which size and other practical concerns influence the appropriate choice of policy.

5 Summary

Our purpose, in this paper, was twofold: to both *specify* and *realize* policies for managing shared ephemeral teleconferencing state. To that end, we first defined a family of policies within which application policy requirements could be expressed. These policies addressed the issues of who could initiate changes to the shared state, what the voting rules were on these changes, and the degree of consistency required in the distributed view of the state. While quite general, the adequacy of this family of policies can only be evaluated relative to the social conventions that arise in future teleconferencing applications.

We then described a mechanism that implemented these policies for two reliable communications models. The mechanism allows applications to combine many different policy requirements while still using only a single implementing mechanism. We will not know whether the mechanism is sufficiently robust and efficient without implementation experience and testing in real application settings. This is the subject of our current work.

We are also studying mechanisms for models with unreliable communications, with particular focus on those applicable to IP multicast on the Internet. While the full range of policies introduced in this paper are not naturally realized in these models, we are discovering mechanisms that support more limited policies and scale well to the large conferences that are enabled by IP multicast over the *MBone* [9].

6 Acknowledgments

We would like to thank many members of the IETF Multiparty Multimedia Session Control Working Group and of the Touring Machine project at Bellcore for valuable discussions and advice.

Eve Schooler was supported in part by ARPA contract DABT63-1-0001, the Airforce Office of Scientific Research grant AFOSR-91-0070, and a grant from the AAUW Educational Foundation. The views and conclusions in this document should not be interpreted as representing the official policies, either expressed or implied, of the AAUW, ARPA, AFOSR, or the U.S. government.

References

1. A. Demers, D. Greene, C. Hauser, W. Irish, J. Larson, H. Sturgis, S. Shenker, D. Swinehart, and D. Terry, "Epidemic Algorithms for Replicated Database Maintenance," *Proc. of 6th ACM Symp. on Principles of Dist. Comp.*, pp. 1-12, 1987.
2. Arango et al., "The Touring Machine System," *Comm. of the ACM*, Vol. 36, pp. 68-77, Jan. 1993.
3. V. Mak, "Session Management for Distributed Multimedia Applications," *Proc. 5th IEEE COMSOCC Int'l Wkshp. on Multimedia '94*, Kyoto, Japan, May 1994.
4. S.R. Ahuja and J.R. Ensor, "Coordination and Control of Multimedia Conferencing," *IEEE Comm.*, Vol. 20, No. 5, pp. 33-43, May 1992.

5. H.M. Vin, D.C. Swinehart, P.T. Zellweger, P.V. Rangan, "Multimedia Conferencing in the Etherphone Environment," *IEEE Computer*, Vol. 24, No. 10, pp. 69-79, Oct. 1991.
6. M. Altenhofen et al., "The BERKOM Multimedia Collaboration Service," *Proc. ACM Multimedia 93*, pp. 457-463, Aug. 1993.
7. E.M. Schooler, "Case Study: Multimedia Conference Control in a Packet-switched Teleconferencing System," *J. of Internetworking: Research and Experience*, Vol. 4, No. 2, pp. 99-120, June 1993.
8. T. Turletti, "H.261 Software Codec for Videoconferencing over the Internet", Research Report 1834, Institut National de Recherche en Informatique et en Automatique, Sophia-Antipolis, France, Jan. 1993.
9. S. Casner, S. Deering, "First IETF Internet Audiocast", *ACM Sigcomm Comp. Comm. Review*, Vol. 22, No. 3, pp. 92-97, July 1992.

Computational Components for Synchronous Cooperation on Multimedia Information

Christophe Logé, Valérie Gay and Eric Horlait
Université Paris VI,
Institut Blaise PASCAL, Laboratoire MASI,
4, place Jussieu,
75252 PARIS cedex 05, France.
E-mail: {loge, gay, horlait}@masi.ibp.fr

Abstract. Nowadays, the design of cooperative applications is becoming more complex due to the introduction of real time cooperation and multimedia aspects. To ease and accelerate the development of these applications it is necessary to build modular and reusable functional components that can be used by application designers.

The objective of this paper is to define some of those functional components. They will be integrated in the cooperative applications to ensure the *synchronous cooperation on multimedia objects*. For openness and possible wide use in an open distributed heterogeneous environment, the specification of those components is based on the computational language of the Reference Model of Open Distributed Processing (RM-ODP).

Keywords. Multimedia cooperative applications, CSCW, Open Distributed Processing, ODP, Distributed Multimedia Application Architecture.

1 Introduction

The fast advancement of computer technology and its decreasing price enable the use of high performance high speed networks and powerful computer systems. In this environment, users expect to work from their desktop on *multimedia information* in a *cooperative way* but the design of cooperative applications handling multimedia information is rather complex. The scope of this paper is to provide indications for the development of such applications. It specifies modular and reusable functional components that may be integrated in the cooperative application to ensure cooperation on multimedia information.

Collaborative applications can be classified in two types: *formal cooperation* and *informal cooperation*. In formal cooperation participants collaborate to reach a common goal that will change some persistent data. For example, a co-authoring system enables the editing of a multimedia document that remains at the end of the session. In contrast, in informal cooperation the produced information is not persistent and is only exchanged between participants. The use of a telepointer to point-out some parts of shared information or the use of a videophone to exchange ideas during a collaborative activity are two examples of informal cooperation. Informal cooperative systems are needed to ease the formal ones and to bring more interactivity between the users. A videophone application is an example of informal cooperative system that can help users of a co-authoring application to elaborate their shared multimedia document. A telepointer tool is an other example of application useful for a teacher during his/her courses in a tele-teaching application.

This paper studies formal and informal cooperation in synchronous applications (e.g. telepointer, videophone). It specifies so-called synchronous cooperation components for synchronous cooperation on multimedia information. As illustrated on figure 1, these components interact with some application components. The synchronous cooperation components enable the cooperation of distributed object handlers over distributed multimedia objects. The application components concern more specifically application designers that deal with information processing without considering cooperative aspects (e.g. text justification). The synchronous cooperation components manage user's interactions over shared *multimedia information*. They ensure that each user has its own view of the shared information and that this information stays consistent also in case of simultaneous multi-user requests. For openness and possible wide use in an open distributed heterogeneous environment, the specification of those components is based on the computational language of the Reference Model of Open Distributed Processing (RM-ODP [1], [2], [3], [4]).

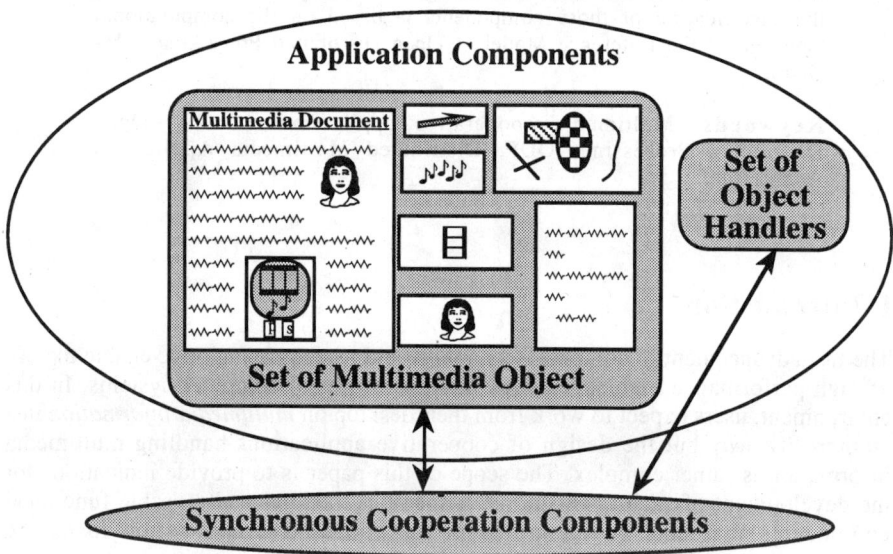

Fig. 1. Application and synchronous cooperation components interactions.

The paper is structured as follows: Section 2 gives an overview of the object modelling concepts based on RM-ODP that will be used to specify the computational components for synchronous cooperation on multimedia objects. Then, it models the handling of multimedia objects in non-cooperative distributed multimedia applications. In section 3, this modelling is completed to introduce the synchronous cooperation aspects. In this process, the computational components enabling the real time cooperation on multimedia object appear and are described. Then a scenario is given illustrating the use and role of the computational components in a cooperative session. In section 4, the computational components for synchronous cooperation on multimedia object are further described and specified. Conclusions are then drawn on the components specified and on their future use in open distributed multimedia applications.

2 From a non-cooperative to a cooperative environment

This section specifies the basic elements used to go from a non-cooperative environment to a cooperative environment. An object-based approach is used to specify our components for modularity and reusability purpose. It is obvious that synchronous cooperation will take place in a distributed environment. This distribution implies the interoperability of various environments and therefore it requires some standard basis.

The specification used in this paper relies on the computational language of the ongoing standard on Open Distributed Processing (ODP). This computational language is object-based and enables the specification of functional components (objects) and their interfaces. The specification in the computational language is distribution and replication transparent.

The objects are characterized by a state and a behaviour. The behaviour is determined by a set of actions that can be invoked by the other objects. An interface is a sub-set of this set of actions. Two interfaces type are used in this paper the *operational interfaces* and the *stream interfaces*. The *operational interface* (⊣ symbol) is used to enable operational interactions between objects. The *stream interface* (⊣ symbol) enables interactions between objects that continue throughout a period of time. To create an application, several objects are chosen and they interact through their interfaces. The complete definition of stream and operational interfaces can be found in [3].

2.1 Non-cooperative environment

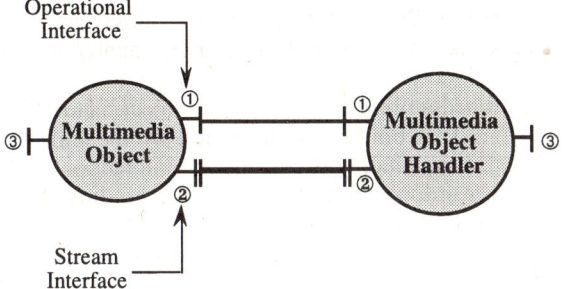

Fig. 2. Modelling concepts shown on a simple multimedia application

As illustrated on figure. 2, we start our study by the modelling of a non-cooperative multimedia application. We consider that a multimedia application can be seen as a set of components that interact to achieve the goal of the application. For example, a single-user multimedia editor is composed of components to edit the multimedia document, to provide and manage an editing graphical environment, to interact with the user. Amongst these components, some of them hold a particular role. They are in charge to interact 'directly' with the multimedia information. In this context, we always consider that an application has some objects called *object handlers* that interact with the multimedia information called *multimedia object*. For example, in a multimedia editor, the editor engine represents the object handler and the multimedia document represents the multimedia object.

A **multimedia object** encapsulates the following characteristics: a state and a set of actions that can be invoked by the other objects. For example, a TextObject (multimedia) object can be characterised by the content of the text and by the set of actions that can be asked to this text like the methods TextObject.DeleteCharacter(Position) and TextObject.InsertWord(Word, Position). A VideoObject (multimedia) object can be characterised by the set of frames that constitute the video with the actions like VideoObject.StartVideo(), VideoObject.StopVideo() and VideoObject.PauseVideo(). Similar characteristics can be found for each kind of media like text, graphics, still pictures, audio and video. In addition a multimedia object can be composed of other multimedia object to constitute a more complex object. For example, the multimedia document on the left hand side in figure 1, is composed of a TextObject, a VideoObject and an AudioObject.

A multimedia object has some interfaces for interaction with its environment. It has an operational interface (①) for operational interactions like TextObject.DeleteCharacter(Position) or TextObject.InsertWord(Word,Position). It may have a stream interface (②) for the interactions that continue throughout a period of time (e.g. VideoObject.StartVideo()). These interfaces (① and/or ②) are sometime referenced in the paper as *Object Interaction Interfaces*. Finally, a multimedia object has an operational interface of type (③), for the other interactions (e.g. to register in a trader).

A **multimedia object handler** is a component of the application that interacts with the multimedia object. Some examples are the editor engine of a multimedia editor or the video presentation component of a complex audio/video application.

To **interact** on a multimedia object, a multimedia object handler must be bound to this multimedia object. This binding is illustrated by the links between the two objects in figure 2. Once the binding is realised the interactions between the bound objects can occur. More information concerning the binding action between two objects can be found in [1], [2] and [3].

2.2 A cooperative environment

This section shows how the cooperation of several object handlers on the same multimedia object is managed.

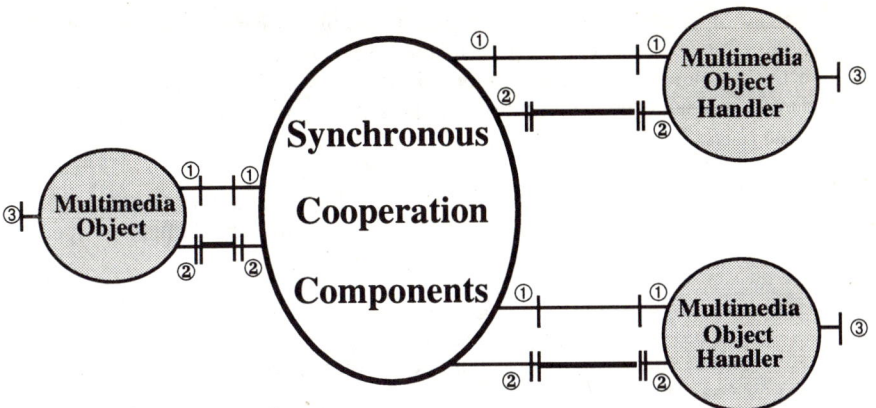

Fig. 3. A multi-user multimedia application

As illustrated in figure 3, we consider now that several object handlers interact simultaneously on the same multimedia object. Each of them can individually be bound to the multimedia object. To manage their cooperative activity a set of computational components is put between them. They will be in charge to analyse and to manage all the interactions that occur between the objects. This implies that interactions initiated by the multimedia object handlers and their result will go through this set of components.

3 The synchronous cooperation components

The specification presented in section 2 is the basis of the study. This section develops the set of components required to complete this object decomposition and introduces the synchronous cooperation aspects in the modelling. The context of the cooperation is the following: several multimedia object handlers expect to handle synchronously the same multimedia object.

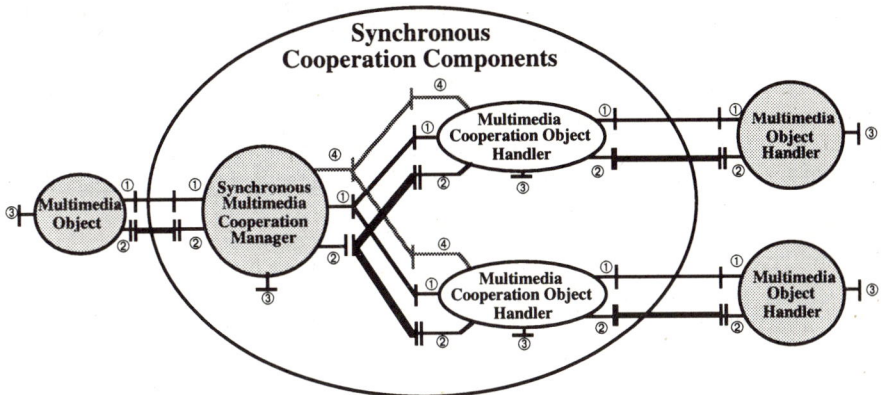

Fig. 4. Synchronous cooperation handling of multimedia object

As illustrated in figure 4, when several multimedia object handlers work on the same multimedia object, it is necessary to manage the interactions they may have. To achieve the management of these interactions, the multimedia cooperation object handler and the synchronous multimedia cooperation manager are inserted between the multimedia object handlers and the handled multimedia object.

The figure 4 shows also a new interface (④ in grey). This interface enables interaction related to the management of the cooperation. An example of operation exchanged on this interface is:

```
ObjectHandlerCooperationRequirements(
        <CooperationRequirements>,
        <MultiMediaCooperativeObjectHandlerId>,
    Result <ResultStatus>).
```

<CooperationRequirements> represents a structure containing the user requirements on a given multimedia object for a given object handler, <MultiMediaCooperativeObjectHandlerId> represents the identifier of the multimedia cooperative object handler that engage the operation and <ResultStatus> carries the results of the operation after its analyses by the synchronous multimedia cooperation manager. It is used by the multimedia cooperation object handler to give its cooperation requirements to the synchronous multimedia cooperation manager.

3.1 Multimedia cooperation object handler

The multimedia cooperation object handler interacts with the multimedia object handler (as presented in section 2.1). It traps the interactions between a multimedia object handler and the multimedia object and manages these interactions in collaboration with the synchronous multimedia cooperation manager. By trapping we mean that the interactions between a multimedia object handler and a multimedia object pass through the cooperative component before being presented to their 'receiver'. The interactions between the multimedia object handlers and the multimedia object are not exchanged by way of a 'direct' link between them but are sent to the cooperation components and then sent to the 'dedicated' object. An example is when a text handler of a multimedia editor invokes the method TextObject.DeleteCharacter(Position), the invocation is trapped by the multimedia cooperation object handler before being 'propagated' on the multimedia object. If the invocation provides some results, then these results pass through the multimedia cooperation object handler before being presented to the text handler of the multimedia editor.

The multimedia cooperation object handler has two sets of *interaction interfaces* (① and ②), and a *cooperation management interface* (④). A first group of *interaction interface* is used to trap the multimedia object interaction that comes from (or dedicated to) a multimedia object handler. The second group is used for the interactions that the multimedia object handler expects to have with a multimedia object. The *cooperation Management Interface* (④), is used for cooperation management interactions.

3.2 Synchronous multimedia cooperation manager

When a cooperative activity occurs on a multimedia object, the synchronous multimedia cooperation manager manages the synchronous cooperative aspects of this activity. Its role is threefold.

One role is to collate the interactions that come from the multimedia cooperation object handlers. By collation we mean that all the interactions generated by the multimedia object handlers are gathered into the synchronous multimedia cooperation manager to be analysed. This collation may concern object interactions. For example, if several users ask simultaneously for the right of modification of a TextObject during a co-authoring session, the collation of their request by the synchronous multimedia cooperation manager will enable it to decide which user can modify the TextObject and which of them cannot. The collation may also concern the management of the cooperation. For example, each user may specify their requirements concerning a cooperation.

An other role is to manage the interactions. This means that the synchronous multimedia cooperation manager decides what it has to do with the interactions that comes from the multimedia cooperation object handlers and from the multimedia object.

The last role is to propagate the results of the interactions to the concerned multimedia cooperation object handlers. For example, in a tele-teaching context, if the teacher starts a video of its course, the synchronous multimedia cooperation manager must send this video to the set of student.

The interactivity evolves with the granularity of the multimedia object. For example, in a joint editing session, they are more simultaneity of work when the concurrence access control and the view sharing are made on part of the document and not on the whole document.

3.3 Illustrating scenario

This section exemplifies the use and role of the computational components during a session of cooperation. This scenario considers two users using their own *object handler* (their multimedia editor) to interact synchronously on a multimedia object called 'Cooperative Paper'. It is a multimedia document with the text and figures that corresponds to this paper and a video that gives a demonstration of its implementation. This document has a table_of_content organising its content.

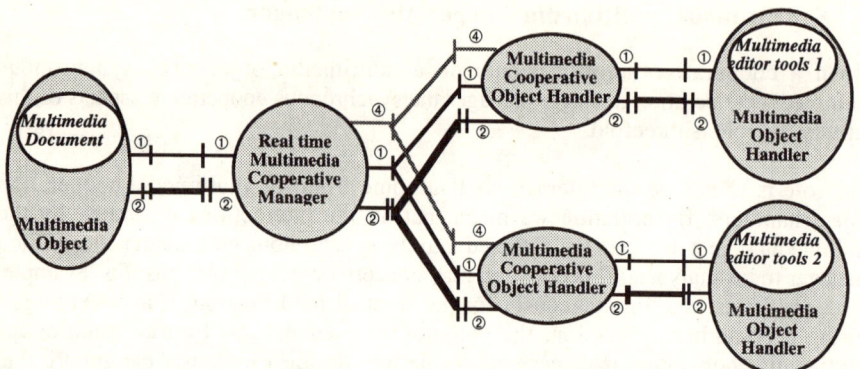

Fig. 5. Synchronous cooperation on a multimedia document.

The *synchronous multimedia cooperation manager* component is already instantiated. Each multimedia editor is bound to its associated *multimedia cooperation object handler* that is bound to the synchronous multimedia cooperation manager. Both types of bindings are established. The interfaces ① & ② are bound between each multimedia editor and each associated multimedia cooperation object handler. Similarly, the bindings are established between interfaces ① & ② of the multimedia cooperation object handler and the synchronous multimedia cooperation manager. In addition, the multimedia cooperation object handlers are already bound to the synchronous multimedia cooperation manager through their *cooperation management interfaces* ④. It is also bound to the 'Cooperation Paper' through the interfaces ① & ②.

Both users decide to set up the condition of their cooperation. They engage the operation *SetUpCooperation* that contains their requirements by selecting an option in a menu. The synchronous multimedia cooperation manager analyses them, and stores them in the *information base* of the synchronous multimedia cooperation manager. This later informs both users on interface ④ that their operation was successful.

Then both users decide to work on the 'Cooperation Paper', but on different parts. One user wants to delete the word 'Multimedia' from the title. He starts engaging an operation *RemoveWord* that will delete the seventh word. The second user wants to play the video that shows the results of the implementation of the ideas presented in the 'Paper'. It pushes on a button near its video and its multimedia editor starts the operation *PlayVideo*. The two operations are trapped by the multimedia cooperation object handler manager. Before propagating them it asks to the synchronous multimedia cooperation manager through the interface ④ for the right to do this. The component analyses the operations and concludes that there is no problem for their execution. It propagates them on the 'Cooperation Paper' object and the two operations are applied on the multimedia document. The word 'Multimedia' is deleted and the video is sent on the stream interface of the second user.

After both users wish simultaneously to work on figure 6 to add operation examples. They click in the figure and their request *SelectFig.* that have been trapped by the multimedia cooperation object handler is sent to the synchronous multimedia cooperation manager. As the operation concerns the same figure, one request is accepted while the other user is informed that there was a conflict. This user decides then to work on the conclusion.

4 synchronous multimedia cooperation manager

The following section presents the synchronous multimedia cooperation manager that manages the interactions of several multimedia cooperation object handlers on a multimedia object on which it is bound. It is the component responsible for the controls of synchronous and cooperative aspects during a cooperative activity. To fulfil its roles it always contains a cooperative information base, a shared view manager and a concurrent access manager as illustrated on figure 6.

The synchronous multimedia cooperation manager collates *object interactions* and *cooperation management interactions*. Then it treats these interactions and possibly multicast the results to the concerned multimedia cooperation object handlers.

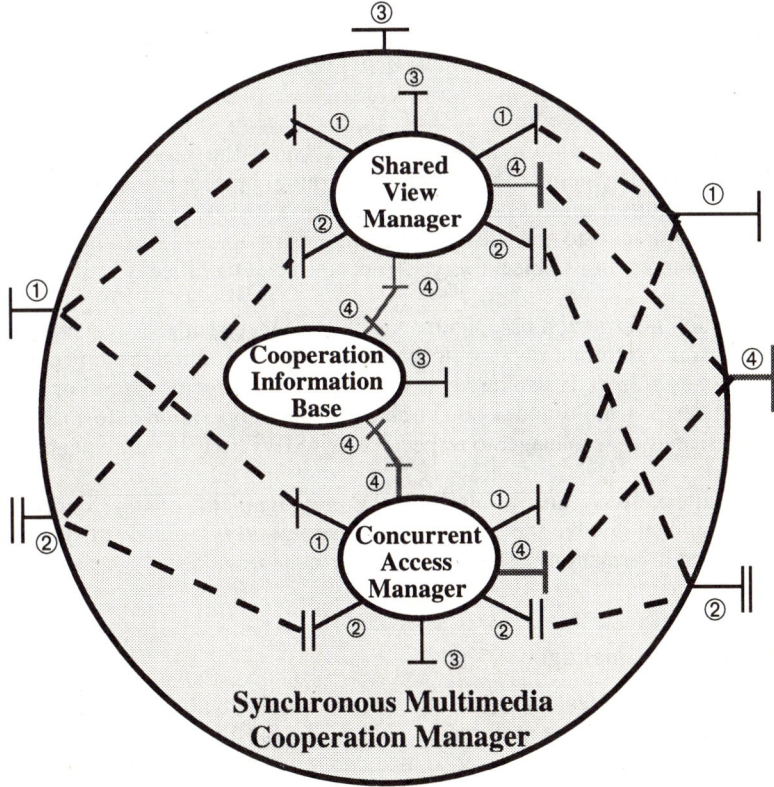

Fig. 6. Synchronous multimedia cooperation manager decomposition

4.1 Cooperation information base

For a given shared multimedia object, the cooperation information base knows a set of information that are needed for the management of the cooperation. It knows the set of object handlers acting on a multimedia object. This refers to a *group notion* that regroups a set of object handlers in a single group. This group may be fixed before a cooperation occurs and/or evolves dynamically during the cooperation. For example, the members of a project correspond to the group of person that are allowed to process a multimedia document.

It knows what are the interactions that each member of the group can perform on a multimedia object. This refers to a *right notion* that gives to each member of a group a set of action it can perform on a multimedia object. These rights may also be fixed before the cooperation or evolve dynamically. An other possible assumption is that only one member of the group can change a multimedia document at the same time.

In addition to the previous notions there is the *role notion* that gives to one or more users some privileged rights during a cooperative activity. One kind of policy can be '*to speak during a multi-user videophone activity, a user must ask to a chairman for the right to speak*'.

The cooperation information base must also know the requirements and the characteristics of each member of a group concerning the cooperation. For example, during a co-authoring session, a user may accept a lost of synchronisation of its shared view compared to the view of the other participants. An other user, may require a refreshment of its object every 10 second (if it is modified).

The cooperation information base interacts with the concurrent access manager when it waits for some information (e.g. to know whether a multimedia object may be change or not by a user). It also interacts with the shared view manager. For example, when the state of a multimedia object changes the cooperation information base informs the shared view manager to respect the WYSIWIS[1] [17] requirements.

All these interactions are related to the management of the cooperation. Consequently, they involve the *cooperative management interfaces* (of type ④) that the cooperation information base shares with the concurrent access manager and the shared view manager.

4.2 Shared view manager

The role of the shared view manager is to guarantee to the multimedia object handlers a correct presentation of the multimedia objects. To fulfil its role it has two groups of *object interaction interfaces* ① & ②, two *cooperation management interfaces* ④ and an interface of type ③. One group of *object interaction interfaces* ① & ② links the shared view manager to the multimedia object. It enables the shared view manager to 'receive' the results of multimedia object interactions. These results are multicasted to the multimedia cooperation object handlers that required them by the way of the second group of *object interaction interfaces* ① & ②. For example, in a tele-teaching

---

[1] WYSIWIS: What You See Is What I See.

context, when a video is started by the teacher, this video must be sent to all the students, it is the role of the shared view manager to ensure that the video is seen by all of them.

The shared view manager is linked to the cooperation information base by a *cooperation management interface* ④ for their interactions. It has also a cooperation management link with the multimedia cooperation object handlers to send them the results of previous interactions or to initiate *cooperation management interactions*. For example, to inform the multimedia cooperation object handlers that a cooperative activity will stop.

This functional component could be used to implement the different *WYSIWIS policies* that are presented in [9]. Basically, it allows several users to see synchronously all the information handled by the others according to their requirement. This information may be directly involved in the cooperative task or be generated by tools helping user's collaboration. GROVE[2] uses the 'cloudburst model and aged text' to propagate the modifications [18]. For the MERMAID[3] system ([14], [15]) where only one user can change the shared document at a time, all statements are sent among the replicated application that can, locally, show modifications. This implies that the application must be replicated with the same hardware and software on all sites. The problem raised by this function is to decide how and when it must be applied. Cooperative participants may expect to have all the modifications of their shared multimedia object at the same time. But what is the 'same time' ? If a user changes a pixel of a picture or a character of a text, must this modification be reflected in real time? On which granule size must the WYSIWIS function be applied?

A strict synchronisation between the shared views is necessary while using telepointing or audio/video tools. Otherwise, the user may accept a loss of synchronisation, excepted if he/she asks explicitly for it. We introduce the notion of *variable WYSIWIS* where the function may evolve between two extremes. The most demanding is the *strict WYSIWIS* function in which a strict synchronisation is asked between shared view (during telepointing and confronting functions). The less demanding is the *lax WYSIWIS* function, which enables a relative loss of synchronisation between users.

Another problem concerns management of continuous media together with the WYSIWIS function. For example, if an object handler plays a video sequence, must this play be perceptible by the other object handlers having the video on their screen? If users ask for telepointing and/or videophone, the play-out must be visible by the subset of participants with a strict synchronisation of different streams of information (telepointers and/or videophone and shared continuous information). But, for an individual projection, this multiplicity of play-out is not to be necessary. So two other notions are introduced: the *absolute WYSIWIS* function in which a play-out of continuous media is perceptible by all users owning this continuous media on their

---

[2]GROVE: GRoup Outline Viewing Editor, a real-time groupeware system proposing a text editor design for use of group of people simultaneously editing.
[3]MERMAID: Multimedia Environment for Remote Multiple Attendee Interactive Decision-making: a system allowing participants to jointly view and process multimedia conference document

shared view, and the *relative WYSIWIS* function in which the projection is not necessary.

4.3 Concurrent access manager

The concurrent access manager collates the interactions of several multimedia cooperation object handlers. This collation is realised by applying the 'collation rules' provided by the collation policy.
Similarly to the shared view manager, it has two groups of *interaction interfaces* ① & ②, two *cooperation management interfaces* ④ and an interface of type ③.

A first group of *object interaction interfaces* ① & ② collates the object interactions arriving from the multimedia cooperation object handler. The other group links the concurrent access manager to the multimedia object. It is used to enable the concurrent access manager to propagate the interactions on the multimedia objects.

It shares with the cooperation information base a *cooperation management interface* ④ for their interactions (e.g. to check user's access right). It has also a *cooperation management interface* ④ for the establishment of links with the multimedia cooperation object handlers, for the collation of the *cooperation management interactions*.

Its collation role must ensure that there is no access conflict on the multimedia objects. If the concurrent access manager does not have functional components that solve such access conflicts, it will execute these interactions in sequence. It can also discard one or both interactions when such conflicts occur. This decision depends on the chosen collating policy.

Once users will raise some level of 'maturity' in the use of cooperative tools, a good definition and specification of the components involved in the presentation of the interactions to the users will reduce the overhead of the mechanisms needed to control concurrent access.

5 Conclusion

In this paper we have described two computational objects that could be integrated in future multimedia cooperative applications to enable synchronous cooperation on multimedia information. The first component, called *multimedia cooperation object handler*, is in charge of the synchronous cooperation functionalities for a particular participant. The second component called *synchronous multimedia cooperation manager* is used to manage in a cooperative way the cooperation of several participants. The synchronous multimedia cooperation manager is in charge for the collation of the interactions of set of users interacting on the shared multimedia object. It presents to some participants the results of the cooperative interactions. It manages an information base containing information that is necessary for cooperation management.

The functionalities of these components can be found in some prototypes, but these prototypes are technology dependent and can hardly inter-operate. The specification and the definition of generic objects for the management of a cooperation activity will ease the development and the maintenance of prototypes and products. For openness and possible wide use in an open distributed heterogeneous environment, the specification of those components is based on the computational language of the Reference Model of Open Distributed Processing (RM-ODP).

6. Classified References

6.1 Standards

1. CCITT X901 and ISO 10746-1.'(CD) Part 1: overview and user model', ISO/IEC/JTC1 SC21/WG7, July 1994.

2. CCITT X902 and ISO 10746-2.'(DIS) Part 2: descriptive model', ISO/IEC/JTC1/SC21/WG7, February 1994.

3. CCITT X903 and ISO 10746-3 '(DIS) Part 3: prescriptive model', ISO/IEC/JTC1/SC21/WG7, July 1994.

4. CCITT X904 and ISO 10746-4.'(CD) Part 4: architectural semantics, specification techniques and formalisms', ISO/IEC/JTC1/SC21/WG7, July 1994.

5. MHEG, 'Coded Representation of Multimedia and Hypermedia Information Objects', ISO/IEC JTC1/SC29/WG12, Juin 1993.

6. Office Document Architecture - Office Document Interchange Format, ISO 8613, April 1988.

7. HyperODA, 'Hypermedia Office Document Architecture', ISO 8613/PDAM 7-10, October 1992.

6.2 Prototypes

8. G. Dermler and K. Froitzheim: 'JVTOS - A Reference Model for a New Multimedia Service'. In the proceedings of 4th IFIP Conference on High Performance Networking 92 (HPN '92), pp. D3/1 - D3/15. Edited by A. Danthine, O. Spaniol. Liege, 1992.

9. M. Weber, H. Biscaia, G. Dermler, P. Drabik, T. Gutekunst, T. Schmidt, E. Ostrowski, N. Pires and Heiner Wolf. 'JVTOS V3.0 - Specification (M34)'. December 31, 1993. Deliverable Number R2060/ETHZ/CIO/DS/P/004/b1.

10. G.Dermler, T. Gutekunst, B. Plattner, E. Ostrowski, F. Ruge and M. Weber, 'Constructing a Distributed Multimedia Joint Viewing and Tele-Operation Service for Heterogeneous Workstation Environments', Proceedings of the Fourth Workshop on Future Trends of Distributed Computing Systems, IEEE Computer Society Press, Lisbon, September 1993.

11. G. Dermler, T. Gutekunst, E. Ostrowski, N. Pires, T. Schmidt, M. Weber and H. Wolf. 'JVTOS - A Multimedia Telecooperation Service Bridging

Heterogeneous'. Proceedings of Broadband Islands '94: Connecting with the End-User. W. Bauerfeld, O. Spaniol and F. Williams (Editors). 1994 Elsevier Science B.V. PP 463-478.

12. S. Cronjaeger, W. Reinhard and J. Schweitzer. 'Functional Components for Multimedia Services'. In proceedings of the IEEE International Conference on Communications ICC'93. Geneva, Swizeland. Nov 23-26, 1993. Pp 1563-1568.

13. M. Altenhofen, J. Dittrich, R. Hammerschmidt, T. Kappner, C. Kruschel, A. Kuckes, T. Steinig. 'The BERKOM Multimedia Collaboration Service'.

14. K. Watabe and S. Sakata, K. Maeno, H. Fukuoka and T. Ohmori, 'Distributed Desktop Conferencing System with Multiuser Multimedia Interface', IEEE Journal on selected areas in communications, Vol. 9, No 4. pp 531-539, May 1991.

15. T. Ohmori, K. Maeno, S. Sakata, H. Fukuoka, and K. Watabe, 'Distributed Cooperative Control for Sharing Applications Based on Multiparty and Multimedia Desktop Conferencing System: MERMAID', C&C Systems Research Laboratories, pp 539-546, 1992.

6.3 Additional Ideas and Concepts

16. H. Lubich and B. Plattner. 'A Proposed Model and Functionality Definition for a Collaborative Editing and Conferencing System'. In Multi-User Interfaces and Applications. S. Gibbs and A.A. Verrijn-Stuart (Editors). Elsevier Science Publishers B.V. (North Holland). IFIP, 1990. PP 215-232.

17. M. Stefik, D.G. Bobrow, G. Foster, S. Lanning, and D. Tatar, 'WYSIWIS Revised: Early Experiences with Multiuser Interfaces', ACM Transaction on Office Information Systems, Vol. 5, No 2. pp 147-167, April 1987.

18. C.A. Ellis, S.J. Gibbs and G.L. Rein, 'Groupware = Some Issues And Experiences', Communications of the ACM, Vol. 34, No 1. pp 39-58, January 1991.

19. I. Greif and S. Sarin, 'Data Sharing in Group Work' In 'Computer-Supported Cooperative Work: A book of readings' Edited by I. Greif. pp 477-508.

20. F. Horn and J.B. Stefani. 'On Programming and Supporting Multimedia Object Synchronization'. In The Computer Journal, Vol. 36, N°1, 1993. pp 4-18.

A Binding Architecture for Multimedia Networks

Aurel A. Lazar*, Shailendra K. Bhonsle** and Koon Seng Lim**

*: Department of Electrical Engineering and Center for Telecommunications Research, Columbia University, New York, NY 10027-6699, email: aurel@ctr.columbia.edu

**: Institute of Systems Science, National University of Singapore, Singapore, 0511.

Abstract: An open architecture that achieves seamless binding between networking and multimedia devices is proposed. The building blocks of the binding architecture consist of a set of interfaces, methods and primitives. The former abstract the functionalities of multimedia networking devices and are organized into a binding interface base. The methods and primitives are invoked for implementing binding applications. The binding architecture is embedded into a reference model for multimedia networking architectures that supports a clean separation between binding interfaces and binding algorithms. Communication between the interfaces of the architecture is supported by CORBA. Public interfaces in the binding interface base are specified using CORBA IDL. The architecture is illustrated with a simple connection management algorithm and an example of computational binding.

1. Introduction

We start by presenting the motivation for our work. This is followed by a review of some of the pertinent literature and a description of the methodology employed for designing a binding architecture for multimedia networks.

1.1 Motivation

Binding is the process of associating (interconnecting) different components of a system. The binding architecture of multimedia networks dictates how its entities are modeled and how these entities are associated with each other in order to provide the user of a service with a "holistic" picture. The architecture itself consists of a binding interface base and binding algorithms. Binding architectures and applications for networking and multimedia computing have been developed for the most part independently. As a result, there is no uniform terminology in these fields: connection management, binding, signalling protocols, etc. are words often used interchangeably.

Connection management in telephone networks is resolved by defining a User/Network Interface (UNI) and a Network/Node Interface (NNI). These interfaces are realized through the Q.93b and CCSS #7 (Common Channel Signalling System), respectively. International standards bodies are considering the CCSS #7 together with the Q.93b interface as the basis for signalling in broadband networks. There are a number of problems with this solution, however.

The UNI and the NNI concepts, introduced in the 60s, rightly recognized that the Customer Premises Equipment (CPE) had a low level of intelligence in comparison with the switching equipment. That has now changed as the customer might possess the latest powerful workstation or parallel machine. In fact, the customer equipment is often at least as intelligent as the switching controllers. Broadband networking requirements for defining and manipulating virtual networks and multicasting are readily modeled as high level objects. It is natural, therefore, to provide higher level language constructs in describing connection management operations. Note that in this context, the UNI/NNI model is akin to a *low level* programming language. Development of signalling protocols based on object-oriented call models does not change this basic assessment.

The Internet community has developed connection management capabilities as part of the TCP/IP suite. Currently, in order to support extensions of the architecture to multimedia services, reservation protocols are being investigated. An evolutionary path towards interworking with future broadband networks is not yet available.

The Interactive Multimedia Association (IMA) is considering proposals for interoperability of distributed multimedia systems. While the networking aspects have not yet been considered, CORBA and IDL have gained wide acceptance for implementing any such systems. The IMA recommended practice is likely to gain wide acceptance in the computer industry.

While there has been considerable work in the individual areas of signalling protocols, object based network architectures and on addressing the issue of interworking of multimedia devices, there has been little work on defining a *seamless architecture* that will integrate all these concerns. It is our belief that such an architecture is needed to support a multitude of applications such as connection management, distributed computing, signalling protocols, etc. The need to design and implement a binding architecture as will be discussed in this paper has been first recognized in [19].

1.2 Related Work

In [11] the facilities required to control and manage multiservice applications in ATM networks are examined in detail. The requirements necessary for dealing with diverse service configurations are defined. As such, issues pertaining to multimedia services and in particular, Quality of Service (QOS), are not addressed.

Another perspective which focuses on specifying the service description of a system rather than its individual components was proposed in [14]. Here, the work concentrates on developing service primitives and their sequence of invocations at service access points for multimedia multiuser services. Architectural and implementation issues were deliberately omitted.

Defining a signalling protocol capable of supporting complex multimedia services is central to the investigations in [26]. Although object-oriented in nature, the model presented limited itself to representing only call related aspects of the network and omits any discussion of the architectural aspects of how these may be realized and integrated within an overall architecture.

The ATM Forum is investigating the applicability of a subset of the Q.93b UNI [7] for broadband applications in the local environment. [12] is representative of this work. Work describing the NNI and the CCSS #7 can be found in [8].

In the areas of protocols that guarantee QOS, one of the more promising developments has been the proposal of two reservation-based protocols for the Internet. The RSVP protocol [29], is a receiver-oriented simplex protocol that can accommodate heterogenous receivers in a multicast group and allow dynamic changes in group membership through maintenance of a 'soft-state' in each node. As opposed to this, the sender-initiated approach is adopted by the ST-II protocol [28]. The protocol uses multiple simplex reservations to create stream-based multicast trees. Since reservation is on a per-tree basis, ST-II cannot accommodate heterogeneous receivers.

The initial version of the Touring Machine project [5] focused on providing a simple, point-to-point desktop video communication service. The second generation system provides APIs for application developers to aid widespread deployment. In both instances, the emphasis was on building a workable prototype so that showcase multimedia applications can be developed rather than on building a generic architecture.

On a different track and much wider in scope, the work by the TINA consortium [3], [4] centers on the development of an Information Networking Architecture that would bring together distributed computing, telecommunications and management standards into a single framework. TINA has not yet addressed the problem of interworking with multimedia devices. The same applies to the substantial contributions put forth by ANSA [2] and the follow up ODP architecture [25].

The methodology developed for binding multimedia objects in Multimedia System Services (MSS) [15] is based on modern foundations of distributed algorithms and software engineering [1], [9]. The MSS proposal is object-oriented. The interfaces are specified in the IDL interface definition language [17]. In order to support interaction among distributed interfaces, MSS depends upon the Object Management Group's (OMG) Common Object Request Broker Architecture (CORBA) [10].

1.3 Methodology

An ad-hoc approach for interconnecting a multimedia system such as MSS to a broadband network specified via an UNI would be to present it with the Q.93b interface. This approach, however, exhibits the limitations already mentioned in section 1.1. We, therefore, advocate a solution based on a different modeling paradigm.

In this paradigm, the binding architecture and binding applications are clearly separated. The architecture itself is *open* and hence possibly subject to standardization efforts. The binding applications, however, might not be. Binding entities in the architecture are modeled as communicating objects. As in MSS, CORBA provides the high level location independent communication facilities. This allows for a seamless binding environment between the network and the multimedia resources. Overall, binding operations exhibit a much lower level of complexity.

This paper is organized as follows. In section 2 the modeling framework provided by the Extended Integrated Reference Model is briefly reviewed and binding within this model discussed. The architectural model of the binding architecture is presented in section 3. The relationship between the binding architecture and other ongoing work is also discussed. The binding interface base, including the interface inheritance diagram and some of the interface definitions are presented in section 4. Section 5 describes the binding methods and primitives. Examples of binding applications are given in section 6. Conclusions and future directions are given in section 7.

2. Modelling Framework

In this section the framework for binding architectures provided by the Extended Integrated Reference Model (XRM) is presented. The XRM is discussed in section 2.1. In section 2.2 the positioning of binding within the XRM is described.

2.1 The Extended Integrated Reference Model (XRM)

In parlance of network architectures, Figure 1 is an abstract representation of the *Extended Integrated Reference Model (XRM)* [19]. The XRM models the communications architecture of broadband networks and multimedia computing platforms. The foundations for the operability of multimedia computing and networking devices is the same. Both classes of devices can be modeled as producers, consumers and processors of media. The only difference appears to be in the overall goal that a group of devices is set to achieve in the network or the multimedia platform.

The restriction of the XRM to broadband networks is called the Integrated Reference Model (IRM) [18]. The IRM incorporates monitoring and real time control, management, communication, and abstraction primitives that are organized into the *Traffic Control Architecture*, the *Management Architecture*, the *Information Transport Architecture* and the *Telebase Architecture*, respectively. The subdivision of the IRM into the Management and the Traffic Control Architectures on the one hand, and the Information Transport Architecture on the other, is based on the principle of separation between controls and communications. The separation between the Management and the Traffic Control Architecture is primarily due to the different time-scales on which these architectures operate.

The Integrated Reference Model is organized into five planes that model the above architectures (Figure 1). The Management Architecture resides in the network management or N- plane, and covers the functional areas of network management, namely, configuration, performance, fault, accounting and security management. Manager and agents, its basic functional components, interact with each other according to the client-server paradigm. The Traffic Control Architecture consists of the resource control, or M-, and the connection management and control, or C-, planes. The M-plane comprises the entities and mechanisms responsible for resource control, such as cell scheduling, call admission, and call routing; the C-plane those for connection management and control. The Information Transport Architecture is located in the user transport or U-plane, and models the protocols and entities for the transport of user information. Finally, the Telebase Architecture resides within the Data Abstraction and Management or D-plane, and implements the principles of data sharing for network monitoring, control and communication primitives, the functional building blocks of the N-, M-, and C- and U-plane mechanisms. (A mechanism is a functional atomic unit that performs a specific task, such as setting up a virtual circuit in the network [24].)

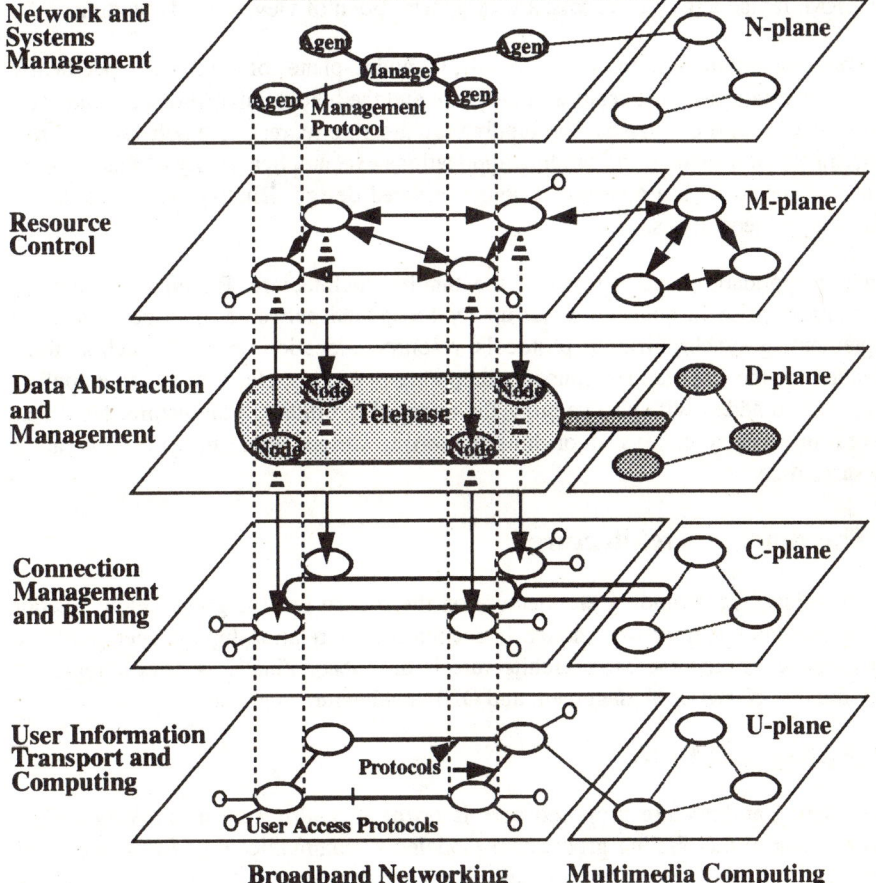

Fig. 1. The Extended Integrated Reference Model (XRM).

The restriction of the XRM to the multimedia computing platform has a similar functionality as the IRM. The N-plane includes system management functionality, and the

M-plane includes process scheduling, memory management, routing (when applicable), admission control and flow control. The D-plane also contains objects modeling multimedia devices, the C-plane binding functionality, and the U-plane transport of user information within the Customer Premises Equipment.

2.2 Binding within the XRM

Binding requirements arise in each of the planes of the XRM. However, dynamic binding requirements are particularly demanding in the C- and M- planes. In order to better understand and fullfil these requirements, a *separation principle* between the binding architecture and applications running on top of it is defined. This separation principle gives a clear focus towards what should be and what needs to be standardized within the XRM. It also allows us to take a very general point of view towards binding.

The binding architecture resides in the M-, D- and C-planes of the XRM. Specifically, the binding interface base resides in D- plane and the binding algorithms execute from within the M- and C- planes. The binding architecture represents a software environment on top of which all the binding applications execute. Scalability of this architecture is achieved with a distributed object-oriented design. Binding interfaces can be added as the need arises.

Binding applications run on top of the binding architecture. Examples of binding applications arise in connection set up for broadband networks, distributed systems implementing synchronization protocols, resource allocation protocols such as those intended for the Internet, multimedia computing platforms, etc. New binding applications can be added without changing the underlying binding architecture. Note that, several proprietary binding algorithms supporting various applications can operate at the same time.

3. The Binding Architecture

An overview of the binding architecture on the system level is given below. Section 3.1 presents the architectural model. In section 3.2 the relationship between our binding architecture and the MSS architecture is described. Finally, a brief comparison with the OSI Network Management and ODP architecture is given.

3.1 Architectural Model

The binding architecture proposed here is open: all multimedia networking entities participating in the binding process are modeled as communicating objects with well defined interfaces that can be externally invoked. Interface methods and some global primitives are used for these invocations. Binding algorithms operate upon these interfaces.

The interfaces are realized as objects modeling resources such as switches, links, multimedia devices, etc. All interfaces reside in a data repository called the Binding Interface Base (BIB). More abstractly, the BIB provides multimedia networking abstractions for producers, consumers, and processors of media. Interfaces in the BIB are defined using the CORBA IDL (Interface Definition Language) specification language. The BIB containing all binding interface instances (called binding objects) reside in the D-plane of the IRM. CORBA provides naming facilities to locate interface implementations and invoke methods. A factory is used to instantiate an interface and one of the embedded methods within the interface can be invoked to delete the interface instance.

Public methods are visible to different "multimedia networking clients" who can invoke them. Multimedia networking clients are "clients" in CORBA sense. For example a binding algorithm that invokes binding interfaces is a multimedia networking client as is the user of a "service" that invokes the BIB interfaces and binding algorithms. Scalability of the binding architecture is readily achieved by adding new interfaces or by upgrading existing ones. The addition of new binding algorithms can also be easily accomplished.

The components of the binding architecture consisting of the binding interface base and the binding algorithms are depicted in Figure 2. This figure shows the distributed nature of BIB interactions, and the distributed interactions amongst different binding algorithms.

3.2 Relationship to Binding for Multimedia

The reader has probably recognized by now a number of similarities between our binding architecture proposal with the Multimedia System Services [15] platform considered by the Interactive Multimedia Association (IMA). Recall that MSS constitutes a framework of "middleware" — system software components lying in the region between the generic operating system and specific applications. Its goal is to provide an infrastructure for building multimedia computing platforms that support interactive multimedia applications dealing with synchronized, time-based media in a heterogeneous distributed environment. It is under evaluation by the IMA and is expected to become a recommended practice within the computer industry.

How does the MSS framework fit into the XRM? Here we distinguish between facilities for creating and removing objects as well as binding operations. In MSS a number of interfaces have been defined to enable both the creation and destruction of objects that participate in the binding process. Creation and destruction operations are D-plane native. Binding algorithms on the other hand are C- and M- plane visible. Sizing of virtual resources derived from QOS requirements is exported through M- and N-plane interfaces.

What are the differences between our architecture and the architecture of the MSS? We believe that only the binding interface base should be standardized although there might be a need to standardize some of the binding algorithms and applications for higher level interoperability. We feel that management and control tasks, such as QOS control and management that the current MSS architecture proposes to fullfil, are best modeled as M- and C- plane binding algorithms. The fundamental abstractions that these algorithms operate upon are modeled as BIB interfaces.

3.3 Relationship to the OSI Network Management Architecture and ODP

There are also important conceptual similarities between our binding architecture and the OSI management architecture [6]. As in OSI management, we propose to have an object-oriented model for the entities of interest, a standard communication support and a well defined set of interfaces that support basic (binding) operations. Both, the OSI management architecture and the binding architecture, are separated from the management and binding applications, respectively. There are of course a number of differences, the main one being that the OSI management architecture is entirely centralized whereas the binding architecture discussed here is entirely decentralized. Another major difference is the time scale on which these architectures operate.

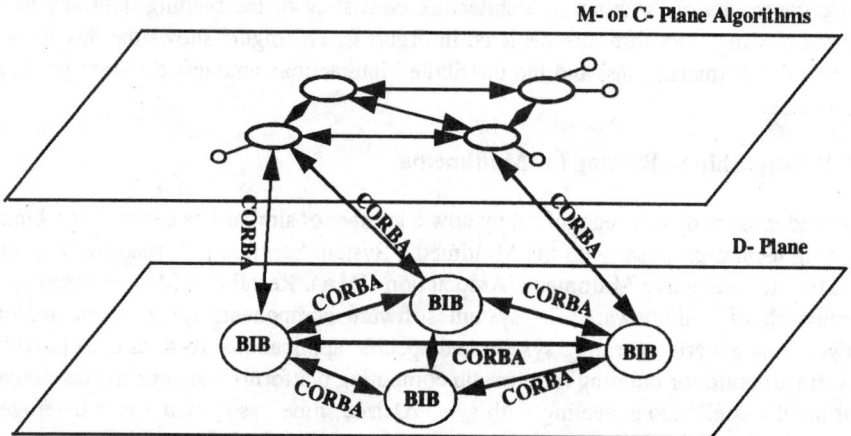

Fig. 2. Distributed Binding Algorithm Interacting with Distributed BIBs

We note a number of similarities with the ODP architecture as well. For example, the ODP information model, for expressing the meaning of information and information processing tasks is analogous to our definition of the D-plane; the computational model for expressing functional decomposition of the systems into distributable units with well-defined interfaces corresponds directly to our standard object-based implementation; the engineering model for describing components and structures needed in support of distribution is expressed in our proposed use of RPC and CORBA both of which are stable and well accepted; and finally the technology model for describing the makeup of a system in terms of components that conform to appropriate standards fits in nicely with our emphasis on CORBA and interworking with MSS.

4. The Binding Interface Base

In this section the inheritance diagram underlying the BIB is discussed. We briefly present the structure and the semantics of some of the BIB interfaces.

4.1 Interface Inheritance Diagram

The interface inheritance diagram for the BIB is depicted in Figure 3. In the future specialized interfaces will inherit from the generic interfaces, thus extending the diagram horizontally. When new interfaces, such as those shown with shaded lines are added to the architecture, the diagram is simply extended vertically. As with the OSI network management MIB, the BIB resides in the D-plane of IRM. Both are integral parts of the architecture of the Telebase.

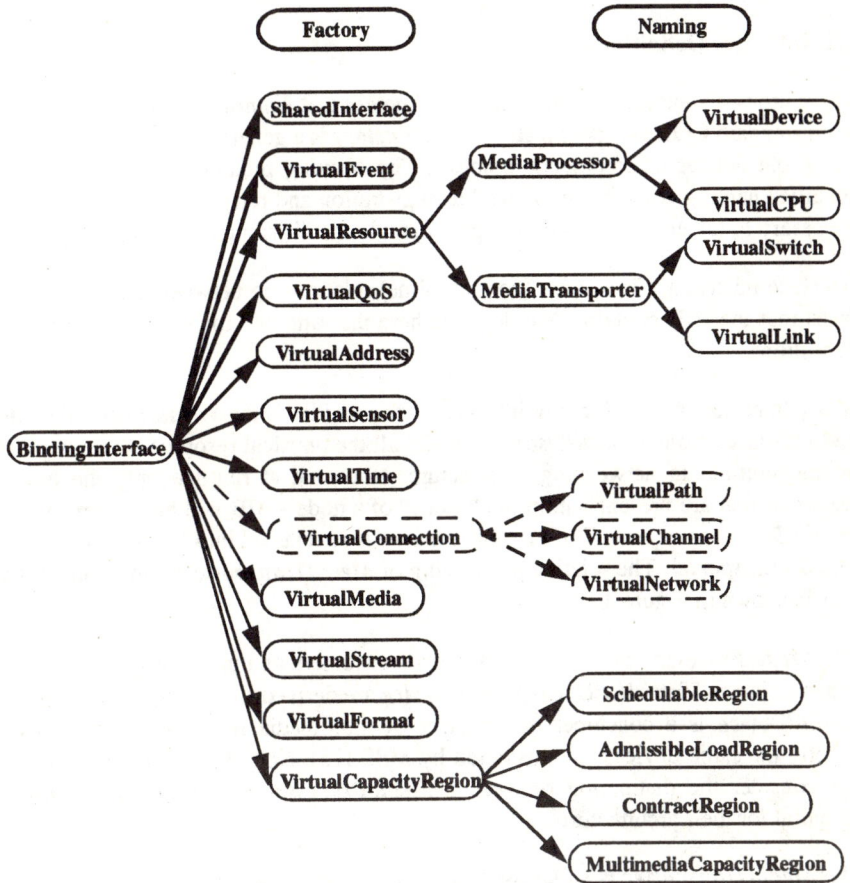

Fig. 3. Interface Inheritance for Binding Interface Base

The BIB interfaces are defined in CORBA's [10] Interface Definition Language [17]. Adopting CORBA terminology, they are realized as object-implementations (servers). An instantiation of an interface is an object within these object-implementations. Since CORBA's view of interfaces is analogous to a "class" in object-oriented programming and since we only employ an object-oriented implementation of interface definitions, the term "class" will be used interchangeably with "interface" and "object" interchangeably with the notion of an "instantiation of an interface".

Interface implementations contain many object management specific public interfaces like creation and deletion of objects. These will be of no concern here. Our *primary goal* is to show what are the various public interfaces pertinent to binding operations and what are the *local states* that they manipulate. The following section briefly describes the essence of our design. A more detailed specification is available in [21].

4.2 Interface Definitions

There are two standard interfaces namely *Factory* and *Naming*. They are both derived from standard CORBA specifications. The *Factory* is a generic component that instantiates and deletes other objects remotely. The *Naming* module deals with "CORBA specific" naming conventions as well as registration and retrieval of interfaces as it is necessary to locate and invoke the right object in the distributed environment.

Interface inheritance is a static (compile-time) relationship between the various interfaces that the BIB contains. We describe here the rationale for some of the interface definitions.

A key interface that the binding interface base provides is the *VirtualResource*. Inherited interfaces from *VirtualResource* model all the physical resources that are present in the multimedia networking architecture. Note that at runtime only the physical resources that are present within the "scope" of a node's BIB will have corresponding *VirtualResource* objects. The *VirtualResource* is subclassed into *MediaProcessor* and *MediaTransporter*. The relative positioning of *MediaTransporters* and *MediaProcessors* is shown in Figure 4.

The *MediaProcessor* models all producers, consumers and processors of media. The generic classes *VirtualCPU* and *VirtualDevice* are derived from *MediaProcessor*. The *VirtualDevice* is a consumer or producer of multimedia information. It represents exactly the same devices as considered by MSS [15]. Note, however, that MSS does not consider the distinction between a *MediaProcessor* and *MediaTransporter* as essential for their architecture.

The *MediaTransporter* is subclassed into *VirtualSwitch* and *VirtualLink*. A physical switch consists of a set of multiplexers interconnected through a switch fabric with an associated control unit. The *VirtualSwitch* represents the "local" control elements of a switch fabric. In our model these control elements are distinct from the control elements and the intelligence associated with the output multiplexers. The multiplexers located at the output ports of the switch are modeled by the *VirtualLink* interface. Thus, the *VirtualLink* models the cell and call resources consisting of output buffers and links whereas the *VirtualSwitch* models the resources associated with the VP/VC routing tables. The distinction between the *VirtualSwitch* and the *VirtualLink* is depicted in Figure 5.

Another important interface contained in the BIB is the *VirtualCapacityRegion*. This interface characterizes the capacity of multimedia networking resources with QOS guarantees. With every *VirtualResource* a *VirtualCapacityRegion* is associated. Its state is typically represented by an *OperatingPoint* within this region. The axes of the *VirtualCapacityRegion* represent the QOS classes (which in turn may be related to QOS parameters). Different resources may have their *VirtualCapacityRegion* axes labeled differently. For example with a *VirtualCPU* a *MultimediaCapacityRegion* is associated whereas a *VirtualLink* is characterized by its *SchedulableRegion*. It is the responsibility of the underlying resource control mechanisms to map "service" class information into appropriate traffic class information and vice versa. The *VirtualQoS* interface is provided to do this translation.

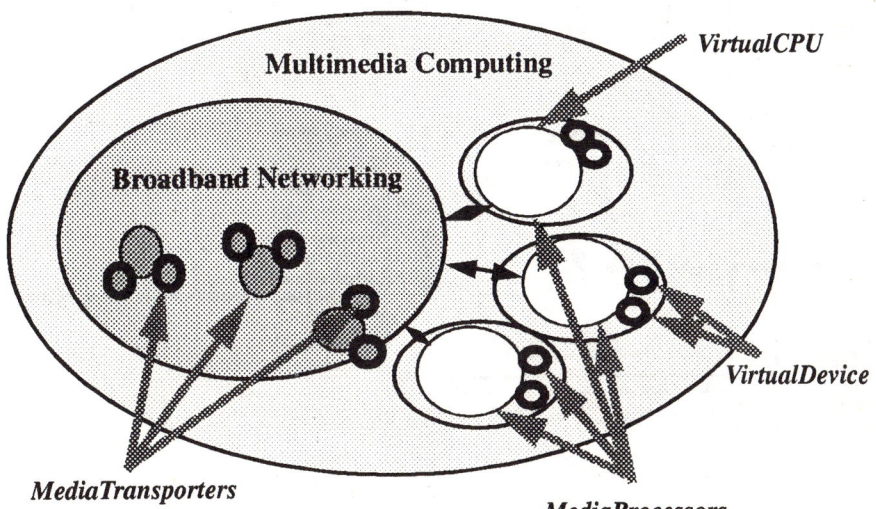

Fig. 4. Relationship between *MediaProcessor and MediaTransporter*

The *VirtualAddress* interface provides translation of various types of addresses (ATM-Forum addresses, Generic addresses, CORBA related addresses) into each other, whenever possible. Different inherited interfaces may add/extend address translation facilities of this interface.

There are two generic interfaces that can be inherited from to provide event forwarding services as well as "monitoring/sensing" parameters of interest. *VirtualEvent* provides for the definition, registration and forwarding of events to any number of client algorithms. The *VirtualSensor* interface provides for monitoring activities of interest within the binding architecture. Both of these interfaces are for "resource" control purposes and hence are "different" in purpose than N-plane specific management event forwarding and monitoring capabilities. Since they are going to be used by fast timescale control algorithms, they must be kept lightweight and efficient and their use must be guarded by some overall real-time constraints.

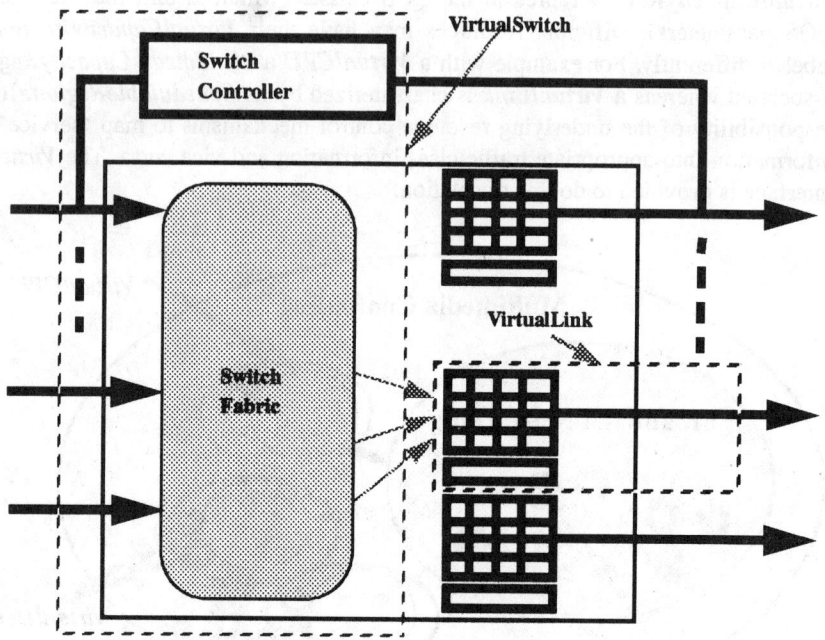

Fig. 5. VirtualSwitch and VirtualLink

Real-time requirements that the various binding algorithms may have are supported by the *VirtualTime* interface that provides a local clock. This interface supports both synchronous and asynchronous timeouts. Time related *VirtualEvents* may be associated with the *VirtualTime* interface.

There are three media related interfaces: *VirtualMedia, VirtualStream, and VirtualFormat*. *VirtualMedia* defines the properties of the various multimedia: video traffic, CD Quality traffic, etc. Associated media format and stream-properties have been abstracted into two related interfaces, *VirtualStream* and *VirtualFormat*, respectively. These interfaces are defined for use by *VirtualResource*, especially *MediaProcessor*, and also for annotating *VirtualCapacityRegion* axes.

SharedInterface is an interface that allows for generically manipulating other interfaces. It is also capable of manipulating "state expressions" to answer certain resource state specific queries. We are currently in the process of defining this interface.

CORBA integrates the normal remote procedure call (RPC) paradigm with "event" handling facilities. It provides standard RPC protocol related events and one can also define events through using its "exception" and "raises" primitives. In [21] it is shown how interface definitions incorporate standard as well as user defined events.

5. Binding Methods and Primitives

In this section binding methods and primitives are discussed. There is a distinction between methods and primitives. The former are interface methods and are invoked by CORBA clients. Apart from methods which are associated with individual interfaces, there are also global "primitives" of the binding architecture. These primitives are implemented as interface methods as depicted in Figure 6. Primitives are global atomic units of execution.

There is a need to identify and classify separately important methods that *must be present* in the BIB interfaces. Specifically since we are controlling certain devices by changing their states, there is a need for imposing consistency of usage by "different" algorithms that coexist in the multimedia networking architecture. In this section we briefly describe the essential elements of these. Details can be found in [21].

---

1. **getState(in BI bObject, out State biState);**
2. **setState(in BI bObject, in State newState);**
3. **getReservationState(in BI bObject, out ReservedState rState);**
4. **setReservationState(in BI bObject, in State newState);**
5. **commitReservedState(in BI bObject);**
6. **evalStateExp(in TypeExp type, in StateExp stateExp, out Boolean result, out State rState);**
7. **parseStateExp(in TypeExp type, in StateExp stateExp, out long result);**
8. **executeMethod(in BI bObject, in FunctionPtr func, out sequence of OutPars outparams, out RetType returnedValues);**
9. **queryBI(in BI bObject, out sequence of BAttributes bAttrs);**

Fig. 6. Primitives Used for Binding

The public methods are used for porting the BIB onto physical devices, bootstrapping, and attaching appropriate transport protocols for supporting CORBA RPC communications. They consist of atomic primitives for getting and setting states.

5.1 Binding Primitives

There is a set of important primitives that must be present either as methods in *VirtualResource* and many other derived interfaces or as methods in *SharedInterface*. These primitives essentially locate and manipulate states, reserved states, and other attributes of an interface.

Primitives for State

The concept of current state has been associated with all interfaces that model physical resources. The primitives *getState()* and *setState()* extract the current state and set the current state, respectively. For example separate state interfaces can be attached in the same BIB for a *VirtualLink*. It is the responsibility of *getState()* and *setState()* primitives to maintain the consistency between various such state interfaces.

Primitives for Reserved State

The reserved state is an important concept in our architecture. It indicates certain resources that have been locked by a "client". Such locking might occur for enforcing consistency requirements on the part of the client. There are two methods for setting and getting reserved states: *getReservationState()* and *setReservationState()*.

EvalStateExp() Primitive

This primitive takes as an input a *StateExp* and evaluates it to either a *Boolean* (true or false) or to another *State*. A *StateExp* could be either a predicate or an expression involving state variables and state operators. It is important to note that the *StateExp* may contain an evaluation of expressions with variables outside the scope of a given BIB. Note also that there is a limit to the expressive power of *StateExp* as it is not intended that binding algorithms evaluate very complex functions.

ExecuteMethod() and QueryBI() Primitives:

These primitives provide means to implement generic "method execution" and getting the values of certain attributes indirectly.

6. Examples of Binding Algorithms and Applications

The example given in section 6.1 describes how the binding architecture can support connection management whereas the example in section 6.2 describes how a distributed computing application can be supported.

6.1 Connection Management with QOS Guarantees

In this section we briefly describe an example of a simple connection management algorithm that can be built on top of the binding architecture of a broadband ATM network. The emphasis of the exercise is not on the connection management algorithm itself, but on illustrating how network binding can be implemented within our architecture.

Figure 7 shows the interface inheritance diagram for this example. We define an interface called *VirtualConnection*. A physical connection exists as a set of interconnected entries in appropriate tables (e.g., routing tables). A corresponding interface in the BIB represents these physical entities. The *VirtualConnection* interface is subclassed into *VirtualPath, VirtualChannel, and VirtualNetwork* modeling the "virtual path", the "virtual channel" and the "virtual network", respectively. These interfaces also provide methods to setup/release connections on ATM networks. The realizations of *VirtualPath, VirtualChannel, and VirtualNetwork* as objects are called network applications. Applications with QOS guarantees are called services.

Fig. 7. Interface Inheritance Diagram for Connection Management.

Connection management algorithms create the above network services (i.e., the *VirtualPath, VirtualChannel, and VirtualNetwork* objects). The algorithms reside on the C-plane, operate on objects residing in the D-plane and interact with algorithms on the M-plane only through the D-plane objects. A "connection manager" in the C-plane can implement any connection management algorithm (in addition to the ones that adhere, e.g., to the Q.93b or SS7 specifications).

Connections with guaranteed QOS (i.e., network services) can be established because the states of local D-plane interfaces are bound to multimedia networking devices with QOS guarantees as discussed in a separate paper [22]. QOS parameters are used to define service classes [7]. As already mentioned, the mapping to traffic classes [23] or other QOS parameter specifications is the prime responsibility of the *VirtualQOS* interface.

Figure 8 shows the M-, D- and C- planes of the XRM and the respective connection management related objects (see also Figures 1, 2 and 3) that reside on them. As indicated in the figure, algorithms that provide call and connection level abstractions are built on top of the "local standardized interfaces" lying on the D-plane of the IRM.

In this example, there is a request for a new connection to be setup between switches A and B. When the request is received by the Connection Manager (the client) on the C-Plane, several actions are performed. At first, the Connection Manager will request from the *Route* object (on the D-plane) a route from switch A to B. After the Connection Manager has obtained the route, it uses the information to poll each intermediate

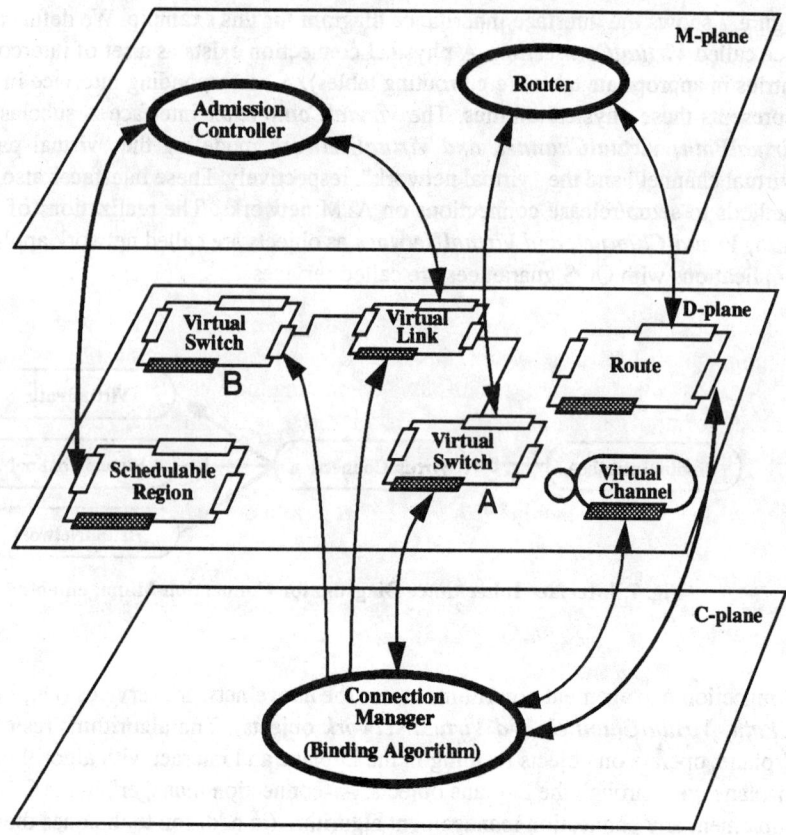

Fig. 8. Binding for Connection Management with QOS Guarantees

node's *Virtual Link* with a link access request. Based on the information provided by the schedulable region associated with the given link, the link admission control method determines whether the call will go through or not. Once the Connection Manager is assured that the call can be accepted by all links along the route, it calls the interface of the *Virtual Switch*es and/or *Virtual Link*s representing the resources along the selected route, requesting a connection to be setup. (For fast call set up this step can be combined with the previous ones.) Because the *Virtual Switch*es and *Virtual Link*s contain abstract representations of real physical objects such as switch controllers, virtual path controllers, virtual circuit controllers and link controllers (not shown for lack of space), they can setup the physical connection by simply modifying the values of some of these objects (in this case, to set up the virtual channel, the connection manager invokes the appropriate method of the *Virtual Switch* and *Virtual Link* object representing the corresponding physical switches and links between points A and B).

At this stage, the call setup is complete and the Connection Manager's role is over. Whenever a new call is accepted into the system, the operating points of the *Virtual*

Switches and *Virtual Link*s along the path of the call changes. After a call disconnect request the connection manager executes the reverse operations.

Once again, note that the example above merely outlines the implementation of a class of possible connection management algorithms. Other schemes supporting, for example, the Q.93b signalling interface can similarly be implemented using this architecture. A general methodology for designing binding applications is given in [27].

6.2 Parallel Virtual Machine

In this section, we further illustrate the generality of our architecture with an example of computational binding. Again, the emphasis is not on the application itself, but on the fact that even generalized distributed computing tools can be easily built on top of our architecture. The Parallel Virtual Machine is an example of a binding application that can be built upon our binding architecture.

The Parallel Virtual Machine (PVM) [13] is a public domain distributed computing software library developed at the Oak Ridge National Laboratory for facilitating the development of general purpose distributed computing applications. Essentially, PVM presents a reliable connectionless data service interface to applications, thereby freeing them from the concerns of the underlying network. PVM also provides a registration facility that allows applications to register themselves with a specified name. This provides flexibility for reconfiguration because applications are identified only through their registered names and not by their locations (such as a network address). Note that the binding between a name and its associated instance is static, i.e., registration is performed only once at application start-up time and cannot be subsequently changed. Several other useful service primitives are also provided for the convenience of distributed system developers. These include facilities for synchronizing parallel applications, primitives for configuration management, process control and group management.

Figure 9 shows the interface inheritance tree for PVM. A *PVMService* interface generic to all PVM services is defined. Within this, three specific interfaces are further specified. The *SynchronizationService* defines an interface of the object providing synchronization to PVM clients. Clients that wish to be synchronized send requests to the *SynchronizationService* object and wait for a reply. When the last client has sent the request, the *SynchronizationService* signals to all the waiting clients. The *ConfigurationService* object allows clients to obtain information about the system configuration (e.g., what other clients are running). Finally, the *ProcessManagementService* object allows PVM processes to manipulate other PVM processes. For example, a client may want to spawn off child processes to perform some task and destroy them after that.

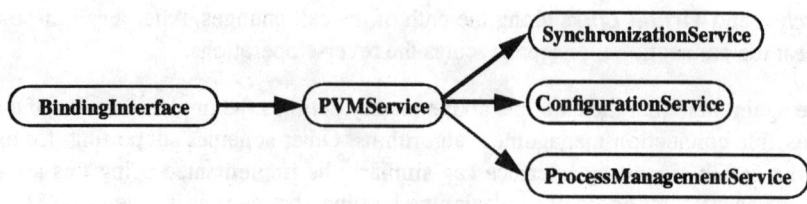

Fig. 9. Interface Inheritance for PVM

Figure 10 shows the implementation of clients, synchronization, process control and configuration management services as objects. Note that PVMs registration service is now not explicitly required because CORBA automatically provides a naming service to all objects. This allows greater flexibility than the PVM name registration because the association is not static and objects may be freely moved. As each client is now an object, special primitives for passing data between themselves are no longer required. Client objects can simply call each other supported by the respective standard interfaces. In a similar manner, one or more client objects can call on a service object for the desired service to be performed. In Figure 10, suppose Client A wants to synchronize with Client B. A first informs the *Synchronization Service* object and then goes to sleep. When B calls the *Synchronization Service* object, A is notified. Similarly, when Client A wants to spawn off a child process, Client C, it requests the *ProcessManagementService* to execute an UNIX 'remote shell' command to start up the child process at an appropriate network node.

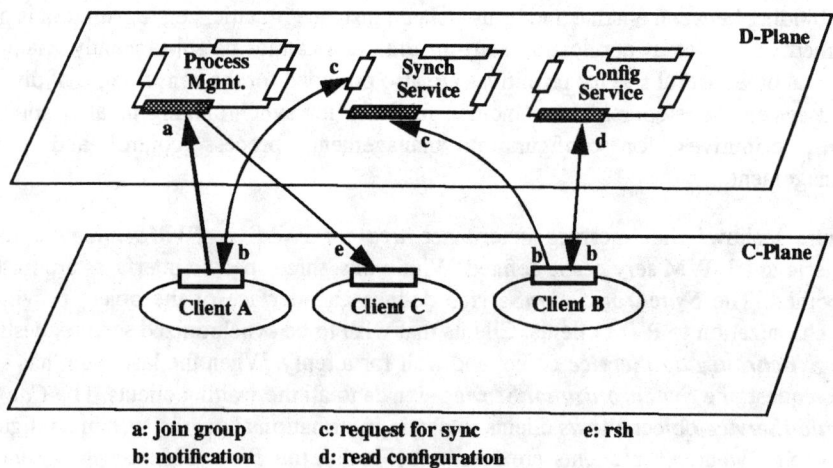

Fig. 10. PVM as an example of Computational Binding

PVM is an example of what we call computational binding. It can be implemented as shown directly above our binding architecture. The various interfaces that implement binding operations in PVM, can be derived directly from the BIB interfaces of the binding architecture. PVM will require certain C- and M- plane specific binding algorithms. The implementer has the flexibility to implement these above the binding interface base.

7. Conclusion and Future Directions

The simplicity of our binding architecture is intentional. The idea was to present the concept of binding in the simplest possible way and illustrate the advantages. In the future more sophisticated BIB interfaces can be added for modeling new broadband network devices and concepts. The design of the binding methods and primitives is also evolutionary and we have only described a bare minimum of these.

We have intentionally left out the description of many "object management" related primitives like interface instantiation and deletion, interface location and interface migration, etc. These are standard activities that many object based systems provide and we have been emphasizing that by using CORBA, we have comfortably bypassed many of these issues and concentrated mainly on multimedia networking requirements.

The advantages of our binding architecture are manifold. Firstly, by providing open interfaces such an architecture "naturally" satisfies the requirements for defining multimedia services both within the network and the multimedia computing platforms. Secondly, our architecture facilitates guaranteeing end-to-end QOS as it seamlessly supports cooperation among distributed algorithms. Thirdly, it provides efficient ways of supporting distributed binding algorithms. For example, a connection establishment algorithm could be carried out either sequentially or in parallel.

Our binding architecture, defined by the BIB and binding algorithms, is conceptually similar to the MIB and CMIS/CMIP protocol of the OSI management architecture. Hence, the integration of the control architecture with the management architecture of multimedia networks becomes greatly simplified. By putting the BIB and the MIB into the Telebase (D-plane), the sharing of data among the C-, M- and N-planes becomes manageable. Using IDL and GDMO for representing information in the BIB and MIB, respectively, further simplifies this task.

As expected, many issues still remain open or have been simply left out because of space limitations. Currently, an implementation for validating our design is underway.

References

[1] Andrews, R.G., "Paradigms for Process Interaction in Distributed Programs", *ACM Computing Surveys*, Vol. 23, No. 1, March1991, pp. 49-90.

[2] ANSAware Version 4.1 Manual, Architecture Projects Management Ltd., Cambridge, UK, May 1992.

[3] Appeldorn, M., Kung, R. and Sarraco, R., "TMN + IN = TINA", *IEEE Communications Magazine*, March 1993, pp. 78-85.

[4] Barr, W.J., Boyd, T. and Inoue, Y., "The TINA Initiative", *IEEE Communications Magazine*, March 1993, pp. 70-76.

[5] Bellcore Information Networking Research Laboratory, "Touring Machine System", CACM, Vol. 36, No. 1, Jan. 1993, pp. 68-77.

[6] Black, U., Network Management Standards, McGraw-Hill Inc., New York, NY, 1992.

[7] CCITT: Recommendation I.413, "B-ISDN User Network Interface", Geneva, 1991.

[8] CCITT: Recommendation Q.761, "Functional Description of the ISDN User Part of Signalling System No. 7," Blue Book, Fascicle VI.8, Geneva, 1989.

[9] Chin, R.S. and Chanson, S.T., "Distributed Object-Based Programming Systems", *ACM Computing Surveys*, Vol. 23, No. 1, March1991, pp. 91-124.

[10] Common Object Request Broker: Architecture and Specification. OMG document 91-12-1.

[11] Crutcher, A. L. and Waters, A. G., "Connection Management for an ATM Network", *IEEE Network*, November 1992, pp. 42-55.

[12] Gaddis, M.E., Bubenick, R. and J.D. DeHart, "A Call Model for Multipoint Communication in Switched Networks", Proceedings of the *International Conference on Communications*, 1992.

[13] Geist, A., Beguelin, A., Dongarra, J., Jiang, W., Manchek, R. and Sunderam, V., PVM 3.0 User's Guide and Reference Manual, Oak Ridge National Laboratory, Tennessee, February 1993.

[14] Heijenk, G. J., Hou, X. and Niemegeers, I. G., "Communication Systems Supporting Multimedia Multi-user Applications", IEEE Network, Vol. 8, No. 1, Jan/Feb 1994, pp. 34- 44.

[15] HP, IBM and SunSoft, "Multimedia System Services Architecture", Response to the *Multimedia System Services Request for Technology of IMA*, June 1, 1993.

[16] Hyman, J.M., Lazar, A.A. and Pacifici, G., "VC, VP and VN Resource Assignment Strategies for Broadband Networks", Proceedings of the *Workshop on Network and Operating Systems Support for Digital Audio and Video*, Lancaster, United Kingdom, November 3-5, 1993, pp. 99-110.

[17] IDL C++ Language Mapping Specification - Joint Submission to the Object Request Broker 2.0 Task Force's C++ Request for Proposals by Hewlett-Packard, IONA, and SunSoft. OMG document number 93-4-4.

[18] Lazar, A.A., "A Real-Time Control, Management and Information Transport Architecture for Broadband Networks", Proceedings of the *1992 International Zurich Seminar on Digital Communications*, March 16-19, 1992, pp. 281-296.

[19] Lazar, A. A., "Challenges in Multimedia Networking", Proceedings of the *International Hi-Tech Forum*, Osaka, Japan, February 24-25, 1994, pp. 24-33.

[20] Lazar, A.A., "A Research Agenda for Multimedia Networking", Proceedings of the *Workshop on Fundamentals and Perspectives on Multimedia Systems*, International Conference Center for Computer Science, Dagstuhl Castle, Germany, July 4-8, 1994.

[21] Lazar, A. A., Bhonsle S., Lim, K. S., "Specification of the Binding Architecture, Version 1.0", July 1994.

[22] Lazar, A.A., Ngoh, L.H. and Sahai, A., "Multimedia Networking Abstractions with QOS Guarantees", *Technical Report # 375-94-22*, Center for Telecommunications Research, Columbia University, New York, NY 10027-6699, August 1994.

[23] Lazar, A.A. and Pacifici. G., "Control of Resources in Broadband Networks with Quality of Service Guarantees", *IEEE Communications Magazine*, Vol. 29, No. 10, October 1991, pp. 66-73.

[24] Lazar, A.A. and Stadler, R., "On Reducing the Complexity of Management and Control of Future Broadband Networks", Proceedings of the *Workshop on Distributed Systems: Operations and Management*, Long Branch, NJ, October 4-6, 1993.

[25] Leopold, H., Coulson, G., Frimpong-Ansah, K., Hutchison, D. and Singer, N., "The Evolving Relationship between OSI and ODP in the New Communications Environment", *Presented at 2nd RACE International Conference on Broadband Islands*, Athens, Greece, June 15-16, 1993.

[26] Minzer, S., "A Signalling Protocol for Complex Multimedia Services", *IEEE Journal of Selected Areas in Communications*, Vol. 9, No. 9, Dec. 1991, pp. 1383-1394.

[27] Pacifici, G. and Stadler, R., "Separating Policy Implementation from Policy Execution: A Paradigm for Resource Management in Broadband Networks", Proceedings of the Network Operations and Management Symposium, Kissimmee, FL, February 1994.

[28] Partridge, C. and Pink, S., "An Implementation of the Revised Internet Stream Protocol (ST-II)", *Journal of Internetworking: Research and Experience*, Vol. 3, No. 1, March 1992.

[29] Zhang, L., Deering, S., Estrin, D., Shenker, S. and Zappala, D., "RSVP: A New Resource ReSerVation Protocol", *IEEE Networks Magazine*, September 1993.

Implementation of a End-to-End Quality of Service Management Scheme

Linda Fédaoui[1], Aruna Seneviratne[2] and Eric Horlait[1]

[1] Laboratoire MASI, Université Pierre et Marie Curie
4, place Jussieu, 75252 Paris Cedex 05, France
e-mail: fedaoui or horlait @masi.ibp.fr
[2] School of Electrical Engineering, University of Technology, Sydney
P.O.Box 123, Broadway, NSW 2007, Australia
e-mail: aps@ee.uts.edu.au

Abstract. In this paper, an implementation framework of a generic end-to-end QoS management scheme is presented. Then, the compliance of the proposed scheme to the ODP (Open Distributing Processing) reference model is verified, by illustrating the correspondence of the various components of the proposed scheme to RM-ODP. The implementation of the end-to-end QoS scheme on a micro kernel based system is then described, with details of all the interfaces and system components. Finally, the applicability of the proposed scheme is demonstrated by applying it to a video on demand application, and carrying out a brief performance analysis.

1 Introduction

It is likely that the next generation of applications will involve multiple media running on general purpose personal computers interconnected via public and private communication networks. Furthermore, the requirements of these applications will change during the life time of application's execution (session), and will frequently utilize data which are distributed over a wide geographical area. For example, in a video on demand application, the video storage server will be located outside the premises of the client. Moreover, during a session, the client will also want to control the viewing by using fast forward, rewind or pause functions. The control operations such as fast forwarding will place different requirements on the system components to the normal playback operation such as requiring more bandwidth and not requiring error correction, and synchronisation between audio and video data streams.

These distributed multimedia systems will rely on three system components for their operation, namely the host operating system for the management of end system resources, host communication subsystem for providing end-to-end data delivery, and the interconnecting communication network for the transportation of data. The correct operation will depend on the availability of adequate resources and functionality within these three system components. However, the determination of the required functionality and resources, and the availability of them is complicated due to (a) the changing requirements of the applications during a session and (b) heterogeneity of the system components. Under these conditions, it is necessary to coordinate the available resources and tailor the functionality of these three system components to suit an individual application before it can be launched.

It is becoming increasingly recognized that this coordination is best achieved through the notion of application Quality of Service (QoS)[2,20,21,22]. Furthermore, numerous schemes for providing guarantees of service from the network [15,1,13,23], for tailoring the host communication subsystem functionality to suit applications [3,24,19,4], and reserving resources within a host operating system [11,12] have been proposed in isolation. Thus far however, the capabilities provided by the individual system components have not been integrated into an end-to-end QoS management scheme as have been suggested.

On an other hand, the Reference Model of Open Distributed Processing (RM-ODP) standard came to enable the federation of distributed systems design by offering an open framework that considers requirements and constraints on communication systems.

In this paper we present an implementation framework for a generic end-to-end QoS management scheme, and describe it in terms of RM-ODP. Then the framework is applied to manage the end-to-end QoS of an experimental video on demand application, XMovie [7], to show the applicability of end-to-end QoS management and highlight the system support that will be required on top of the services provided by the individual system components.

2 Reference Model of ODP

Open Distributed Processing (ODP) systems enable the integration of distribution, interworking, interoperability and portability. The Reference Model of ODP [18] standardizes such Open Distributed Processing systems. The standard defines how ODP systems are specified and identifies the features of systems to qualify them to be characterized as ODP systems. RM ODP consists of 4 parts: an "overview and guide to use of the Reference Model", a "descriptive model", a "prescriptive model" and a "architectural semantics".

Part 1 is an overview and guide to use of the Reference Model, it gives a general introduction and tutorial on the ODP Reference Model, but is not normative. Part 2 is the descriptive model. It contains the definition of the concepts, the analytical framework, and notation for formalised description of distributing processing systems. Part 3 is the prescriptive model. It contains the specifications of the required system characteristics which test the compliance and conformance to open distributed processing requirements, the constraints to which ODP standards must conform. Part 4 concerns architectural semantics and contains a formalisation of the ODP modelling concepts.

In order to deal with complexity of systems, RM-ODP provides a framework of abstraction which defines five viewpoints: enterprise, information, computational, engineering and technology viewpoints.

The enterprise viewpoint allows the description of roles and activities, including human interactions, security and management policies. The information viewpoint describes the information structures, the rules and what is of concern for information. In the computational viewpoint, the processing functions and all the data types are visible, in other words, the description of the system as a set of interacting objects. The engineering viewpoint concerns mainly the communication mechanisms to support distribution. The technology viewpoint includes the components and links from which the distributed system is constructed (including the operating system).

There is a language associated to each viewpoint. It consists of concepts, rules and structures appropriate for the specification of that viewpoint. The modelling in each viewpoint is object oriented. Each object interface comprises a set of operations that may be invoked by other objects.

In the engineering part of ODP, an established connection for continuous media corresponds to a channel (which is an agreed specified level of QoS) that supports distributed transparent interaction between communicating engineering objects.

In the following sections of this paper, we will focus on the communication aspect of distributed multimedia applications. We will first describe how QoS can be managed and what are the computational and engineering objects that are involved and how they interact.

3 QoS management frameworks

Quality of service management is identified in ODP as a part of the global application management and is considered by RM-ODP as a major issue. Quality of service management should be specified in terms of:

- objectives and policies at the enterprise viewpoint: what actions should be taken in order to establish connection at a given level (negotiation, re-negotiation, monitoring...) and thus, the objects that are involved in QoS management.
- requirements concerning information (e.g. MIB use and information maintenance)
- activities that occur and the interactions between objects (at computational viewpoint).
- defining the role of the different objects at engineering viewpoint (e.g.: resource reservation scheme, monitoring).

– how specifications for management are implemented (e.g. mechanisms for resource reservation) at technology viewpoint.

The requirements and policies described in the enterprise viewpoint affect the specification in all the other viewpoints. Thus QoS management frameworks that have been reported in the literature [2,21,20] that essentially advocate negotiation of an acceptable level of service at session establishment time, with provisions for re-negotiation of the service level corresponds to the enterprise viewpoint.

This requires mechanisms to be specified in the technology viewpoint of ODP, for reserving the necessary resources at the end systems and the interconnecting networks and configuring the host communication subsystem to suit the application as schematically shown in figure 1. In such a system, the application reserves a percentage of each of the system resources to satisfy its requirements.

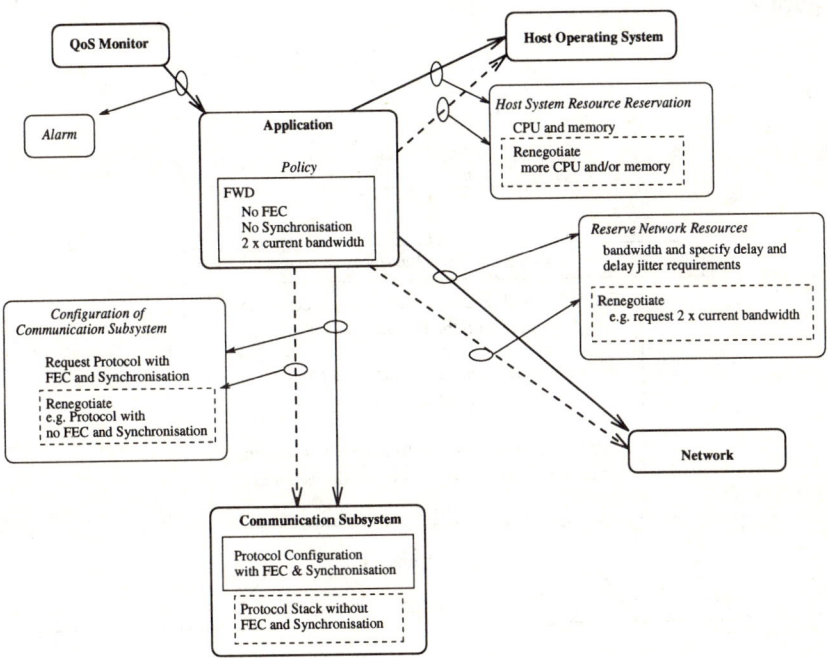

Fig. 1. Operation of a QoS Managed System

RM-ODP identifies the *communications domain function*, it has being particularly adapted to dealing with the variations of application requirements and services provided by the system components. A QoS management framework also requires a policy, and a mechanism for re-negotiating the level of service when the initially negotiated QoS is no longer appropriate. The policy will stipulate what actions are to be taken when operational conditions change. This is illustrated in figure 1 by the dashed lines.

Finally, the QoS management system will need a facility for monitoring the QoS provided to the application, and raise an alarm if sustained QoS violation occurs. These activities can be viewed as part of the *communications domain function*.

Therefore, a QoS management system can be viewed as an ODP system. However, in this type of ODP system, if explicit support for QoS management is not provided by the system, the handling of the interaction of the system components will become the responsibility of the application developer making the development process immensely more complex. Moreover, the negotiation and renegotiation will be difficult if not impossible because of the differences in capabilities of the systems involved and as the state of the system components will be distributed across the entire system. The obvious way of overcoming this is to introduce a QoS manager object which will manage all the interactions, maintain state and export a standardised interface to the application, as shown in figure 2.

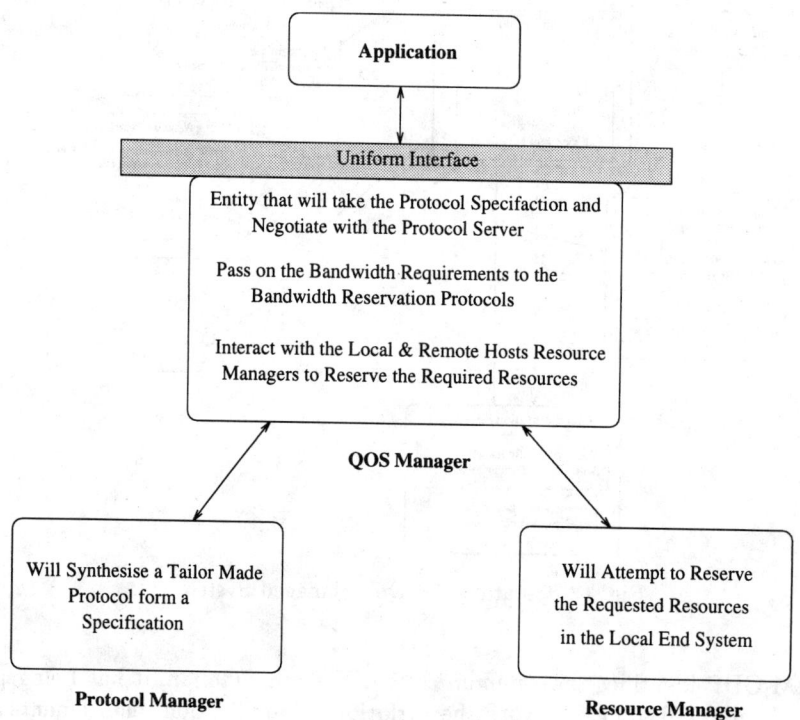

Fig. 2. Overview of the System

3.1 A generic QoS management system

This section presents the scheme for establishing connection at a given quality level. Four main objects are involved, namely QoS manager, resource manager, protocol server and performance monitor, the interactions between objects are performed by invocations on their interfaces (e.g. Q.Request, R.Request, P.Request, Q.Monitor).

In the QoS management scheme outlined above, the sequence events and invocations that will take place will be as follows. The steps referred to correspond to the numbers in figures 3 and 4.

Once activated, the QoS managed application will interact with the *QoS Manager* to obtain a specified level of service - step **1**. The *QoS Manager* will then interact with the *Resource Manager* and the *Protocol Server* to determine whether there are sufficient resources and to synthesize the most appropriate protocol stack for the application - step **2** and **2'**. The *Resource Manager* will either grant the resource request and reserve the resources or signal a failure and indicate the maximum available resources - step **4**.

The *Protocol Server* will synthesize a protocol stack with the appropriate functionality and signal back [3] - steps **3** and **3'**. If a failure occurs, the application has the option of lowering the QoS specification and attempting to reserve the resources as before.

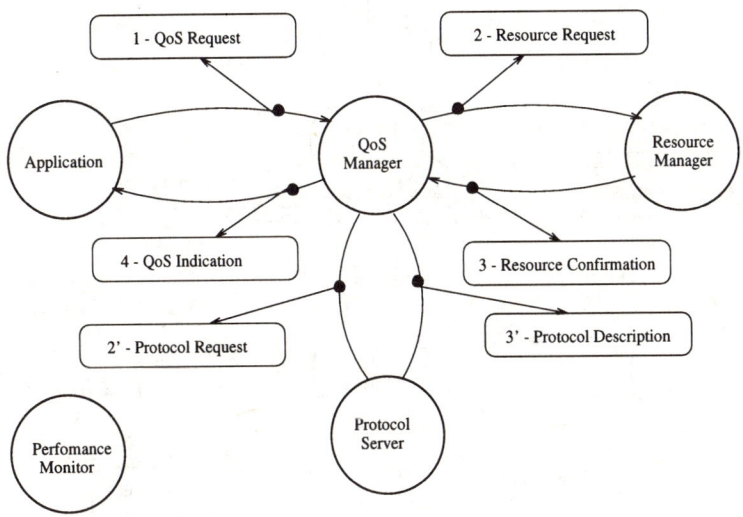

Fig. 3. Schematic Diagram of the Interaction of QoS Management System

Once an acceptable level QoS has been negotiated, the resources necessary to provide that level of service will be provisionally reserved. Then the peer

---

[3] The method of reserving resources and synthesizing the protocol stack will be discussed in section 4.2 and 4.3 respectively

QoS Manager will be contacted to determine the level of service that can be supported end-to-end - step **5** shown in figure 4.

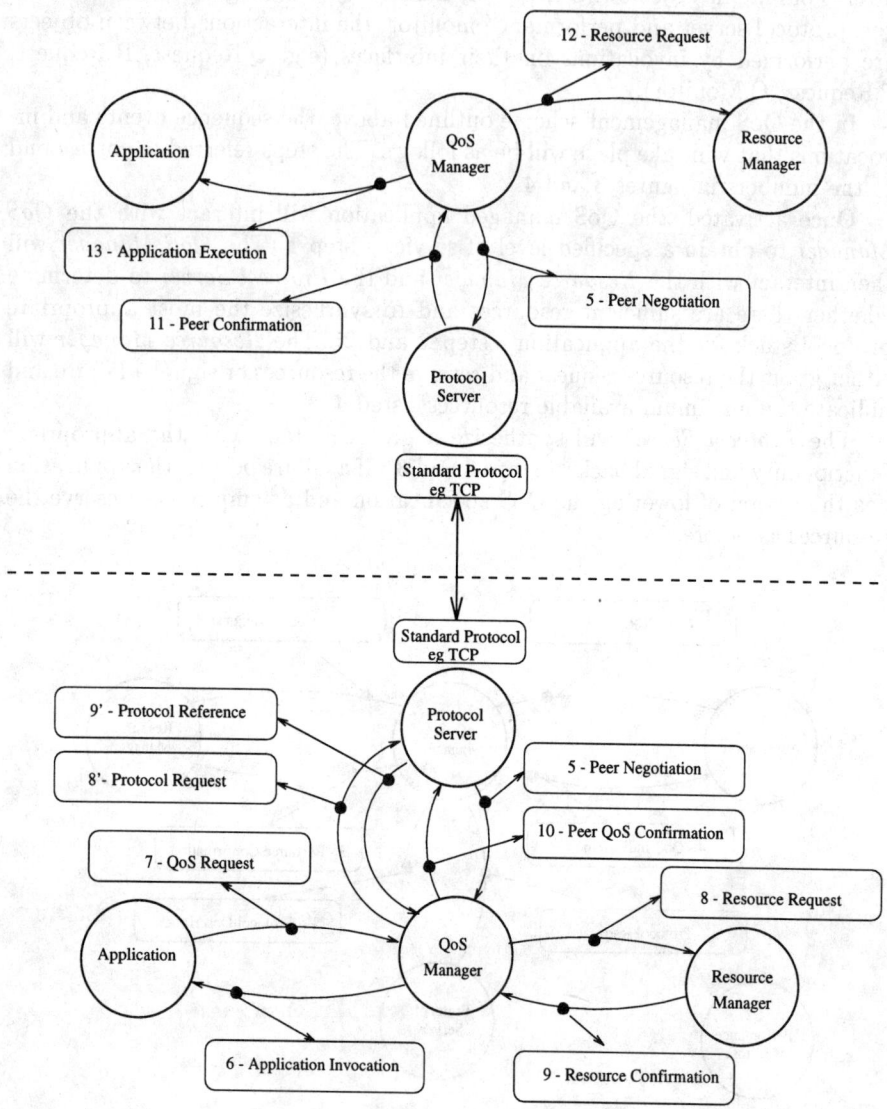

Fig. 4. Schematic Diagram of the Interaction of QoS Management System

While the connection is being established to the peer *QoS manager*, the network resources necessary to support the session will be reserved.

When this message is received, the remote *QoS Manager* will activate the specified application with the appropriate parameters - step **6**. The parameters to be used will be discussed in section 4.1. The remote application will in turn interact with its *Resource Manager* to determine the level of service that can be provided - step **7**. The *QoS Manager* will then negotiate with the resource reservation similarly to the initiating system - steps **8** and **9**. Further it will indicate to the *Protocol Server* the required protocol - step **9'**. Again, the application may lower the QoS requirements up to the minimum acceptable level, indicated at the invocation, if a failure occurs. If the lowest specified level of QoS cannot be provided, the initiating *QoS Manager* will be informed to terminate the application, and release all reserved system resources.

If the application is supportable, the *QoS Manager* will reserve the necessary resources and it will indicate to the initiating *QoS Manager* the level of support that can be provided in the form of a confirmation - steps **10** and **11**.

The initiating *QoS Manager* will then confirm the initial reservations releasing any unnecessary resources that might result form a lowering of the service by the remote system - step **12**. Finally, the initiating *QoS Manager* will inform the *QoS managed application* of the level of service - step **13**.

Once the application is executing, its performance will be monitored by a *performance monitor*. If sustained QoS violations occur, these will be notified to the *QoS Manager*. The *QoS Manager* in turn will notify the application which will attempt to re-negotiate the level of QoS.

In summary, the QoS manager needs to perform the following functions:

— Take the QoS specification provided by the application and map it into a set of end system resource requirements and a protocol specification;
— Request resources from the resource manager and keep account of resource reservation of all QoS managed applications;
— Request a protocol with the appropriate functionality;
— And finally, negotiate the end-to-end QoS with the remote QoS manager.

The protocol server needs to take the specification provided by the QoS manager and synthesize a protocol stack with the appropriate functionality. The resource manager need to carry out a schedulerbility analysis to determine whether the requested resources can be provided without jeopardizing the already established sessions.

Once the application is running, the performance monitor needs to monitor a specified entity, average the observations to eliminate transient effects, and raise an alarm if the value reaches a predetermined level. The QoS manager in turn will inform the application of the QoS violation.

3.2 Specification of application requirements

With the proposed system, the application designer will be responsible for defining its host resource requirements, the network characteristics and communication subsystem characteristics. This is an extension of the philosophy used for current network resource reservation [15,23], i.e. the inclusion of the specification of protocol functionality and resource requirements.

There is no established methodology for defining protocol functionality, and this is currently an open research issue. However, a number of methods have been proposed in the literature [24,3,4]. As this paper is primarily concerned with QoS management, for simplicity, it will be assumed that a protocol will be chosen for a set of pre-defined protocol implementations.

Estimation of local resource requirements is also an open research issue. As shown by [11], good estimates can be achieved through simple tests. Hence it is assumed that the application designer will be able to provide estimates of host system resource requirements as described in 4.1.

Finally the application designer will have to provide the network resource requirements. Currently, in resource reservation network protocols it is necessary to provide a flow specification [23,15]. This flow specification can be directly used in the proposed scheme.

4 Objects involved in QoS management

A QoS managed system consists of the following objects, they are described in the next sections:

- QoS manager
- Resource manager
- Protocol server
- Performance monitor

4.1 Structure of the QoS manager

The primary tasks of the QoS manager are the determination of total system resource requirements of the application, maintaining state of all local QoS managed applications, and negotiation of QoS with the remote QoS manager.

QoS manager object invocation. Apart from providing its system resource and protocol functionality requirements, and a flow-specification, where appropriate, the application needs to implement the re-negotiation options. This will be facilitated through a standard interface to the QoS manager. The interface will be as follows:

Q.Request($operation, syst\_resources, syst\_names, prot\_spec,$
$flow\_spec, app\_info, perf\_monitor, result$)

The structures *syst_resources*, *syst_names*, *prot_spec*, *flow_spec*, and *perf_monitor* will be discussed in the following sections. *operation* indicates the whether it is a request from an initiator or a responder of a QoS negotiation session. *app_info* structure provides the name of the remote application to be invoked, the number of parameters to be passed to it, and their values.

```
struct app_info {
            string  name;
            int     no_parameters;
            string  parameters[];
}
```

The *result* will indicate success or failure. In the case of failure the current system status will be returned via the *syst_resources*.

For example, if CPU and network bandwidth are the only system resources that are considered, the QoS renegotiation options can be implemented as described below. In the description, for brevity of representation, the *syst_resources* structure can be represented by *des_CPU* and *min_CPU*, *prot_spec* is represented by *TCP* and *flow_spec* is represented by *BW*. The remote application is *test* and is assumed to have no invocation parameter. BW_{ava} and CPU_{ava} are the currently available CPU and bandwidth capacity that will re returned by the QoS manager when a *Q.Request()* fails. And finally, CPU_{min} and BW_{min} are the minimum CPU capacity and network bandwidth that is required to run the application.

```
Q.Request( remote, {des_CPU, min_CPU}, TCP, {des_BW, min_BW},
           {test}, result)
begin
      if (result not OK)
      if ( CPU_ava < CPU_min or BW_ava < BW_min )
          terminate;
      else
      begin
            if (CPU_ava ≥ CPU_min)
            begin
                  alternate strategy;
                  des_CPU = new value;
            end
            if ( BW_ava ≥ BW_min )
            begin
                  alternate strategy;
                  des_BW = new value:
            end
            /* for all possible negotiation options */
      end
end
```

Q.Request(*remote*, {*des_CPU*, *min_CPU*}, *TCP*, {*des_BW*, *min_BW*}, {*test*}, *result*)

Estimation of resource requirements. The application designer will be able to provide a list of all system function libraries the application relies on via the structure *syst_names*.

```
struct syst_names {
            int no_libraries;
            string library_names[];
}
```

Then the minimum and desirable total resource requirements for each type of resource will be estimated as follows. Two separate variables for each resource type, which will contain the sum of the resource requirements of all the system support libraries the application relies on [4]. The applications "stand alone" resource requirements will be provided by the application via *syst_resources*. The minimum and desirable resource requirements of the support libraries will be obtained from a QoS MIB within the QoS manager.

The QoS MIB will be maintained by the system administrator. The QoS manager will provide facilities for the system administrator to register system support libraries and their desirable and minimum resource requirements.

The two major obstacles in implementing the above scheme is associated with the determination of the system resources, and the assumption of independence between the resource requirements of individual system utilities. It is becoming clear that in future multimedia systems, processor capacity, memory, and bus bandwidth will be the most critical [5]. Therefore, it will be possible to describe the system requirements in terms of a finite set of system resources. For interoperability, the set of system resources to be used needs to be standardized.

The construction of system through re-usable software modules have been shown to be viable [15]. The validity of the assumption of independence between them however has not been proven. Despite this, since as only an estimate of the requirements is required because the system can tune the requirements at run time [11], we contend that the above scheme is viable.

The *syst_resources* structure will indicate the resources which need to be reserved to provide the required level of QoS. The resources that can be reserved will be specified by the system, and thus it will have the following structure.

```
struct syst_resources {
            int des_ CPU;
            int min_ CPU;
            int des_ memory;
            int min_ memory;
```

---
[4] e.g. system calls

```
                    int des_ buss;
                    int min_ buss;
}
```

Non zero entries indicate the requirement for that resource to be reserved and the value indicates the application's requirement.

Maintaining state. The requirement for maintaining state arises out of the providing of renegotiation facilities. If a request is received which could not be accommodated with the existing distribution of resources, it is necessary to determine whether it will be possible to accommodate it by lowering the service levels of all the currently active applications to their minimum acceptable level. To enable this, the QoS manager keeps information about each active application's current status, i.e. whether it is operating with its minimum resource requirements, and its minimum resource requirements.

When the QoS manager makes a resource reservation request to the resource manager, it will provide the desirable and the minimum resource requirements. The resource manager, as it will be described in 4.2, will indicate whether (i) the resources can be reserved at the desired level, (ii) by reducing the resource reservations of all applications to the minimum level, thus the request can be satisfied, or (iii) the request cannot be satisfied.

If the request can be satisfied, the QoS manager will initiate negotiations with the peer QoS manager. If the current request can be satisfied under minimum resource conditions, the QoS manager will go renegotiate the resource reservations of all active applications to their minimum levels while negotiating with the peer QoS manager. If the resources request cannot be satisfied, the QoS manager will indicate a failure to the application and provide the current resource availability information.

Remote negotiation. As described in section 3.1, the QoS manager needs to determine whether the remote machine can support the requested QoS. This essentially requires facilities for invoking the remote entity with some application specific parameters which the application will use to translate into its own local resource requirements, and indicating the protocol configuration. Thus the QoS negotiation protocol will be simple, and will have the frame format in figure 5.

The version number enables the evolution of the protocol. The number of channels will indicate the number required by the application. If more than one channel is involved, the protocol will wait for the necessary information before invoking the application. The channel priority field will be used to establish the priority among the different channels of an application. The remote application name length and name indicate the name of the application to be invoked. The number of parameters, and parameters specify the arguments to be used with the remote invocation. Finally the Protocol Specification length and Specification parameters will indicate the protocol to be used on top of the channel on which the message was received.

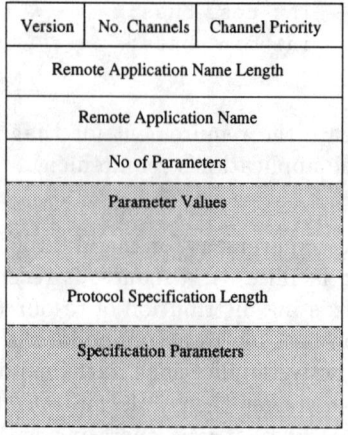

Fig. 5. The frame structure of the peer QoS Manager Interaction Protocol

4.2 Resource manager

The resource manager will be responsible for reserving the host system resources. To enable this, the QoS manager will indicate whether the application is periodic or aperiodic. If it is periodic, the QoS manager will provide the period and if it is aperiodic it will provide the maximum response time, *time*, that can be tolerated. In addition, it will provide an estimate of the total host system resource required to support the desired QoS and the minimum acceptable level of QoS in the structure *resource* will be similar to the *syst_resources* structure described in section 4.1. The only difference is that the values will indicate the total resource requirements of the application computed by the QoS manager.

The resource manager and the QoS manager will interact via the following invocation on the resource manager interface:

$$R.Request(time, resource, result)$$

result will indicate the success or failure, and in the case of failure signal the availability of resources.

```
struct result {
            int outcome;
            int CPU;
            int memory;
            int buss;
}
```

Once a R.Request() is received, the resource manager will perform schedulerbility analysis using the desirable as well as the minimum QoS levels. If the desirable level can be supported, the requested resources will be provisionally reserved. If not, the analysis will be carried out with the minimum levels for all active connections, and an indication will be provided to the QoS manager whether under these conditions the request could be supported. If neither of the above proves successful, the maximum availability of resources will be returned to the QoS manager.

In this paper we will only consider the reservation of CPU. The scheme scales to include other resources as methods for reserving these resources become available.

CPU reservation strategy. The CPU reservation will depend on the scheduling policy used by the system. In this paper we will assume a rate monotonic scheduling (RM) policy. To facilitate RM scheduling, the resource manager will determine the number of host system scheduling time intervals t_H the period corresponds to. Then it will determine the rate of the requested application by dividing the CPUtime requested by the number of time intervals.

$$rate = \frac{CPUtime * t_H}{period}$$

Then either the Liu and Layland [9] approximate analysis or Lehoczky [10] exact analysis can be used to determine the schedulerbility.

Apart form the above functions, the resource manager will measure the CPU usage by the applications and inform the scheduler. The measuring mechanism will take into account the cost of all servers the application uses and also, the time consumed in receiving packets. This can be done as proposed by Mercer [11].

4.3 Protocol server

The function of the protocol server will be to provide standardized protocol implementations for initial negotiations and network resource reservation, and then a protocol that is suited to the given application to be used in subsequent communications.

The specification of the protocol requirements is an open research issue as was mentioned earlier. In this paper it is assumed that the applications will choose from a set of predefined protocols such as RTP, TCP, and UDP.

The interface with the QoS manager is thus given by:

$$P.Request(flow\_spec, prot\_spec, protID)$$

The *flow_spec* will specify the flow specification to reserve the network resources. The flow specification will depend on the bandwidth reservation protocol that will be used, and does not influence the operation of the proposed system. The protocol specification will indicate the protocol to be used. As techniques for synthesizing application specific protocol become available this will contain the specification of application requirements that the protocol sever could use to synthesize an optimized protocol for the given application. The *protID* will provide a reference to the port associated with the requested protocol.

4.4 Performance monitor

The *Performance Monitor* will measure the invocation frequency of specified function, *func_monit*, average it over the defined period, *time_monit* . If the value falls below the specified threshold, *threshold_monit*, will raise an alarm to the *QoS Manager*.

Thus the performance monitor, whose role is part of the *communications domain management function* identified in RM-ODP, will interact with the QoS manager via the following interface:

$$Q.Monitor(perf\_monitor)$$

The structure *perf_ monitor* will thus be

```
struct perf_ monitor { string   func_ monit;
                       int      time_monit;
                       int      threshold_monit;
}
```

5 Application to Xmovie

The previous sections have described a framework for implementing end-to-end QoS management. In the descriptions, abstract interfaces and parameters were defined. The following sections examine how these could be used to manage the QoS of an experimental system, WMovie [8].

5.1 Xmovie

Xmovie enables the transmission of stored movies via networks and displaying in windows of the X window system. It consists of a extended *X server* which enables it to directly display the images sent by a *X client*. Furthermore, the *Movie client* and the *Movie server* exchange data using a specialized protocol referred to as MTP. The basic structure of the Xmovie system is shown in figure 6 from [6].

In this system, the playback application will interact with the users control of the display windows, start the movie server and make the link between the *Movie server* and the *Movie client*. The interaction with the users will involve the obtaining of the movie information, and starting, stopping pausing and the placing of the movie window.

The *X server* will process the movie orders simultaneously and will generate events in response to user action, such as pressing a mouse button. The *Movie client* will be an extended version standard server. The extensions will enable the displaying of a movie in a given window directly from the network. Further, it will retrieve the movie from the *Movie server*, while the *Movie server* will perform the actions necessary to obtain the movie form the file system.

6 Addition of QoS management to Xmovie

For this example the structure shown in figure 6, where the playback application and the x server/movie client is on one workstation and the movie server on a remote workstation will be used. Further, CPU will be the only host system resource, and bandwidth will be the only network resource that will be used.

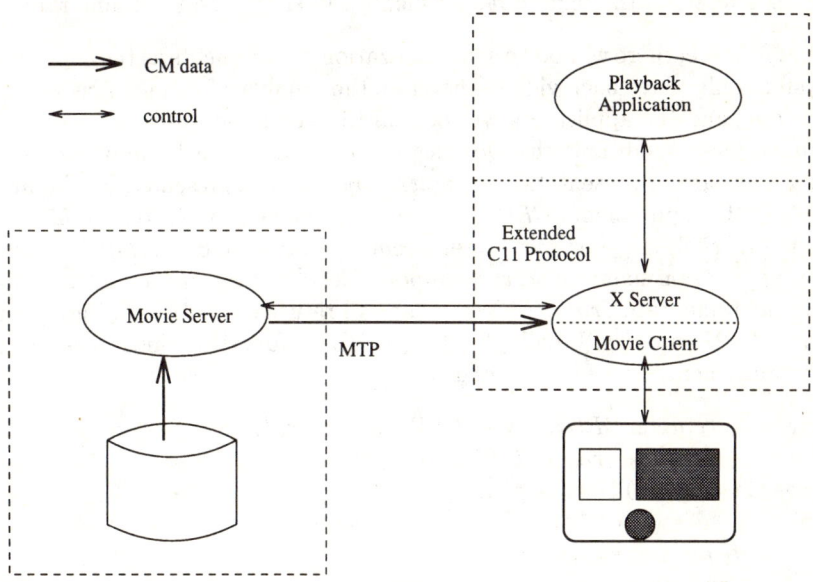

Fig. 6. Xmovie System Architecture

6.1 Playback application

To add QoS management capability to XMovie using the *QoS Manager* and the *Resource Manager* described above, *playback application* and the *movie server* will need to be modified. In the case of the *playback application*, it will be necessary to supply the necessary information to the *QoS Manager* via the **Q.Request()** primitive. The information required can be obtained as follows:

- The bandwidth requirements between the X server/movie client, and the movie server will have to be determined. This will be possible as the average frame size, and the frame rate is known. Also, it will be possible to determine the minimum bandwidth required as the minimum frame rate that will be acceptable will be known [5].

---

[5] Further, the maximum delay that can be tolerated will be similar for most broadcast type applications. With this and the average frame size and rate, it will be possible to determine the memory requirements

- By examining the structure of the system, it can be easily determined that the playback application will use, the X server/movie client continuously, and itself sporadically. Estimates of CPU time required for the X server/movie client and the playback application itself can be determined though some simple experimental runs as suggested in [11].
- Finally, the XMovie operation will depend on the movie server being able to retrieve and transmit the required information at the minimum acceptable rate. Therefore, the remote requirements will simply be the frame rate.

Furthermore, there will be no synchronization requirements as the experimental system only plays back video. Therefore the number of connections required will be one, and the application will be real time periodic.

Under these conditions the QoS negotiation policy can be implemented as follows. For ease of presentation let *syst_resources* be represented by the application's CPU requirements CPU, and the *prot_spec* be represented by MTP the XMovie protocol, the *flow_spec* be represented by BW, and *app_info* be presented by application name $m.server$, supportable frame rate $frate$ and minimum acceptable frame rate $frate_{min}$. When the call returns, let CPU_{ava} indicate the available CPU capacity. Finally, let $f_{CPU}(x)$ be a function which generate the CPU requirements as a function of x.

```
Q.Request( remote, CPU_app, MTP, BW, m.server, frate,
           frate_min, result, CPU_ava)
if (result not OK)
begin
    if (f_CPU(frate_min) ≥ CPU_ava)
        terminate;
    else
    begin
    frate = frate_min;
        do
            frate++;
        while (f_CPU(frate) < CPU_ava)
        CPU_app = f_CPU(frate);
    end
end
Q.Request( remote, CPU_app, MTP, BW, app.name, frate,
           frate_min, result, CPU_ava)
```

The application will start executing in the normal fashion once a positive Q.Indication is received.

6.2 Movie server

The modifications required to the movie server will again be minimal. It will have to be able to be invoked with three parameters, namely the frame rate supportable by the initiating system $frate$, the minimum acceptable frame rate $frame_{min}$, and the band width that can be reserved for the connection BW_{ava}. Then the desirable re-negotiation options have to be programmed.

The modification required to invoke the movie server with the required parameters is trivial.

The re-negotiation options can be implemented as follows. Let $f_{bw}(x)$ be a function which generates the bandwidth requirement for a given framerate x [6].

```
if ( BW_ava < f_bw(frate_min))
     terminate;
if ( BW_ava < f_bw(frate))
begin
     fratenew = frate_min ;
     do
         fratenew++;
     while (f_bw(fratenew) < BW_ava) ;
     CPU_ava = f_CPU(fratenew);
end
else
begin
     fratenew = frate;
     CPU_app = f_CPU(fratenew);
end
Q.Request( local, CPU_app, MTP, BW, app.name,
           frate, frate_min, result, CPU_ava)
if( CPU_ava ≥ f_CPU(frate_min))
begin
     fratenew = frate_min;
     do
         fratenew++;
     while(f_CPU(fratenew) < CPU_ava)
     CPU_app = f_CPU(fratenew);
end
Q.Request( local, CPU_app, MTP, BW, app.name,
           frate, frate_min, result, CPU_ava)
```

7 Performance implications

With all the interactions that are taking place, it might seem that the performance of the proposed system would be poor. However, a closer examination of the interactions shown in figures 3 and 4 indicates that the performance penalty is minimal, and will mainly be confined to the connection establishment phase.

---

[6] It will be assumed that the file access will be guaranteed using a scheme such as the one presented in [17]

The extra overhead are due to the initiating system and remote systems resource reservation and protocol configuration activities, steps 1 to 3 and 7 to 9. This will depend on the execution overheads of the QoS manager, and the protocol server or resource manager. It is likely that in future systems delay associated with these will be minimal compared to the end to end propagation delay. Under these conditions it can easily be shown that the performance of QoS managed applications are not significantly affected by the proposed scheme [7].

8 Conclusions

A framework for implementing an end-to-end QoS management scheme was presented in this paper.

Firstly, it was shown that this type of QoS management schemes can be viewed as ODP systems. The compliance was demonstrated by mapping the various functions into the appropriate viewpoints and languages of the RM-ODP.

Secondly, it was shown that with the provision of such a scheme at the system level will enable QoS multimedia applications to be developed without increasing the complexity of the application developers task, by applying it to an experimental video on demand application.

Thirdly, a methodology for estimation of resources and their reservation applied on one system resource namely the CPU, was presented.

In all the above cases appropriate interfaces between the various system components we presented. The applicability of the proposed scheme was then demonstrated by applying it to an experimental video on demand application. The analysis showed that proposed scheme enable the "easy" development of QoS managed applications.

Finally, a brief analysis of the performance implications of the proposed scheme was presented to illustrate that it does not significantly affect the overall performance of the system.

The viability of the proposed system hinges firstly on the assumption that multimedia applications can be viewed as collection of composite software modules and secondly on the fact that the resource requirements of these modules can be estimated. And finally, that the total resource requirements of an application can be obtained by adding together the requirements of its composite modules. This has not yet been proven, as it is currently being evaluated though an experimental implementation.

9 Acknowledgements

Aruna Seneviratne wishes to acknowledge the support from MASI Laboratory.

---

[7] no re-re-negotiation is assumed as under those conditions applications with no QoS management will fail

References

1. David P. Anderson and Ralf Guido Herrtwich and Carl Schaefer, "SRP: A Resource Reservation Protocol for Guaranteed-Performance Communication in the Internet", Technical report, University of California, Berkeley, 1990.
2. A. Campbell, G. Coulson, F. Garcia , D. Hutchinson and H. Leopold,"Integrated Quality of Service for Multimedia Communications", IEEE INFOCOM'93, 1993.
3. M. Fry and A. Seneviratne and A Richards, "Framework for the Implementation of the Next Generation of Communication Protocols", In 4th International Workshop on Network and Operating Systems Support for Digital Audio and Video, University of Lancaster, November 1993.
4. P. Hoschka,"Towards Tailoring Protocols to Application Specific Requirements", IEEE INFOCOM'93, 1993.
5. M. Jones, "Adaptive Real Time Resource Management Supporting Modular Composition of Digital Multimedia Services", 4th International Workshop on Network and Operating Systems Support for Digital Audio and Video, University of Lancaster, November 1993.
6. R. Keller and W. Effelsberg and B. Lamparter, "Performance Bottlenecks in Digital Movie Systems", 4th International Workshop on Network and Operating System Support for Digital Audio and Video, Lancaster, November 1993.
7. B. Lamparter and W. Effelsberg, "X-MOVIE: Transmission and Presentation of Digital Movies under X", Second International Workshop on Network and Operating System Support for Digital Audio and Video, Heidelberg, November 1991.
8. B. Lamparter and W. Effelsberg and N. Michl, "MTP: A Movie Transmission Protocol for Multimedia Applications", Multimedia92, 4th IEEE ComSoc International Workshop on Multimedia Communications, Monterey, California, April 1992.
9. C. Liu and J. Layland, "Scheduling Algorithms for Multiprogramming in Hard Real Time Environment", Journal of the ACM, 1973.
10. J. P Lehoczky and L. Sha and Y. Ding, "The Rate Monotonic Scheduling Algorithm: Exact Characterisation and Average Case Behaviour", 10th IEEE Real-Time Symposium, 1989.
11. C. Mercer and S. Savage and H. Tokuda, "Processor Capacity Reserves for Multimedia Operating Systems", CMU Technical Report - CMU-CS-93-157, 1993.
12. C. Mercer and S. Savage and H. Tokuda, "An Abstraction for Processor Capacity Reservation", Workshop on Worstation Operating Systems, 1993.
13. C. Paris and G. Ventre and H. Zhang,"Graceful Adapatation of Guaranteed Performance Service Connections", Technical report, University of California, Berkeley, 1993.
14. C. Partridge and S. Pink, "A faster UDP", IEEE Transaction on Networking, vol 1, n.4, 1993
15. C. Partridge and S. Pink, "An Implementation of the Revised Internet Stream Protocol (ST-2)", Internetworking: research and experience, vol 3, n.1, 1993.
16. R. K. Raj and H. M. Levy, "A Compositional Model for Software Re-use",The Computer Journal, n.4, vol 32, dec 1989.
17. P. V. Rangan and H. Vin, "Designing File Systems for Digital Audio and Video", 13th ACM Symposium on Operating System Principles, October 1991.
18. Basic Reference Model of Open Distributed Processing

CCITT X901 and ISO 10746-1. "Part 1: overview and user model", (WD) ISO/IEC/JTC1/SC21/WG7, November 1993.
CCITT X902 and ISO 10746-2. "Part 2: descriptive model ", (CD) ISO/IEC/JTC1/SC21/WG7, November 1993.
CCITT X903 and ISO 10746-3. "Part 3: prescriptive model", (CD) ISO/IEC/JTC1/SC21/WG7, June 1993.
CCITT X905 and ISO 10746-4."Part 4: architectural semantics, specification techniques and formalisms", (WD) ISO/IEC/JTC1/SC21/WG7, June 1993.

19. Douglas C. Schmidt, Donald F. Box and T. Suda, "ADAPTIVE: A Dynamically Assembled Protocol Transformation; Integration; and Validation Environment", Concurrency: Practice and Experience Journal, 1993.
20. A. Seneviratne, M. Fry, V. Withana, V. Saparamadu and A. Richards,"A Methodology for Managing End-to-End QoS of Multimedia Applications", 10th IEEE Phoenix Computer Communication Conference, Phoenix, Arizona, 1994.
21. W. Tawbi and L. Fedaoui and E. Horlait," Dynamic QoS Issues in Distributed Multimedia Systems", 2nd International Conference on Broadband Islands, Athens, June 1993.
22. A. Vogel, G. v. Bochmann, R. Dssouli, J. Gecsei, A. Hafid, B. Kerherve, "On QOS Negotiation in Distributed Multimedia Application", University of Montreal, April 1993.
23. L. Zhang, S. Deering, D. Estrin, S. Shenker and D. Zappala,"RSVP: A New Resource Reservation Protocol", IEEE Network Magazine, Sept 1993.
24. Martina Zitterbart, Burkhard Stiller and Ahmed Tantawy, "A Model for Flexible High-Performance Communication Subsystems", IEEE JSAC, May 1992

A Formal Description Technique Supporting Expression of Quality of Service and Media Synchronization

Howard Bowman[1], Lynne Blair[2], Gordon S. Blair[2] and Amanda G. Chetwynd[2]

[1] Computing Laboratory, University of Kent at Canterbury, Canterbury, Kent, CT2 7NZ, United Kingdom.

[2] Department of Computing, Lancaster University, Bailrigg, Lancaster, LA1 4YR, United Kingdom.

Abstract. Formal description techniques have been applied successfully to the fields of communications and distributed systems. We argue, however, that the recent emergence of multimedia computing will have a significant impact on this work. In particular, existing formal description techniques do not satisfactorily model the real-time behaviour exhibited by distributed multimedia systems. This paper considers the impact of multimedia on formal description techniques and proposes an approach in which functional behaviour is expressed in the language LOTOS and non-functional quality of service is expressed in a real-time temporal logic. This dual language approach to formal description is demonstrated through a number of multimedia examples, culminating in the specification of a lip-synchronization algorithm.

1 Introduction

Formal specification languages are becoming increasingly important in the development of large, complex systems. This trend is particularly evident in the field of communications protocols. Indeed, the importance of formal specification in communicating systems was recognised as long ago as 1979 when the ISO organisation founded an ad-hoc group to enable the formal description of OSI protocols. Three formal description techniques have emerged out of this work: LOTOS [ISO88], Estelle [ISO89] and SDL [CCITT88]. The OSI standardisation initiative has also been extended, with some success, to the field of Open Distributed Processing (ODP) [Linington91].

Recently, however, multimedia systems have emerged; such systems involve the integration of a variety of media types including static media (text, graphics, etc) and continuous media (audio, video, animation, etc). Furthermore, this emergence has been accompanied by a rapid proliferation of multimedia technology, suggesting that standardisation will also be of considerable importance in the multimedia field. Formal description techniques which can express distributed multimedia systems are essential to such a process, since they enable the unambigous and precise expression of standards. This has significance in all areas of standardisation involving multimedia, for example in ODP or in the definition of multimedia teleservices architectures. However, existing formal description techniques for distributed systems do not satisfy the real-time requirements of multimedia systems.

In this paper we will highlight a new approach which is specifically tailored to the requirements of distributed multimedia computing. Our approach has been to reuse existing technology, rather than to develop a completely new formal language. Specifically, our approach is based upon LOTOS and known technology in the field of real-time temporal logic.

It should be pointed out that not only is this a very demanding area (due to the real-time requirements that multimedia impose on the formal description of distributed systems) it is also a very new area. In particular, there is very little published work that takes a formal approach to distributed multimedia computing ([Berra90] is a rare exception to this). However, if distributed systems standardisation activities (such as ODP) are going to successfully handle multimedia, formal techniques which can express such structures are essential. This paper highlights work which has been performed as a first step in the development of suitable formal techniques.

The paper is structured as follows. Section 2 highlights the basic requirements for formal description of multimedia, with particular emphasis on the real-time requirements imposed on system development. In section 3 we consider to what extent timed formal description techniques can realise these requirements and we present our new approach. In particular, we present the specification of a multimedia stream in this section, which indicates the suitability of our approach for formal description of quality of service. The expression of real-time synchronization using our approach is then demonstrated in section 4, through the specification of a lip-synchronization algorithm. In section 5 we consider the development of validation techniques for this approach. Section 6 discusses some related work and finally section 7 presents some concluding remarks. We have avoided discussion of the theoretical aspects of our approach in the main body of the paper. However, we present the major semantic principles which underly our approach (and in particular, the validation process with our approach) in the appendices.

2 Requirements for Formal Description of Multimedia

2.1 The Nature of Multimedia

The most fundamental characteristic of multimedia systems is that they incorporate continuous media [Anderson90] such as voice, video and animated graphics. The use of such media in distributed systems implies the need for continuous data transfers over relatively long periods of time, e.g. playout of video from a remote surveillance camera. Furthermore, the timeliness of such media transmissions must be maintained and this must represent an ongoing commitment for the duration of the continuous media presentation. There are three aspects to such a maintenance of timeliness and each represents a fundamental characteristics of distributed multimedia computing:

(i) Support for Continuous Media

It is necessary to provide *abstractions* for continuous media. For example, a number of projects have extended distributed systems with *streams* as abstractions over continuous media transmissions [Coulson92] [Nicolaou90].

(ii) Quality of Service

The timeliness of media transmissions is maintained through *quality* of service (QOS) parameters. Examples of QOS parameters are throughput, end-to-end delay (or latency), delay variance (jitter), packet error rates and bit error rates [Hehmann90].

(ii) Real-Time Synchronization

Existing distributed systems provide a range of mechanisms to support synchronization between processes. However, in distributed multimedia systems, there is an added requirement to provide *real-time synchronization* to maintain the temporal integrity of media transmissions. The classic example of real-time synchronization is maintaining the lip synchronization between an audio and a video channel.

2.2 Implications for Formal Description

Basic Requirements

It is clear that there are a number of basic requirements for the formal description of multimedia. Firstly, we are interested in formal techniques with explicit representations of *concurrency* and *interaction*, such as classically arising from the concurrency theory community. Techniques such as Petri Nets [Reisig85], Process Algebras [Milner89] and Finite State Machines [Budkowski87] all satisfy these explicit concurrency requirements.

In addition to these standard distribution features, multimedia imposes a number of significantly more stringent requirements on formal description, i.e.,

(i) Real-time

The representation of timing properties is a fundamental requirement for the formal description of multimedia systems. Specifically, multimedia systems belong to the class of soft real-time systems; if timing constraints are lost (such as if lip synchronization is lost), the results, although not optimal, are not catastrophic.

(ii) Probabilities

Timeliness properties are often expressed probabilistically, e.g. throughput may be expressed in terms of averages and jitter in terms of acceptable delay variance.

(iii) Dynamically changing real-time constraints

In multimedia systems real-time constraints may be altered interactively and renegotiated.

The incorporation of time, probabilities and dynamic behaviour into formal techniques are all major areas of ongoing research. Thus, due to the immaturity of formal techniques in these directions, the emphasis in this paper will be on the first of these requirements, that of expressing real-time.

The Functional/Non-functional Distinction

Quality of service has proved a powerful conceptual device in the development of multimedia systems and the central role of this concept has become well accepted [Horn92] [Campbell93] [Keshav93]. The acceptance of quality of service implies that a division is maintained during multimedia system development between the *functional* and the *non-functional* elements of a system. There is some confusion as to the exact distinction between functional and non-functional behaviour. So, by way of clarification, in our work functional is taken to mean that part of a system which can be expressed as simple ordering of events (also sometimes called the qualitative temporal behaviour of the system). In contrast, non-functional behaviour is that part of a system which cannot be expressed as simple ordering of events. The non-functional category includes a large number of properties such as: *dependability*, *criticality*, *security properties* as well as timeliness properties. However, as just indicated we will concentrate on the expression of timeliness properties.

We believe that the functional/non-functional distinction is so pervasive in multimedia system development that it must be reflected in the formal description of such systems. In addition, we believe that maintaining this distinction offers additional benefits in terms of the level of abstraction in the specification; we will return to this issue in our conclusions (section 7).

3 Timed Formal Description and Multimedia

The standard approach to timed formal description is *single language based*, i.e. notations are used which incorporate a notion of quantitative time (e.g. a delay operator or timestamps) directly into a standard untimed formal notation. Numerous such enhancements of standard formal notations exist, for example, timed enhancements of Petri Nets (such as [Walter83]), timed specification logics (such as TAM [Scholefield92]), timed Process Algebras ([Nicollin91] contains an overview of such notations), etc. It should also be pointed out that time is incorporated directly into the two standardised formal description languages Estelle and SDL.

We believe that such approaches are fundamentally unsuitable for formal expression of multimedia since they do not reflect the functional/non-functional distinction. Specifically, what is required is an approach in which untimed functional behaviour is expressed separately from the non-functional properties. For an in-depth discussion of the limitations of formal description of multimedia with single language based approaches the interested reader is referred to [Bowman93b].

In response to this limitation, we propose a dual language approach where one language is used to capture functional behaviour and a second language is used to model non-functional quality of service. This approach has been targeted specifically towards formal description of multimedia systems and is being developed as part of the Tempo project at Lancaster University. We will initially illustrate the approach through the specification of a multimedia stream (data channel) between a data source and a data sink and some associated quality of service statements.

3.1 Modelling Functional Behaviour

In our approach, the language LOTOS is used for the specification of functional behaviour. Briefly, LOTOS is a process algebra building on the experiences from the development of CCS [Milner89], CSP [Hoare85] and CIRCAL [Milne85]. The language has also been supplemented with an abstract data type notation based on ACT-ONE [Ehrig85]. [Drayton92] contains a straightforward introduction to the language, with more advanced treatment given in [Bolognesi88].

LOTOS was chosen for four main reasons:

(i) Importance in standardisation

> LOTOS was developed as a language for specifying OSI protocols and is now a recognised ISO standard [ISO88]. It has been used extensively in OSI standardisation and is becoming accepted in ODP [Sinderen91].

(ii) Fundamental benefits of the process algebra approach

> Process algebras feature an elegant set of operators for developing concurrent systems. Thus, succinct expressions of communicating concurrent processes can be made. Similarly, the emphasis on non-determinism in process algebras encourages elegant specification and abstraction from implementation details. Furthermore, rich and tractable mathematical models of process algebras have been developed. These are based upon concepts of equivalence through observation of the external behaviour of a specification, as derived from the seminal work of Robin Milner [Milner89].

(iii) Availability of tools

> Due to the standardisation of LOTOS and the application oriented nature of the language, a large number of support tools have been developed; the Lite toolkit developed in the Lotosphere project [Mañas92] is a good example of such work.

(iv) Suitability for expression of functional behaviour

> LOTOS enables the (untimed) order of event execution to be elegantly defined. Typical timing structures such as timeouts, watchdog timers, exception handlers and periodic timing behaviour can be expressed without associating the behaviour with actual concrete timing values. LOTOS events can thus be used as *placeholders* for concrete timing values, which can be *ground* in real-time during later stages of system development.

As an illustration of the use of LOTOS, the behaviour of the source process in a multimedia stream could be expressed as follows:

```
process Source[send]:noexit:=
    send; Source[send]
endproc (* Source *)
```

This behaviour simply states that the source process makes repeated transmissions of frames. In addition, behaviour of the sink process could be expressed as:

```
process Sink[arrive,play,timeout,error]:noexit:=
    (   arrive;
        play; Sink[arrive,play,timeout,error]
    []
        timeout; error; stop
    )
endproc (* Sink *)
```

i.e., the sink process merely receives frames and plays them or times out if the frame arrives with the wrong timing characteristics. Critically though, only qualitative ordering of events has been expressed; no expression of the quantitative time at which events will occur has been made.

3.2 Modelling Real-time Quality of Service Properties

Quality of service, by it's nature, expresses *requirements* for distributed systems to satisfy. In this sense QOS is akin to standard distributed systems requirements, such as freedom from deadlock, mutual exclusion and freedom from starvation. This characteristic of QOS is in contrast to the functional part of a distributed multimedia system, which is inherently behavioural. Where QOS differs from the traditional distributed systems requirements is that it expresses requirements for the *performance* of a system to satisfy. Thus, there must be a real-time aspect to the expression of QOS.

We have chosen to express QoS in a *real-time temporal logic*, since it has been demonstrated that mathematical logics both enable the abstract expression of requirements and facilitate rigororous reasoning about these requirements. In addition, we have chosen a real-time temporal logic in preference to other timed mathematical logics, such as RTL [Jahanian86], for two main reasons. Firstly, we wish to utilise the extensive understanding of verification with real-time temporal logics, in particular, model checking techniques [Clarke88] [Ostroff91], offer a powerful mechanism for verification of behaviour against requirements. Secondly, since Quality of Service in it's broadest sense encompasses a range of (non-timed) general distributed systems requirements, the added expressiveness of temporal logic over first order logics [Pnueli77] in modelling concurrent behaviour is needed.

We call the logic we use QTL (for quality of service temporal logic). QTL is based upon J. Ostroff's RTTL language, which has already been demonstrated to be highly suitable for the expression of a wide range of real-time properties in the area of real-time control [Ostroff89] [Ostroff92]. In addition, it should be noted that Alur and Henzinger's insights into explicit clock temporal logics [Alur91] were also influential. QTL is designed specifically to be compatible with LOTOS, hence the logic is *event based*, employing a *trace interleaving semantics*. Additional, features of the logic are that it is *linear time, explicit clock* and employs a *discrete time domain*. We will next consider the syntax of QTL.

The Syntax of QTL

Abstract Syntax. QTL has the following syntax (where ϕ is an arbitrary QTL formula):

$$\phi ::= \text{true} \mid a \mid \chi \mid \neg\phi \mid \phi_1 \wedge \phi_2 \mid \bigcirc\phi \mid \phi_1 U \phi_2 \mid \ominus\phi \mid \phi_1 S \phi_2 \mid \exists T.\phi$$
$$\chi ::= E_1 \theta E_2$$
$$\theta ::= < \mid \leq \mid = \mid > \mid \geq$$
$$E ::= T \mid t \mid x \mid E_1 + E_2 \mid E_1 - E_2$$

where,

$a \in A$ (the set of all possible LOTOS actions). Thus, atomic propositions in QTL are LOTOS actions.

$T \in \mathbf{T}$ (the set of *global time variables*).

$x \in \mathbf{N}$ (the set of natural numbers). Thus, the language is discrete time.

$\chi \in \mathbf{\chi}$, where χ is the domain of *timing constraints*. χ is totally general, i.e. arbitrary combinations of T, t, x, <, ≤, =, >, ≥, + and - are allowed.

$\theta \in \mathbf{\theta}$, where θ is the domain of *timing relations*.

$E \in \mathbf{E}$, where E is the domain of *timing expressions*.

In this language, t is a terminal symbol denoting the explicit clock variable, $\bigcirc$ is the temporal future operator *next*, U is the temporal future operator *until*, $\ominus$ is the temporal past operator *previous* and S is the temporal past operator *since* [Manna92].

Other propositional, existential and temporal operators can be defined as usual, e.g.

False:	false $\equiv \neg$true	Eventually:	$\Diamond\phi \equiv \text{true} U \phi$
Or:	$\phi_1 \vee \phi_2 \equiv \neg(\neg\phi_1 \wedge \neg\phi_2)$	Henceforth:	$\Box\phi \equiv \neg\Diamond\neg\phi$
Implication:	$\phi_1 \rightarrow \phi_2 \equiv \neg\phi_1 \vee \phi_2$	Once:	$\blacklozenge\phi \equiv \text{true} S \phi$
For all:	$\forall T.\phi \equiv \neg(\exists T.\neg\phi)$	Has-always-been:	$\blacksquare\phi \equiv \neg\blacklozenge\neg\phi$

Quality of Service Examples in QTL

Quality of service constraints (written in QTL) can be placed over the LOTOS multimedia stream behaviour presented in section 3.1. For the sake of simplicity in the logic statements, it is assumed that the medium is completely reliable (i.e. messages are never lost, corrupted or re-ordered). All data variables in the logic statements below are flexible variables, expect T and u which are both rigid variables.

1.a $\Diamond$ (arrive $\wedge$ t < 20)

1.b $\forall T. \Box$ ((arrive $\wedge$ t = T) $\rightarrow$ $\Diamond$ ((arrive $\wedge$ T+10 $\leq$ t < T+12)
$\vee$ (timeout $\wedge$ t = T+12)))

Statement 1.a ensures that at least one message arrives at the sink before 20 time units have elapsed and hence places a bound on the *latency* of the first transmission. Statement 1.b ensures that if a message arrives at time T, then at some time in the future another message should arrive within the time interval [T+10, T+12). If a message does not arrive, then a timeout must occur at time T+12. This statement constrains both the *throughput* and *jitter* of the transmission.

More complicated quality of service statements can be constructed, for example based on the mean and variance of message arrival times. As an illustration, the QTL statements below define throughput and jitter based on a mean arrival time of messages. In the following propositions we use Θ to refer to the initial state of the system.

2.a Θ : no_msgs = 0

2.b $\forall u. \Box$ ((arrive $\wedge$ no_msgs = u) $\rightarrow$ $\bigcirc$ (no_msgs = u+1))

3.a Θ : sum_arrivals = 0

3.b $\forall T \forall u. \Box$ ((arrive $\wedge$ t=T $\wedge$ sum_arrivals = u) $\rightarrow$
$\Diamond$ (arrive $\wedge$ sum_arrivals = u + t - T))

Statements 2.a and 2.b define a counter for the number of messages that have arrived (the no_msgs variable) and statements 3.a and 3.b calculate the sum of message arrival times (the sum_arrivals variable) in a similar way. Using these variables we can impose a throughput requirement of a mean of 20 frames per time unit with allowed variance of ±5 frames per time unit:

4. $\Box$ (15 $\leq$ (no_msgs ÷ sum_arrivals) $\leq$ 25)

4 A Real-time Synchronization Example

The examples in section 3.2.2 indicate that quality of service can be expressed in QTL. In this section we will demonstrate the use of QTL in expressing real-time synchronization requirements. This will be done through the presentation of a lip synchronization algorithm. This example was first written in the real-time programming language ESTEREL [Stefani92] and has also recently been written in Temporal LOTOS [Regan93], which is a single language based timed LOTOS. The requirements of the lip-synchronization algorithm are summarized below (we will take one time unit to be equal to a millisecond).

R1. *Sound Intervals*
   Sound packets must be presented every 30 milliseconds (ms).

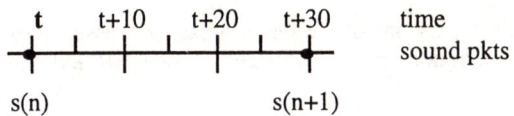

R2. *Video intervals*
   Video packets *should* be presented every 40ms, but a jitter of ±5ms is permitted (i.e. video packets must be presented between 35ms and 45ms after the previous packet).

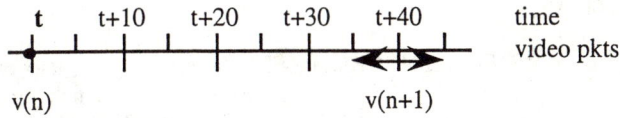

R3. *Synchronization of Sound and Video*
   (i) Video must not lag the associated sound packet by more than 150ms.
   (ii) Video must not precede the associated sound packet by more than 15ms.

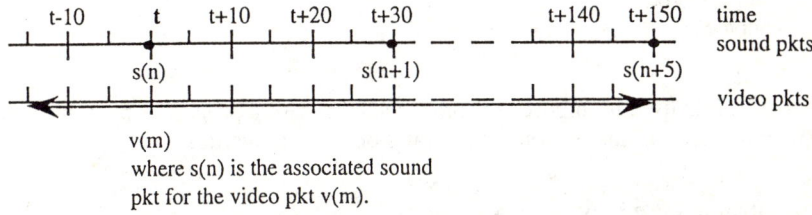

where s(n) is the associated sound
pkt for the video pkt v(m).

The overall structure of the lip-synchronization algorithm is given in figure 1. This figure shows two data sources, each of which communicate with their own manager when a data item is available (s_avail, v_avail). The managers then signal their readiness to a central controller (s_ready, v_ready) and wait for an ok signal (ok_s, ok_v) before asking the presentation device to present the data (s_present, v_present). Finally the presentation device replies with an acknowledgement (s_presented, v_presented) when the data item has been successfully presented.

It is the controller in this system that has the job of maintaining lip-synchronization between sound and video. The LOTOS specification of the controller's behaviour is:

process CONTROLLER[s_ready, v_ready, ok_s, ok_v, s_late_error,
                                    v_late_error,synch_error] : exit :=
   (s_ready; ok_s; exit
   []
   v_ready; ok_v; exit)
   >>

SOUND_SYNCHRONIZER[s_ready, ok_s, s_late_error]
|||
VIDEO_SYNCHRONIZER[v_ready, ok_v, v_late_error, synch_error]
where
(* NOTE: only one of the synchronizer processes needs to raise a synchronization error. The VIDEO_SYNCHRONIZER process does this below.
*)

process SOUND_SYNCHRONIZER[s_ready, ok_s, s_late_error] : exit :=
    s_ready; (ok_s; SOUND_SYNCHRONIZER[...]
    []
    s_late_error; exit)
endproc

process VIDEO_SYNCHRONIZER[v_ready, ok_v, v_late_error, synch_error] : exit :=
    v_ready; (ok_v; VIDEO_SYNCHRONIZER[...]
    []
    v_late_error; exit
    []
    synch_error; exit)
endproc
endproc

Then QTL is used to impose real-time synchronization properties over this behaviour. The remainder of this section concentrates on these properties.

Requirement R1: Sound Intervals

Firstly, we consider the requirement R1 that states sound packets must be presented every 30ms. To achieve this, we specify that the Controller must issue an ok_s action every 30ms. If this is late, an s_late_error must occur at the next time instance. For this first formula we provide a detailed explanation underneath the formula. For other formulae that follow a similar pattern, a detailed description will be omitted.

$$S1. \quad \forall T. \Box ((ok\_s \wedge t=T) \rightarrow \Diamond ((ok\_s \wedge t=T+30) \vee (s\_late\_error \wedge t=T+31)))$$

In this formula we specify that for all future states, if an ok_s occurs at some time T then, at some state in the future, either an ok_s will occur at time T+30 or an s_late_error will occur at time T+31. Note that in this logic formula we are not concerned about the time at which the first ok_s action occurs.

Requirement R2: Video intervals

Secondly, we consider the arrival of video packets. We specify that the controller must issue an ok_v action between 35ms and 45ms after the previous ok_v action. If this is late, a v_late_error must occur at the next time instance.

S2. $\forall T. \Box ((ok\_v \wedge t=T) \rightarrow \Diamond ((ok\_v \wedge T+35 \leq t \leq T+45) \vee$
$(v\_late\_error \wedge t=T+46)))$

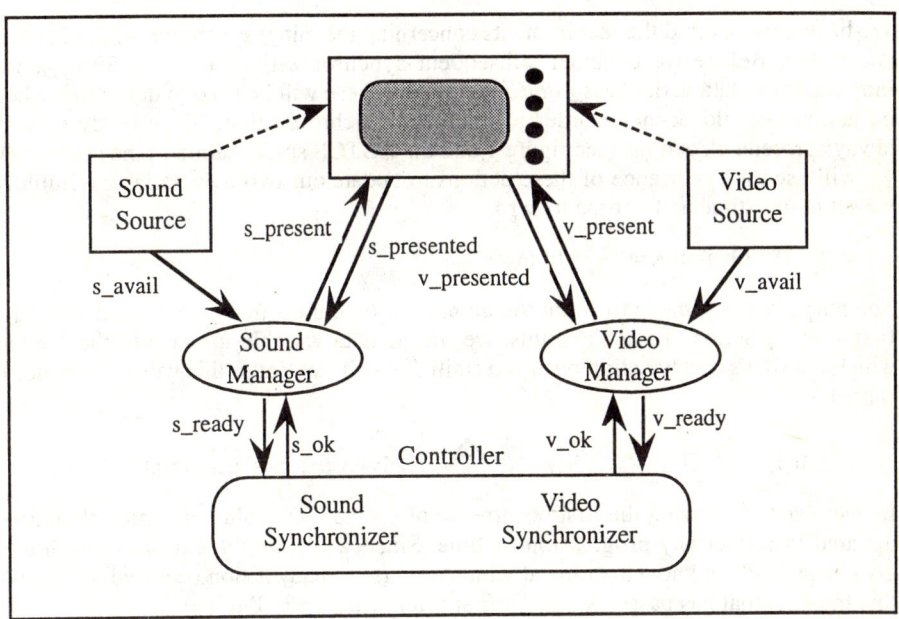

Fig. 1. The overall structure of the lip-synchronization algorithm.

Requirement R3: Synchronization of Sound and Video

Finally we must consider the synchronization between sound and video packets. To start with, we consider initial synchronization errors, i.e. when a sound packet has arrived, but the initial video packet fails to arrive and vice-versa.

In R3(i) we state that video must not lag the associated sound packet by more than 150ms. Hence after the first ok_s action, either an ok_v must follow within 150ms or else a synch_error must occur.

S3(i) $\Diamond (\exists T. ((ok\_s \wedge (\neg \Diamond ok\_s) \wedge t=T) \rightarrow \Diamond ((ok\_v \wedge t \leq T+150) \vee$
$(synch\_error \wedge t=T+151))))$

The premiss of this formula states that an ok_s occurs at some time T and it is not the case that an ok_s has occurred in a previous state (i.e. it is the first ok_s). It is sufficient for the formula to hold in just one future state (since you can only ever get one initial ok_s action); therefore we use the eventualy temporal operator at the start.

Similarly in R3(ii) we state that video must not precede the associated sound packet by more than 15ms. Hence after the first ok_v action, either an ok_s must follow within 15ms or else a synch_error must occur.

S3(ii) $\Diamond(\exists T. ((ok\_v \land (\neg \Diamond ok\_v) \land t=T) \rightarrow \Diamond((ok\_s \land t \leq T+15) \lor (synch\_error \land t=T+16))))$

We have now covered the requirements concerning the initial synchronization of sound and video. Before we consider subsequent synchronization (formula S3(iii)), we introduce two data variables s_time and v_time. These will be used to determine when ok actions should occur in order to maintain synchronization. Since ready actions always precede ok actions (see figure 1 and the LOTOS specification in the appendix), we will use the occurrence of these actions to update our two new variables. Initially we set both s_time and v_time to zero.

S4(i) $\Theta: s\_time=0 \land v\_time=0$

The purpose of s_time is to count the number of time units that have passed since the first s_ready action. To achieve this, we set the data variable initial_s to the time at which the first s_ready action occurred (initial_s will stay set to this value in all future states).

S4(ii) $\Diamond(\exists T. ((s\_ready \land (\neg \Diamond s\_ready) \land t=T) \rightarrow \Box(initial\_s=T)))$

In *every* state following the first occurrence of s_ready, the value of s_time should be updated to reflect any progression of time. Since we know the current time in any given state and we know the time at which the first s_ready action occurred, it follows that the time that has passed since the first s_ready action is T-initial_s.

S4(iii) $\forall T. \Box((\Diamond s\_ready \land t=T) \rightarrow O(s\_time=T-initial\_s))$

The second variable, v_time, will be used slightly differently. Instead of counting time that has passed, v_time will hold the value of the time at which the next ok_v is *expected* (the first ok_v is expected at time 0, the second at time 40 and so on). Therefore, with every occurrence of a v_ready action *except* the first, the value of v_time should be increased by 40.

S4(iv) $\forall u. \Box((v\_ready \land \Diamond v\_ready \land v\_time=u) \rightarrow O (v\_time=u+40))$

We can now return to the final synchronization formula. The formulae above have ensured that ok_s actions occur regularly every time unit or else an error is reported. All we are left to check is that the video presentation does not gradually drift out of synchronization from this regular sound presentation (for example, when video packets repeatedly arrive late). This is where we make use of the two variables s_time and v_time. For an ok_v action to occur (except the first ok_v), the variable s_time must lie within the bounds [v_time-15, v_time+150]. If the upper bound is exceeded, a synch_error must occur.

S3(iii) $\Box(ok\_v \to \Diamond(\,(ok\_v \land v\_time-15 \leq s\_time \leq v\_time+150)$
$\lor (synch\_error \land s\_time > v\_time+150)\,))$

We have shown that the LOTOS/QTL approach is powerful enough to express the real-time synchronization properties of lip-synchronization as defined by [Stefani92]. Furthermore, an initial comparison to the lip-synchronization specifications presented in [Stefani92] and [Regan93] indicate that our separated functional and non-functional behaviour approach leads to a significantly more elegant and abstract specification.

5 Validation

Validation typically involves refining a single specification and verifying that this refinement satisfies the specification. However, such an approach is not sufficient for validation with separated functional and non-functional behaviour. Verification with such separated behaviour involves showing not only that the refinement satisfies the previous specification, but also that the refinement satisfies the non-functional requirements.

It has become clear to use that there are two classes of timeliness properties: *timing requirements* (quality of service in the multimedia domain) and *performance assumptions*. The former express the desired behaviour of a system, while the latter define implementation specific timing, i.e. timing which results from the execution speed of system components. In the multimedia stream example a property stating that the sink process transmits frames at one second intervals would represent a performance assumption. These properties state the timing constraints enforced by the implementation, rather than the timing characteristics which are desired of the developed system. It is in using these performance assumptions that an understanding of the quantitative time validity of system behaviour can be determined. Specifically, performance assumptions must be combined with functional behaviour in order to yield a timed behaviour. Broadly speaking, performance assumptions can be seen to ground functional behaviour in real-time. Then it can be determined whether the QOS requirements are satisfied. We call this principle of dividing timing properties separating timing concerns; the prinicple is discussed in more depth in [Bowman93a].

As shown in figure 2 system development with separated timing concerns suggests that two forms of validation can be undertaken: *untimed validation* and *timed validation*. We will consider each in turn.

1. Untimed Validation

The untimed validation of refinement involves checking (vertically) that the refined specification satisfies the previous specification. Standard validation techniques for untimed formal system development can be used in this process. During untimed validation it is also possible to check that the functional behaviour is consistent with the QOS requirements in an untimed sense, i.e. that the two specifications do not express conflicting event ordering. The semantic basis for such a process is expressed in section (b) of appendix ii as the untimed semantic link between LOTOS and QTL.

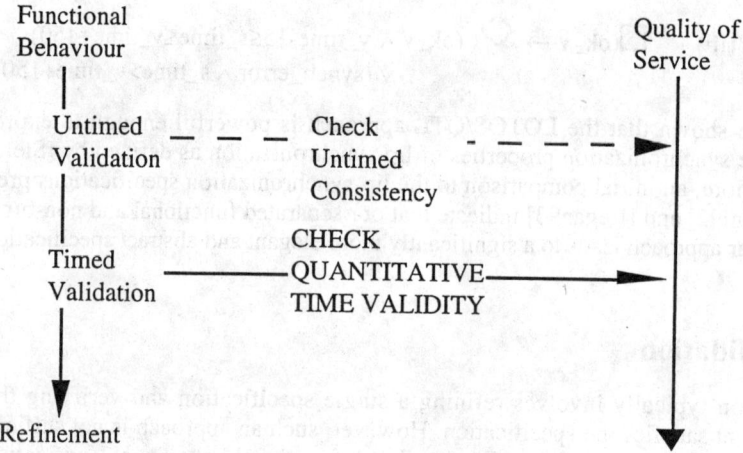

Fig. 2. System Development with Separated Timing Concerns

2. Timed Validation

An understanding of the quantitative time validity of a formal description can only be obtained through timed validation. Thus, while checking vertically that refined specifications satisfy previous specifications would still be undertaken, the major part of timed validation is checking quantitative time validity horizontally. During this process, the functional behaviour must be ground in concrete time. This is done by resolving performance assumptions and behaviour, to yield a representation of timed behaviour for those particular performance assumptions. The quantitative time validity of the behaviour with respect to the QOS requirements can then be determined. The semantic basis for such a process is expressed in section (b) of appendix ii as the timed semantic link between LOTOS and QTL.

It is important to note that the grounding of behaviour in performance assumptions does not embed these properties into the behaviour for further system development. The system developer will work exclusively with separated functional behaviour and the grounding of this behaviour in time is a mechanism completely internal to the validation process.

The Tempo project has investigated a number of means of realising these classes of validation. Unfortunately, a full discussion of these techniques is beyond the scope of this paper. Briefly though, we are presently involved in the implementation of a model checking technique for determining the quantitative time validity of a LOTOS specification with regard to QOS requirements and performance assumptions expressed in QTL. Once again, the emphasis in this work is on reuse of existing techniques by basing our implementation upon J. Ostroff's RTTL tools [Ostroff92].

6 Related Work

There has been little work specifically addressing the impact of multimedia on the formal description of distributed systems. Researchers at CNET, however, have developed an approach based on the real-time language Esterel and are currently developing a separate language to enable the specification of quality of service requirements [Stefani92]. However, this work is not as formally based as our approach.

A number of researchers have considered the more general impact of real-time on the formal description of distributed systems, however, as indicated earlier, the majority of this work has been single language based (which we criticise on the grounds that it produces formal descriptions in which functional and non-functional behaviour are entwined). Some workers have though considered techniques which are more closely related to our approach. Specifically, Timed CSP [Schneider91], Ostroff's TTM/RTTL approach [Ostroff89] and the Z/RTL approach [Fidge92] all employ some form of separated timing concerns. However, despite the fact that all these approaches use a timed logic in order to express timing requirements, none of the techniques go as far as we do in maintaining a complete separation between functional and non-functional behaviour throughout system development. The decision to employ such a complete separation of time and behaviour arises specifically from the multimedia application domain we are considering and the role of QOS in that domain.

Probably the approach most closely related to ours is due to J. Schot [Schot92]. He proposes a method for the formal design of distributed systems which provides facilities for the specification of real-time requirements. As in our approach, LOTOS is used to specify the qualitative temporal ordering of events in a system. In addition, he introduces a new language (based on a timestamped LOTOS-like notation) to separately specify the real-time requirements. However, no formal syntax or semantics are provided for this new language.

7 Concluding Remarks

The formal specification of communication protocols has attracted a great deal of research and a number of techniques, such as LOTOS and Estelle, have been developed. Such techniques are approaching a stage of maturity with comprehensive toolkits now available. However, the advent of distributed multimedia computing adds new challenges to the field of formal description. It is our belief that existing techniques do not adequately address these new requirements, particularly in terms of representing real-time behaviour. Furthermore, those approaches which do enable a representation of time generally encourage functional and non-functional behaviour to be entwined in a single specification.

The approach presented in this paper is based on a complete separation of timing concerns between functional behaviour and non-functional QOS requirements. Specifically, the functional behaviour of a system is specified in LOTOS with QOS expressed in QTL. There are a number of additional advantages which we believe arise when timing concerns are separated:

1. *horses for courses*

   The most suitable technique can be used for each part of the specification, i.e. a process algebra can be used for abstract behaviour and a temporal logic for system requirements.

2. *maintenance of standards*

   With this approach, there is no need to alter the LOTOS language to enable the specification of real-time behaviour. This is important given the standardisation of LOTOS. This also means that existing LOTOS tools can be used with the new approach.

3. *level of abstraction*

   It is our view that the embedding of timing values into specification of behaviour does not lead to abstract formal description. Specifically, it leads to the hard-wiring of performance assumptions into specifications. Critically, our approach enables system behaviour, timing requirements and performance assumptions to be treated separately and thus, the abstractness of the system behaviour does not need to be compromised. We believe that this principle has great relevance to the formal description of real-time systems in general. This aspect of the LOTOS/QTL approach is discussed in full in [Bowman93a].

4. *implementation non-specific*

   With the separation of all performance assumptions, the behavioural specification remains valid if these assumptions change, e.g. the round-trip delay in a network changes. Thus, the specification of behaviour is implementation non-specific and hence is portable.

As highlighted at the start of this paper, the emphasis in this project has been on reuse of known technology from the field of formal specification of distributed systems. Thus, the originality of the work presented is not in terms of the basic techniques used (i.e. LOTOS and real-time temporal logic), but rather in the novel use of this known technology. In particular, we regard the application of these techniques to the multimedia field and the combining of the two approaches in order to facilitate abstract and elegant specification of multimedia systems as the major contribution of the presented research.

Acknowledgements. Tempo is a SERC/DTI funded project under the IED programme (GR/G 01362).

References

[Alur91] Alur, R., and T.A. Henzinger. "Logics and Models of Real-Time: A Survey." REX Workshop. Real-Time: Theory in Practice, Editor: J.W. de Bakker, C. Huizing, W.P. de Roever and G. Rozenberg, Springer-Verlag, pp: 74-106, 1991.

[Anderson90] D.P. Anderson, S.Y. Tzou, R. Wahbe, R. Govindan and M. Andrews, "Support for Continuous Media in the DASH System", Proceedings of the 10th International Conference on Distributed Computing Systems, Paris, May 1990.

[Berra90] P.B. Berra, C.Y.R. Chen, A. Ghafoor, C.C. Lin, T.D.C. Little, and D. Shin. "Architecture for distributed multimedia database systems." Computer Communications Vol. 13 No. 4, pp 217-231, 1990.

[Bolognesi88] T. Bolognesi and E. Brinksma, "Introduction to the ISO Specification Language LOTOS", Computer Networks and ISDN Systems, Vol. 14, No. 1, pp 25-59, North-Holland, Amsterdam, 1988.

[Bowman93a] Bowman, H., G.S. Blair, L. Blair, and A.G. Chetwynd. "Time Versus Abstraction in Formal Description." FORTE'93, Sixth International Conference on Formal Description Techniques (also, available as Lancaster University Computing Department Report MPG-93-09), October1993.

[Bowman93b] Bowman, H., L. Blair, G.S. Blair, and A.G. Chetwynd. "Formal Description of Multimedia Systems: An Assessment of Potential Techniques", Internal Report MPG-93-05, Lancaster University. (also, to appear in Computer Communications), April 1993.

[Budkowski87] Budkowski, S., and P. Dembinski. "An Introduction to Estelle: A Specification Language for Distributed Systems." Computer Networks and ISDN Systems Vol. 4, pp: 3-23, 1987.

[Campbell93] Campbell, A., G. Coulson, F. Garcia, D. Hutchison, and H. Leopold. "Integrated Quality of Service for Multimedia Communications." Infocom '93, 1993.

[CCITT88] CCITT, "Recommendation Z.100:Specification and Description Language SDL", AP IX-35, 1988.

[Clarke87] Clarke, E.M. and O. Grumberg "Research on Automatic Verification of Finite State Concurrent Systems." Annual Review of Computer Science, pp: 269-290, 1987.

[Coulson92] Coulson, G., G.S. Blair, N. Davies and N. Williams, "Extensions to ANSA for Multimedia Computing", Internal Report MPG-90-11, available from the Computing Department, Lancaster University, Bailrigg, Lancaster, U.K., June 1992.

[Drayton92] L. Drayton, G.S. Blair and A.G. Chetwynd, "An Introduction to LOTOS through a Worked Example", Computer Communications, Vol. 15, No. 2, Butterworth-Heinemann, March 1992.

[Ehrig85] H. Ehrig and B. Mahr, "Fundamentals of Algebraic Specification", Springer-Verlag, 1985.

[Fidge92] Fidge, C.J. "Specification and Verification of Real-Time Behaviour Using Z and RTL." Formal Techniques in Real-time and Fault Tolerant Systems, Editor: J. Vytopil, Springer-Verlag, Pages: 393-410, 1992.

[Hehmann90] D.B. Hehmann, M.G. Salmony and H.J. Stüttgen, "Transport Services for Multimedia Applications on Broadband Networks", Computer

Communications, Vol 13, No 4, pp 197-203, Butterworth-Heinemann, May 1990.

[Henzinger90] Henzinger, T.A., Z. Manna, and A. Pnueli. "An Interleaving Model for Real-time." Fifth Jerusalem Conference on Information Technology, IEEE Computer Society Press, pp 717-730, 1990.

[Hoare85] C.A.R. Hoare, "Communicating Sequential Processes", Prentice-Hall, 1985.

[Horn92] Horn, F., L. Hazard, J.B. Stefani, G. Coulson, and G.S. Blair. "An Integrated Computational Model and Programming Platform for Open Distributed Multimedia Applications." 3rd International Workshop on Network and Operating System Support for Digital Audio and Video, 1992.

[ISO88] ISO, "Information Processing Systems - Open Systems Interconnection - LOTOS - A Formal Description Technique Based on the Temporal Ordering of Observational Behaviour", ISO /IEC 8807, Geneva, 1988.

[ISO89] ISO, "Information Processing Systems - Open Systems Interconnection - ESTELLE - A Formal DescriptionTechnique Based on an Extended State Transition Model", ISO/IEC 9074, Geneva, 1989.

[Jahanian86] Jahanian, F., and A.K. Mok. "Safety Analysis of Timing Properties in Real-time Systems." IEEE Transactions on Software Engineering 1986, pp: 890-904.

[Keshav93] Keshav, S. "Report on the Workshop on Quality of Service in High Speed Networks." Computer Communications Review 1993, pp: 74-85.

[Linington91] Linington, P. "Introduction to the ODP Reference Model." *International IFIP Workshop on Open Distributed Processing,* Editor: J. de Meer, V. Heymer and R. Roth, North-Holland, Pages: 3-13, 1991.

[Mañas92] J.A. Mañas, "Getting to use LITE", Proceedings of the Third LOTOSPHERE Workshop, Edited by T.Bolognesi, E. Brinksma, C.A. Vissers, Pisa, Italy, 1992.

[Manna92] Z. Manna, and A. Pnueli. The Temporal Logic of Reactive and Concurrent Systems. Springer-Verlag. New York. 1992.

[Miguel92] C. Miguel, A. Fernandez, and L. Vidaller. "Extending LOTOS towards performance evaluation." FORTE' 92: 5th International Conference on Formal Description Techniques, Editor: M. Diaz and R. Groz, 1992.

[Milner89] R. Milner, "Communication and Concurrency", Prentice-Hall, ISBN 0-13 115007-3, 1989.

[Nicolaou90] Nicolaou, C., "Architecture for Real-Time Multimedia Communication Systems", IEEE Journal on Selected Areas in Communication, Vol. 8, No. 3, pp 391-401, 1990.

[Nicollin91] Nicollin, X., and J. Sifakis. "An Overview and Synthesis on Timed Process Algebras." REX Workshop. Real-Time: Theory in Practice, Editor: J.W. de Bakker, C. Huizing, W.P. de Roever and G. Rozenberg, Springer-Verlag, pp: 74-106, 1991.

[Ostroff89] J.S. Ostroff, "Temporal Logic for Real-TimeSystems", Research Studies Press Ltd, ISBN 0-86380-086-6, 1989.

[Ostroff91] Ostroff, J.S. "Verification of Safety Critical Systems Using TTM/RTTL." REX Workshop. Real-Time: Theory in Practice, Editor: J.W. de Bakker, C. Huizing, W.P. de Roever and G. Rozenberg, Springer-Verlag, pp: 74-106, 1991.

[Ostroff92] Ostroff, J.S. "StateTime - a Diagrammatic Toolset for the Design and Verification of Real-time Systems", Technical Report CS-92-07, Department of Computer Science, York University, Ontario, Canada. July 1992.

[Pnueli77] Pnueli, A. "The Temporal Logic of Programs." Foundations of Computer Science 18, pp: 46-57, 1977.

[Regan93] Regan, T. "Multimedia in Temporal LOTOS: A Lip Synchronisation Algorithm." To appear at PSTV XIII, Protocol Specification, Testing and Verification, Liege, Belgium, May, 1993.

[Reisig85] Reisig, W. Petri Nets. Springer-Verlag. 1985.

[Schneider91] Schneider, S., J. Davies, D.M. Jackson, G.M. Reed, J.N. Reed, and A.W. Roscoe. "Timed CSP: Theory and Practice." Real-Time: Theory in Practice, Editor: J.W. de Bakker, C. Huizing, W.P. de Roever and G. Rozenberg, Springer Verlag, Pages: 640-675, 1991.

[Scholefield92] Scholefield, D.J., and H.S.M. Zedan. "TAM: A Formal Framework for the Development of Distributed Real-Time Systems." Formal Techniques in Real-Time and Fault-Tolerant Systems, Editor: J. Vytopil, Springer-Verlay, pp: 411-428, 1992.

[Schot92] J. Schot, "The Role of Architectural Semantics in the Formal Approach of Distributed Systems Design", ISBN 90-9004877-4, 1992.

[Sinderen91] Sinderen, M.v., and J. Schot. "An Engineering Approach to ODP Design." International IFIP Workshop on Open Distributed Processing, Editor: Jan de Meer and Volker Heymer, 1991.

[Stefani92] J-B. Stefani, L. Hazard and F. Horn, "Computational Model for Distributed Multimedia Applications based on a Synchronous Programming Language", Computer Communications (Special Issue on FDTs), Vol 15, Number 2, March 1992.

[Walter83] B. Walter, "Timed Petri Nets for Modelling and Analyzing Protocols with Real-Time Characteristics", In H. Rudin (ed), Protocol Specification, Testing and Verification III, pp 149-159, North-Holland, 1983.

Appendix i: LOTOS Semantics

LOTOS semantics without time. The LOTOS semantics that we work from are the standard *untimed* semantics from [ISO88], where *labelled transition systems* are used as a model of process behaviour and *many sorted algebras* are used as a model of abstract data typing. In this document, we are interested in the former of these two concepts. So, by way of clarification, each element of the set of all labelled transition system (LTS) is a 4-tuple Sys = <S,Act,Tr,B_0> where:

(i) S is a non-empty set of states. These states represent the set of behaviours that are derivable from the initial state of the specification.

(ii) Act is a set containing all the possible actions that arise from a LOTOS specification (including the internal action i).

(iii) Tr is a set of *transition relations*. There is a transition relation, denoted -a-> for each a∈ Act and -a-> is a set of *transitions* of the form B_1 -a-> B_2 where $B_1, B_2 \in S$.

(iv) $B_0 \in S$ is the *initial state* of Sys.

LOTOS semantics with time. As is indicated in section 5, in order to perform timed validation, functional behaviour must be ground in performance assumptions. In our approach this process will yield a timed semantic model for LOTOS. The notation of this semantic model is described below.

The standard LOTOS labelled transition system can be enhanced with quantitative time to produce a *timed labelled transition system*. The set of all such transition systems is denoted LTS^t and each element of LTS^t has the form: $Sys^t = <S, Act, Tr^t, B_0>$ where:

(i) S, Act and B_0 are of the form defined above.

(ii) Tr^t is a set of *timed transition relations*. Each timed transition relation is denoted -ax-> for a∈ Act and x∈ **N** (the set of natural numbers, which we use as our time domain) and each -ax-> is a set of *timed transitions* of the form B_1 -ax-> B_2 where $B_1, B_2 \in S$. In addition, Tr^t satisfies the property that $\forall a' \in$ Act, $\exists x' \in$ **N** such that -a'x'-> ∈ Tr^t (i.e. there is at least one timed transition relation for each action in Act).

The time value associated with the transition expresses the execution of the event relative to the previous event or to the initial instant of a behaviour, if there is no previous event. Thus, timed transitions express relative event timing rather than a a global time instant. The semantic model of LOTOS-T [Miguel92] has influenced this structure.

Appendix ii: Semantics for QTL

(a) Basic Semantics

QTL is defined in the usual way over a model. The notion of a *timed action trace* is used as the model for QTL. These correspond to a timed trace of the execution of a LOTOS process. The set of all timed action traces is denoted TAT and σ∈ TAT is of the form:

$$\sigma = \varepsilon \quad \text{(the empty trace) or}$$
$$\sigma = \tau_0, \tau_1, \tau_2, \tau_3, \tau_4, \ldots$$

where timed actions τ_i (we may also use τ when the position in the sequence is not significant) have the form $< a_i, x_i >$ for $a_i \in A$ and $x_i \in N$, i.e. they are a pair associating a discrete time instance x_i at which the action a_i occurs. With timed action pairs, we use $\alpha(\tau_i)$ to access the action in the pair and $\beta(\tau_i)$ to access the associated time instance. We use $l(\sigma)$ to denote the number of timed actions in σ and concatenation of traces is expressed using the infix operator ●.

In timed action traces the firing of actions is assumed to be instantaneous. In addition, we assume that time progresses validly through the trace. Specifically, we assume that $\forall i, 0 \leq i \leq l(\sigma)-1, \beta(\tau_{i+1}) \geq \beta(\tau_i)$, i.e. time cannot go backwards. However, time does not have to progress from timed action to timed action; more than one action is allowed to occur at the same time. This corresponds to micro time in [Henzinger90], where events are only distinguished by event ordering, and is consistent with an interleaved trace semantics for QTL. In addition, time gaps can occur in timed action traces, i.e. there can exist $x \in N$, $0 \leq x \leq \beta(\tau_{l(\sigma)})$ such that there does not exist an i such that $\beta(\tau_i)=x$. In particular, the first action does not have to occur at time 0. One reason for these time gaps is that the actions offered by a LOTOS process could be offered to the environment, which may not take them immediately.

Notation:

(i) The notation $(\sigma,i) \models_\varepsilon \phi$ should be read as the formula ϕ holds at position i of the timed action trace σ for a given environment ε.

(ii) The environment function $\varepsilon : T \rightarrow N$ maps global time variables to timing values, i.e. $\varepsilon(T)$ yields the timing value associated with T. By default, $\varepsilon(T)=\bot$ $\forall T \in T$, i.e. timing variables are undefined until a value has been assigned to them. The notation $\varepsilon \vdash \chi$ states that the timing constraint χ holds in the environment ε.

(iii) $\varepsilon[T/x]$ denotes the environment that is the same as the environment ε on all variables except T, which is mapped to x.

The formulae of QTL are interpreted over timed action traces. Specifically, given a QTL formula ϕ and a timed action trace σ with respect to the labelled transition system Sys, then the satisfaction relation $(\sigma,i) \models_\varepsilon \phi$ for $i \in N$ is defined inductively with respect to an environment function ε, as follows:

$(\sigma,i) \models_\varepsilon$ true = <u>true</u>

$(\sigma,i) \models_\varepsilon a$ iff $\alpha(\tau_i)=a$

$(\sigma,i) \models_\varepsilon \chi$ iff $\varepsilon[t/\beta(\tau_i)] \vdash \chi$

$(\sigma,i) \models_\varepsilon \neg\phi$ iff <u>not</u> $((\sigma,i) \models_\varepsilon \phi)$

$(\sigma,i) \models_\varepsilon \phi_1 \wedge \phi_2$ iff $(\sigma,i) \models_\varepsilon \phi_1$ <u>and</u> $(\sigma,i) \models_\varepsilon \phi_2$

$(\sigma,i) \models_\varepsilon \bigcirc\phi$ iff $(\sigma,i+1) \models_\varepsilon \phi$

$(\sigma,i) \models_\varepsilon \phi_1 U \phi_2$ iff $\exists k, k \geq i$, <u>s.t.</u> $(\sigma,k) \models_\varepsilon \phi_2$ <u>and</u> $\forall j, k > j \geq i, (\sigma,j) \models_\varepsilon \phi_1$

$(\sigma,i) \models_\varepsilon \ominus\phi$ iff $(i>0)$ <u>and</u> $(\sigma,i-1) \models_\varepsilon \phi$

$(\sigma,i) \models_\varepsilon \phi_1 S \phi_2$ iff $\exists k, 0 \leq k \leq i,$ <u>s.t.</u> $(\sigma,k) \models_\varepsilon \phi_2$ <u>and</u> $\forall j, k<j\leq i, (\sigma,j) \models_\varepsilon \phi_1$

$(\sigma,i) \models_\varepsilon \exists T.\phi$ iff $(\sigma,i) \models_{\varepsilon[T/x]} \phi$ <u>for some</u> $x \in N$

For simplicity in the presentation of these semantics, only the modelling of timing variables has been presented. Standard data variables can be modelled in a similar way.

(b) Linking LOTOS and QTL Using Legal Timed Action Traces

The previous section has defined QTL in terms of the general notion of timed action trace models. In order to facilitate formal reasoning about the QTL/LOTOS approach the relationship between the timed action trace model for QTL and the semantic structure for LOTOS must be defined. In this task we must differentiate between the untimed relating of LOTOS and QTL semantics and the timed relating of LOTOS and QTL semantics. The former is used to express the satisfaction relationship required for determining untimed consistency and the latter is used to express the satisfaction relationship for timed validation. We will deal with each separately:

Untimed Semantic Link. In this section we will define the satisfaction relation $\vdash_u \subseteq \text{LTS} \times \text{QTL}$ for untimed behaviour. Firstly then, assuming σ is such that $\forall i, 0 \leq i \leq l(\sigma), \alpha(\tau_i) \in \text{Act}$, we define the notion of σ being a *legally ordered execution* with respect to a behaviour $B \in S$ of a labelled transition system $\text{Sys}=<S,\text{Act},\text{Tr},B_0>$, as:

$$B \cdots \!\!> \sigma \text{ iff } \sigma=\varepsilon \vee (\sigma=\tau \bullet \sigma' \wedge \exists B' \in S \text{ s.t. } B-\alpha(\tau)\text{->}B' \wedge B' \cdots \!\!> \sigma')$$

This property guarantees that σ is consistent with the given labelled transition system.

We now define, for $B \in S$, the set of all legally ordered timed action traces with respect to Sys as:

$$\$\text{TAT}^{lo}(B) = \{\, \sigma \mid B \cdots \!\!> \sigma \,\}$$

Then the satisfaction relation can be defined as:

$$B \vdash_u \phi \text{ iff } \forall \sigma \in \$\text{TAT}^{lo}(B), (\sigma,0) \models_\varepsilon \phi$$

In particular, a QTL formulae holding (untimed) over the whole of a LOTOS specification represented by the labelled transition system Sys, is expressed as $B_0 \vdash_u \phi$.

Timed Semantic Link. In a similar way, we can define the satisfaction relation $\vdash_t \subseteq \text{LTS}^t \times \text{QTL}$ for timed behaviour. Firstly then, we define the notion of a trace σ being a *legally timed execution* with repect to a behaviour $B \in S$ of a timed labelled transition system $\text{Sys}^t=<S,\text{Act},\text{Tr}^t,B_0>$, as:

(~) $\quad B \!\rightarrow\! \sigma$ iff $\sigma=\varepsilon \vee ((\exists B' \in S \text{ s.t. } B-\alpha(\tau_0)\beta(\tau_0)\text{->}B') \wedge B' \!\rightarrow\! \sigma)$

where assuming that $\sigma \neq \varepsilon$, we define:

$\quad\quad B \!\rightarrow\! \sigma$ iff $\sigma=\tau \bullet \varepsilon \vee$
(*) $\quad\quad\quad (\sigma=\tau \bullet \tau' \bullet \sigma' \wedge (\exists x \in N, \exists B' \in S \text{ s.t. } B-\alpha(\tau')x\text{->}B' \wedge \beta(\tau')-\beta(\tau)=x$

$$\wedge \; B' \xrightarrow{\tau' \bullet \sigma'})\,)$$

The complexity in these conditions is required in order to make sure that a timed transition exists in Sys^t which satisfies the timing between each timed action in σ. In particular, we are relating TATs, in which the timing of each action execution is expressed in terms of a global clock, to LTS^ts in which the timing of each action execution is expressed relative to previous and succeeding behaviours (i.e. locally). Subclause (*) is the general case; it compares timed action global times to timed transition relative times. Subclause (~) determines that the first timed action is correctly timed. This is an exception to the general rule of subclause (*), since the first timed action will define a relative timing (relative to the initial instant of the behaviour B).

In a similar way to above we define, for $B \in S$, the set of all legally timed TATs with respect to Sys^t as:

$$\$TAT^{lt}(B) = \{\, \sigma \mid B \xrightarrow{\sigma} \,\}$$

The satisfaction relation can be defined as:

$$B \vdash_t \phi \;\; \text{iff} \;\; \forall \sigma \in \$TAT^{lt}(B), \; (\sigma,0) \models_\varepsilon \phi$$

and $B_0 \vdash_t \phi$ expresses the property that a QTL formula holds (timed) over the whole of a LOTOS specification as represented by the timed labelled transition system Sys^t.

The satisfaction relations $\vdash_u$ and $\vdash_t$ are the semantic basis of validation techniques being developed in the Tempo project for the LOTOS/QTL approach.

On the Synchronization Mechanisms for Multimedia Integrated Services Networks

Wei Yen and Ian F. Akyildiz

School of Electrical and Computer Engineering
Georgia Institute of Technology; Atlanta, GA 30332
Tel: 1-404-894-5141; Fax: 1-404-853-9410
E-mail: wei@eecom.gatech.edu; ian@armani.gatech.edu

Abstract. With the advance of communication technology, integrated services networks make possible real-time, multimedia applications. Unlike traditional data traffic, real-time multimedia traffic requires synchronization; temporal relationships among media must be maintained. Yet delay jitter and the absence of a global clock may disrupt these temporal relationships. This paper introduces new group synchronization protocols for real-time, multimedia applications, including teleconference, teleorchestration and multimedia on-demand services. The proposed protocols achieve synchronization for all configurations (one-to-one, one-to-many, many-to-one, and many-to-many), and it does so without prior knowledge of the end-to-end delay distribution, or the distribution of the clock drift. The only a-priori knowledge the protocols require is an upper bound on the end-to-end delay. The paper concludes with simulation experiments showing that the mechanisms work effectively in both LAN and WAN environments.

Key Words: High Speed Networks, Multimedia Services, Synchronization, Asynchrony, Delay Jitter, Local Clock Drift, Initial Collection Time, Initial Playback Times, Performance Evaluation.

1 Introduction

Real-time[1], multimedia applications present a critical problem to integrated services networks – synchronization. Such applications require both *intramedia* and *intermedia* synchronization. *Intramedia* synchronization defines the timing relationship of a single medium over a single connection. *Intermedia* synchronization defines those relationships for multiple connections, or for multiple media interleaved in a single connection.

Multimedia applications require synchronization in four configurations. The first configuration, known as *unicast*, is a one-to-one relation. In the unicast configuration, one user transmits temporally related media units and another user receives the media units and plays them out via different display devices.

---
[1] End-to-end delay must be less than a few hundred milliseconds for real-time applications.

For example, video/voice mail has the unicast configuration. The second configuration, *multicast*, is a one-to-many relation. One sender multicasts temporally related media units to many receivers. For example, some teleconferencing applications such as tele-lectures have the multicast configuration. The third case is the *retrieval* configuration which is a many-to-one relation. In this configuration, a receiver retrieves media from different senders. A user may, for example, get video and voice from two different databases. Finally, there is the *group* configuration which is a many-to-many relation. Many users in a communication group, each play the role of sender and receiver. Teleconferencing examplifies the group configuration. Note that the unicast, multicast, and retrieval configurations are special cases of the group configuration.

In all four configurations, four sources of asynchrony can disrupt synchronization: (*a*) Delay jitter, (*b*) Local clock drift, (*c*) Different initial collection times and (*d*) Different initial playback times.

(a) Delay Jitter. Delay jitter is the variation of end-to-end delay which consists of three parts [6]. First is the collection delay, the time needed for the sender to collect media units and prepare them for transmission. Collection may occur directly from media recorders such as camera or from databases, or both. The second part is the network delay from the network boundary at the sender to the boundary at the receiver. Finally, delivery delay adds to the end-to-end delay. Delivery delay is the time the receiver needs to process the media units and prepare them for playback. Note that none of these delay components are necessarily constant. For example, the network delay in ATM networks varies because different queueing delays caused by the unpredictable burstiness in the network and by the variable transmission bit rates. The collection delay and the delivery delay also vary from time to time due to different processing time (coding/decoding, segmentation/assembling etc.) of media units.

To see the effect of delay jitter, consider several temporally related media units sent from a sender to a receiver via different connections. These media units must be played back with appropriate temporal relationships. The presence of delay jitter, however, may destroy the temporal relationships that exist at the sender. When the receiver is ready to play a media unit, that unit may not be available yet because it experienced greater end-to-end delay than preceeding units. In such a case, a playback *discontinuity* occurs at the receiver. Delay jitter[2] becomes so significant in WAN environments that playback discontinuity is perceptible to humans.

(b) Local Clock Drift. Local clock drift arises when clocks at users run at different rates. Without a synchronization mechanism, the asynchrony gradually will become more and more serious. If the playback rate of the receiver is faster than the collection rate of the sender, the receiver may suffer *starvation*. As with delay jitter, playback discontinuity may result. Alternately, the receiver can become *flooded* with media units if its playback rate is slower than the collection rate of the sender. Both of these problems can be considered to be aspects of the *asynchrony*.

---

[2] End-to-end delay is also an important factor in WAN environments.

(c) Different Initial Collection Times. In both of the retrieval and the group configurations, there are more than one sender in the communications. These senders must collect and transmit synchronously; otherwise, the temporal relationships among media units might be destroyed. For example, consider two media sources, one providing voice and the other video. If they start to collect their media units at different times, then playback of the media units of voice and video from two sources at the destination loses semantic meaning. Media units with no temporal correlation are replayed simultaneously resulting in the so-called "lip-sync" problem. Note that the different end-to-end delays and clock drift can also cause the "lip-sync" problem.

(d) Different Initial Playback Times. As receivers, users in a group must start to play temporally related media units simultaneously, so that each user initially perceives media units synchronously. If the initial playback times are different for each user, then asynchrony will arise. For some multicast applications in which fairness is the major concern, the playback times of media units of all receivers should be the same; otherwise, the earlier a receiver gets media units, the earlier he can react.

Note that only delay jitter and local clock drift arise in the unicast configuration. Multicast configurations must also deal with different initial playback times, while retrieval configurations may experience different initial collection times in addition to delay jitter and local clock drift. All four sources of asynchrony occur in the group configuration.

In addition to the classification by configuration, intermedia synchronization may be classified based on application. Such a classification results in *Lip* synchronization and *Synthetic* synchronization [12, 17]. Lip synchronization applications produce, transmit and play media units in real time. Synthetic synchronization applications synchronize media units retrieved from databases. Typically, the temporal relationships among media units are static in lip-sync applications and are variable in synthetic-sync ones. For some synthetic-sync applications, data can be retrieved well ahead of their playout time so that the system can have the flexibility to schedule the communication task. Teleconferencing and on-line multimedia inquiry services are two examples of lip and synthetic synchronizations, respectively. All four sources of asynchrony affect both lip and synthetic synchronization applications.

2 Related Work

Recently there has been a growing interest in the development of synchronization protocols to solve asynchrony problems. Several papers have been published dealing with various types of asynchrony problems[3].

The adaptive feedback protocol [15] solves the asynchrony in multimedia on-demand services, i.e., 1-n configuration. It requires an additional connection for each sender and receiver pair to transmit feedback units. In the model [15], a

---

[3] A study on the sensitivity of human perception for various multimedia applications has been done in [13].

multimedia server transmits media units to several mediaphones which are playback devices such as audiophones and videophones. Among the mediaphones, one is assigned to be the master mediaphone and others are slave mediaphones. The protocol synchronizes slaves to the master by collecting and comparing the feedback units at a multimedia server. The multimedia server stores media units and provides them to mediaphones as they request them. The multimedia server estimates the asynchrony and instructs mediaphones to skip or pause media units accordingly. Adaptive feedback performs well in LAN environments but is not suitable for WANs because its synchronization guarantees are limited by the maximum network delay.

Pre-compiled scheduling protocol [16, 17] is designed for synthetic-sync applications. They assume media units are retrieved well ahead of transmission so that they can use an *Object Composition Petri Net* (OCPN) to capture the temporal relationships among media units. According to end-to-end delays and the OCPN, they calculate the playout and transmission schedules for receivers and senders, respectively. Senders and receivers will then follow the computed schedules to transmit and play the media units. As long as the network itself is synchronized, and the computed schedules are followed precisely, pre-compiled scheduling solves delay jitter, initial collection times and initial playback times. This method [16, 17] adds nearly all overhead to the beginning of the communication; it requires almost no overhead when the actual transmission starts. The critical problem with this approach is that the schedules cannot be calculated in advance if the temporal relationships among media are not known or predictable. In addition, the scheme does not consider the clock drift problem.

Although not specifically devised for multimedia synchronization, delay jitter control [5, 14] can be used to correct delay jitter problems. It distributes the synchronization responsibility among intermediate nodes between senders and receivers. However, the intermediate nodes which perform synchronization are unlikely to be work-conserving. The resulting average network delay will become longer while the variance becomes smaller.

The flow synchronization protocol [6] can, in some cases, correct delay jitter, initial collection times, and initial playback times. The protocol assumes that a clock synchronization scheme exists to handle the local clock drift problem. The protocol lets users [4] in a synchronization group exchange flow information periodically so that the common synchronization delay for the group can be calculated. The main problem of this protocol is the assumption of the underlying clock synchronization scheme. In contrast, our protocol corrects local clock drift explicitly.

A real-time *Stream Synchronization Protocol* (SSP) aiming to provide multimedia news services was proposed in [8]. This protocol, first, determines the upper bound of tolerant asynchrony among media streams according to the QoS requirements of applications. In order to synchronize media streams, this protocol adds "intentional delay" to those media streams which experience less end-to-end delay without violating the tolerant asynchrony bound. User interac-

---

[4] "users" are identical to "processors" in [6].

tions, including pausing, skipping, and scanning forward or backward, are also addressed in this paper. However, the local clock drift problem is not explicitly mentioned in this paper nor is the different initial playback times problem.

In addition to these five protocols, a media mixing algorithm was proposed in [9] to support multimedia conferences. The algorithm performs hierarchical mixing in a set of mixers, the root mixer multicasts the mixed media streams to all participants. This algorithm corrects delay jitter and local clock drift, but it is not clear how a multimedia conference is initiated synchronously. Media mixing leaves unresolved the problems of different initial collection times and different initial playback times. Furthermore, mixers add the additional end-to-end delay. The additional delay may be particularly critical in WAN environments. Furthermore, Steinmetz and Engler conducted experiments to explore the sensitivities of human perceptions to asynchrony in multimedia applications [13]. Their results define synchronization requirements for different types of multimedia applications.

In this paper, we develop synchronization protocols proposed in section 3 to eliminate the four sources of asynchrony for the unicast, multicast, retrieval and group multimedia configurations in real-time broadband integrated services networks. Our protocols should be placed at orchestration/sychronization layer to provide synchronization services to application layer [1, 7]. However, application layer needs to inform our protocols of the description of multimedia objects including the synchronous flows and their collection and playback periods. In section 4, we demonstrate that our protocols perform equally well in LANs and WANs[5] without a global clock. In section 5 we conclude the paper.

3 A Protocol for Group Synchronization

A group in the network consists of a set of users of an application and of a set of connections among them. Each user in the group has full duplex connection to all other users in the group. Moreover, each user has its own processor and may play roles of a sender or a receiver or both as Figure 1 shows. A sender collects media units from multimedia sources such as media recorders or databases while a receiver plays out media units via media display devices such as a speaker or terminal. The users in the same group need to collect and playback media units synchronously.

In teleconference, a set of users at different sites attempt to have a virtual meeting via an integrated services network. Each user transmits his own media units while receiving media units from the other participants. Media units collected by users in the conference at approximately the same time should be played by all users at the same time. Thus, all participants of the teleconference are in one group. For a teleorchestra, a conductor and a group of musicians at distributed locations play a symphony and broadcast the music to some audience

---

[5] In some cases, LAN delay can be greater than WAN delay. However, in this paper, we assume end-to-end delay is less than $20ms$ in LANs and greater than $100ms$ in WANs.

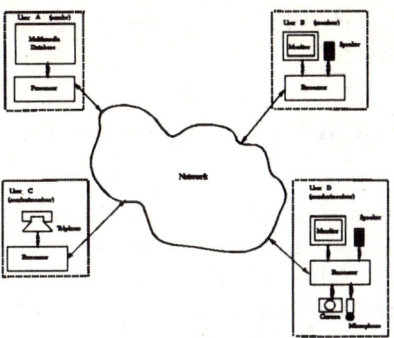

Fig. 1. User Model

in real time. Media units collected from the conductor must be played out to all musicians simultaneously; the playback time of the media units from the conductor must be synchronized at each musician. Failure to meet the requirement would cause the teleorchestra to become discordant and cacophonous. However, the playback times of the temporally related media units at each audience need not to be the same because fairness is not the concern in this case.

Distributed multimedia databases provide on-demand multimedia services to an audience. The distributed databases playing the roles of senders need to collect media units sychronously. Although each audience needs to play out the media units from the databases synchronously, it is not necessary to synchronize the playback time of media units among the audiences, even when they order the same program. Thus, each audience playing the role of receiver belongs to a different group while the sending databases belong to all groups.

3.1 Compensating the Delay Jitter Effect

The asynchrony caused by delay jitter between sender i and receiver j can be solved by prefetched buffering with buffer size B_j given by the following formula [15].

$$B_j = \lceil \frac{J_{i,j}}{\mu_j} \rceil \tag{1}$$

where $J_{i,j}$ is the delay jitter between i and j and is computed by

$$J_{i,j} = \Delta_{i,j} - \delta_{i,j} \tag{2}$$

with $\Delta_{i,j}$ is the maximum end-to-end delay between i and j; $\delta_{i,j}$ is the minimum end-to-end delay between i and j; μ_j is the playback period of a medium at receiver j; and $\lceil x \rceil$ is the smallest integer greater than x. If $J_{i,j}$ can be kept small, then the required prefetched buffer size B_j will decrease as can be seen in equation (1).

Since the values of $\Delta_{i,j}$, $\delta_{i,j}$ and μ_j can be determined in the negotiation phase for multimedia communication between users i and j, the delay jitter effect can be solved by selecting an appropriate size of prefetched buffer at the receiver as given in equation (1).

3.2 Dynamic Period Adjustment Protocol for Local Clock Drift Problem

To compensate for local clock drift our scheme synchronizes the clocks of all users in the same group according to a time reference [11] which we call *virtual global time* (VGT). The value of VGT is equal to the value of the local clock of the *chairman* elected by the distributed election algorithm presented below. The chairman distributes the time information, i.e., the value of the VGT, based on its local clock to all other users in the group (Section 3.3). This time information is then used to synchronize the local clocks of all users to the VGT. First we present the algorithm for the election of the chairman.

Distributed Election Algorithm. In the negotiation phase, each user of a group generates a one-byte random number between 0 and 2^8 and broadcasts it to all other users. Each user then sorts these random numbers in a descending order. The user generating the largest random number is chairman. If more than one user generates the same largest number, then the algorithm ignores that number and selects the next largest number from the remaining set. The search for a chairman continues until a unique maximum is found. If the algorithm fails to pick a chairman by using the set of random numbers generated by the users in the first round, then it proceeds with another round in the same fashion. Once the chairman is selected, it creates a queue which contains all users in the group. The order of the users in the queue represents their priority to become the next chairman. The value of the local clock of the first user in the queue (the chairman) is used as the VGT. This queue is broadcast by the chairman to all other users in the group. If a user wants to leave the group, it is removed from the queue by all other users. If a user wants to enter the group, it is added to the end of the queue by all other users. Note that another election algorithm for a distributed clock synchronization system is given in [10]; there the existence of a master clock[6] is assumed and the algorithm simply elects the next master clock when the original one fails.

Dynamic Period Adjustment Protocol. Once the group has elected a chairman, a mechanism must synchronize the playback period at user j to the collection period at user i, given i is the chairman. We denote the initial playback period decided by i and j in the negotiation phase by $\mu_j^{(init)}$ and the adjusted playback period which we want to achieve by μ_j. Note that the initial playback period $\mu_j^{(init)}$ at j and the collection period η_i at i have the same value. User j needs to obtain the time information of user i so that he can use it to adjust its playback period μ_j.

We stamp the first packet of each media unit k from i to j with the timestamp $\phi_{i,j}(k)$ at which i began to collect the media unit. The delaystamp $\psi_{i,j}(k)$ is the sum of all times spent by the packet in the nodes (switches) of the network plus

---

[6] "Master clock" and "chairman" are equivalent in [10].

the collection delay at i and the delivery delay at j [18]. The k-th media unit sent from user i is available to be played by j at time $S_{i,j}(k)$ which is

$$S_{i,j}(k) = \phi_{i,j}(k) + \psi_{i,j}(k) + \theta_{i,j} \tag{3}$$

where $\theta_{i,j}$ is the propagation delay between users i and j.

Let

$$\sigma_i(k) = S_{i,j}(k+1) - S_{i,j}(k) \tag{4}$$

represent the value of the interarrival time between media units k and $k+1$ at user j referring to local clock of i.

Substituting equation (3) into equation (4) we obtain

$$\sigma_i(k) = \phi_{i,j}(k+1) - \phi_{i,j}(k) + \psi_{i,j}(k+1) - \psi_{i,j}(k) \tag{5}$$

Ideally, the *intercollection* time $[\phi_{i,j}(k+1) - \phi_{i,j}(k)]$ in equation (5) is constant for all $k = 1, 2, \ldots$ media units and is equal to the collection period η_i at user i, i.e,

$$\eta_i = \phi_{i,j}(k+1) - \phi_{i,j}(k) \tag{6}$$

In this case equation (5) can be rewritten as:

$$\sigma_i(k) = \eta_i + \psi_{i,j}(k+1) - \psi_{i,j}(k) \tag{7}$$

If the *intercollection* time is not constant, then we can use equation (5) to estimate $\sigma_i(k)$.

In order for user j to find out the local clock drift between i and himself, our protocol requires a timer at user j. Let $\nu_j(k)$ be the value of this timer which counts the interarrival times between two consecutive media units, say k and $k+1$, using the local clock of j.

Now we want to adjust the negotiated playback period $\mu_j^{(init)}$ at j to μ_j so that the time user i spends from zero to the collection period η_i is the same as the time user j spends from zero to the adjusted playback period μ_j. Therefore, the ratio of the interarrival time $\nu_j(k)$ referring to user j's clock and the interarrival time $\sigma_i(k)$ referring to user i's clock is equal to the ratio of the adjusted playback period μ_j at user j and the collection period η_i at user i:

$$\frac{\nu_j(k)}{\sigma_i(k)} = \frac{\mu_j}{\eta_i} \tag{8}$$

By substituting equation (7) into equation (8) we obtain

$$\mu_j = \frac{\eta_i \nu_j(k)}{\eta_i + \psi_{i,j}(k+1) - \psi_{i,j}(k)} \tag{9}$$

User j plays media units according to the adjusted playback period μ_j obtained from equation (9) referring to its own local clock. Note that equation (9) assumes that the intercollection time is constant. However, equation (9) can easily be modified in the case of variable intercollection time using equation (5).

Let $A_{i,j}(k)$ define the dynamic period adjustment factor as:

$$A_{i,j}(k) = \frac{\mu_j - \mu_j^{(init)}}{\mu_j^{(init)}} \qquad (10)$$

where $\mu_j^{(init)}$ is the negotiated playback period at user j and μ_j is the adjusted playback period, equation (9). $A_{i,j}(k)$ measures the drift between the local clocks of users i and j.

By substituting equation (9) into equation (10) and using the fact that the value of $\mu_j^{(init)}$ is equal to η_i, then $A_{i,j}(k)$ can be computed from:

$$A_{i,j}(k) = \frac{\nu_j(k)}{\eta_i + \psi_{i,j}(k+1) - \psi_{i,j}(k)} - 1 \qquad (11)$$

The local clock of each user in the group can be synchronized to the chairman's clock by periodically applying the dynamic period adjustment protocol to all users pair $(chairman, j)$ where j is any user in the group except the chairman. The period of adjusting μ_j depends on the synchronization requirements of applications and the level of clock drift. The performance of the dynamic period synchronization mechanism with different adjustment periods will be shown in section 4.

3.3 Synchronization Protocol for Initial Collection Time

In some applications, such as teleconference, it is important that users start to play and collect at the same time to maintain the temporal relationships among media units, i.e., the semantic meaning of communication. In this subsection, we present the mechanism to synchronize the initial collection time of all users in a group. The combination of the initial collection time synchronization mechanism and the period synchronization mechanism introduced in section 3.2 assures that users provide media units in a synchronous manner.

Initial collection time synchronization is either explicitly stated or implicitly assumed in some of the related works [6, 9, 17]. However, the methods devised to guarantee initial collection time synchronization require some sort of initiator who sends the initialization message to users in the group [6, 17]. The initiator may or may not be one of the users in the synchronization group. The initiator will not send the initialization message until all users in the group are ready so users in the group need to inform the initiator when they are ready to collect the media units. Those methods delay the global initial collection time of the group because of the message exchange among the users and the initiator. In our approach the initial sending time is computed and decided in a distributed manner and it is the earliest time that users can initiate their collection of media units.

Our method requires the recovery of VGT at all users in the group. If the propagation delay $\theta_{i,j}$ between the chairman and every user in the group is known, then every user can compute the VGT value by using the equation (3). In the following we explain how to measure the propagation delay.

In the negotiation phase, user i transmits a packet α to user j with the timestamp $\phi_{i,j}(\alpha)$. The delaystamp $\psi_i(\alpha)$ is the total time spent by the packet in the network plus the collection and delivery delays as we described in section 3.2. After receiving the packet α, user j adds the time that α spends there to the delaystamp $\psi_{i,j}(\alpha)$ and sends α back to user i. So, the packet α comes back to user i at time $T_{i,j}(\alpha)$ which is:

$$T_{i,j}(\alpha) = \phi_{i,j}(\alpha) + \psi_{i,j}(\alpha) + 2\theta_{i,j} \tag{12}$$

where $\psi_{i,j}(\alpha)$ is the accumulated delaystamp for packet α; $2\theta_{i,j}$ is the round-trip propagation delay between i and j. Since $T_{i,j}(\alpha)$, $\phi_{i,j}(\alpha)$, and $\psi_{i,j}(\alpha)$ are known, the propagation delay $\theta_{i,j}$ of the packet α for the connection can be obtained by rewriting equation (12).

$$\theta_{i,j} = 0.5[T_{i,j}(\alpha) - \phi_{i,j}(\alpha) - \psi_{i,j}(\alpha)] \tag{13}$$

In the negotiation phase, each user measures the propagation delay $\theta_{i,j}$ between itself and all other users in the group. Once the chairman is appointed, user i can use the estimated propagation delays $\theta_{ch,i}$ between itself and the chairman to compute the value of VGT using equation (3). According to equation (3), the estimated propagation delay $\theta_{ch,i}$ and delaystamp $\psi_{ch,i}$ affect the accuracy of computed VGT. If intermediate nodes provide precise measurement on $\theta_{ch,i}$ and $\psi_{ch,i}$, then we can expect the accuracy is within millisecond range. The users (excluding the chairman) increase the VGT with the speed of their local clocks and confirm the VGT values with the time information carried implicitly in some packets sent from the chairman. After the chairman starts to distribute VGT, each user i can inform all other users in the group of its initial collection time I_i in terms of VGT.

Note that the initial collection time I_i of user i must be bounded as:

$$I_i \geq \max_j\{\Delta_{i,j}\} + \tau_i \tag{14}$$

where $\max_j\{\Delta_{i,j}\}$, is the largest among the maximum end-to-end delays of the user pairs (i,j) in the network and τ_i is the time in terms of VGT at which user i broadcasts its initial collection time I_i to all other users. If I_i does not satisfy the inequality (14), then there exists a user j with the maximum end-to-end delay $\Delta_{i,j}$ such that $\Delta_{i,j} + \tau_i > I_i$. Namely, it is possible that j is still waiting for I_i at the moment other users start to collect media units.

After each user receives the initial collection times I_i for all users i in the group, the largest initial collection time is picked by users in the group as the global initial collection time G_c.

$$G_c = \max_i\{I_i\} \tag{15}$$

where I_i is obtained from equation (14). Each user begins to collect his media units at G_c.

3.4 Synchronization Protocol for Initial Playback Time

Once the initial playback times are synchronized given by the protocol in the previous section, the dynamic period synchronization mechanism described in section 3.2 will adjust the collection/playback periods according to the VGT. The combination of initial playback time synchronization protocol and the period synchronization protocol assures that users play media units in a synchronous manner.

Now we show the derivation of the global initial playback time G_p. Without loss of generality, we assume that the sending times of a media unit from user i to all other users in the group are different. In the negotiation phase of the connections, the maximum end-to-end delay $\Delta_{i,j}$ is decided by the QoS requirements of applications and available resources [1, 3].

By assuming the maximum collection and delivery delay are known, we can determine the maximum end-to-end delay $\Delta_{i,j}$. The minimum end-to-end delay $\delta_{i,j}$ is equivalent to the propagation delay $\theta_{i,j}$ between users i and j plus the minimum collection delay and minimum delivery delay. We can estimate the propagation delay $\theta_{i,j}$ estimated in the negotiation phase by the mechanism we introduced in section 3.3.

With the knowledge of the two parameters $\delta_{i,j}$ and $\Delta_{i,j}$, we can calculate the prefetched buffer sizes $B_j(m)$ for any medium m of user j using equation (1). For the connections between i and j, the latest time $L_{i,j}$ that j could start to playback media units is equal to

$$L_{i,j} = I_i + \max_m \{B_j(m)\eta_i(m) + \Delta_{i,j}(m)\} \qquad (16)$$

where I_i is the initial collection time of the media units from i; $B_j(m)$ is the prefetched buffer size of medium m of receiver j (equation (1)); $\eta_i(m)$ is the collection period of a medium m of user i; $\Delta_{i,j}(m)$ is the maximum end-to-end delay between users i and j for the medium m.

In equation (16), we see that $L_{i,j}$ is the latest time that the $B_j(m)$-th media unit is ready to be played out at user j for each medium m sent from user i. It is obvious that receiver j gets at least $B_j(m)$ media units for all media by the time $L_{i,j}$. If receiver j starts to playback media units before $L_{i,j}$, then discontinuity may occur due to delay jitter and insufficient prefetched buffer size. On the other hand, if j starts to play media units from i after the latest playback time $L_{i,j}$, then j must buffer more media units than it needs, a wastes of network resources.

User i can compute the latest playback time $L_{i,j}$ for any user j in the communication group from equation (16). However, if we want to synchronize the initial playback times M_i of media units from user i at all other users in the group, then we must select a time instant based on VGT. We can select the synchronous initial playback time of the users as:

$$M_i = \max_j \{L_{i,j}\} \qquad (17)$$

where M_i represents the maximum $L_{i,j}$ among the connections associated with user i. Assume that user i is aware of the initial collection time I_i in the negotiation phase, it can calculate the latest initial playback time $L_{i,j}$, equation (16), for user j in the communication group. Then, i can get M_i from equation (17) and distribute it to every user in the group.

We can compute M_i for each user i in the group. However, in order to get the global initial playback time, the largest M_i will be selected by all users in the group. The initial collection time of each user i in the group can be synchronized by using equation (15). Note that each user in the group plays both roles of a sender and a receiver. From the senders' point of view, M_i can be obtained from the global initial collection time G_c, equation (15). Thus, each user i in the group distributes its playback waiting period λ_i in terms of VGT to all users in the group where

$$\lambda_i = M_i - I_i \qquad (18)$$

In the playback waiting period λ_i, all users fill their prefetched buffer with media units from user i and wait for the synchronous playback of the media units. After receiving the playback waiting period λ_i from all users, each user in the group will pick up the largest playback waiting period $\max_i\{\lambda_i\}$. The global initial playback time G_p is then computed from

$$G_p = G_c + \max_i\{\lambda_i\} \qquad (19)$$

where G_c is the global initial collection time obtained from equation (15), and $[\max_i\{\lambda_i\}]$ is the largest playback waiting period.

For each user j in the group, a timer ω_j is set up with the initial value given by equation (20). The timer ω_j starts to countdown when the first media unit from the *chairman* is received.

$$\omega_j = G_p - \rho_j \qquad (20)$$

where ρ_j is the time in terms of VGT that the first media unit from the *chairman* is ready to be played at j.

Afterwards, the timer keeps decreasing and corrects its value according to the dynamic period adjustment factor $A_{i,j}(k)$, given i is the chairman. When the timer ω_j reaches zero, user j starts to playback. Once the receivers begin playback, only the dynamic period synchronization mechanism of section 3.2 is involved.

4 Performance

In this section we present simulation results to evaluate the performance of these protocols.

4.1 Performance of the Prefetched Buffering Protocol

Our simulation model consists of one sender and one receiver where the sender transmits 1000 video frames with the collection period $\eta_i = 25ms$ and the receiver consumes media units with the playback period $\mu_j = 25ms$. Each frame contains 0.2 ~ 1.5Mb data. Suppose that the speeds of the local clocks of the sender and the receiver are perfectly matched, i.e., no local clock drift exists. The maximum and minimum end-to-end delays are used as input variables.

In Table 1 we show the number of discontinuities occuring based on different maximum and minimum end-to-end delays and the total discontinuity time. The number of discontinuities counts the times that j has no media units to play while the the total discontinuity time represents the sum of the persistent time of all discontinuities j suffers. As it is clear in Table 1, both the number and the total time of discontinuities increase monotonically with the increasing value of $J_{i,j}$. It can also be seen in Table 1, that no discontinuity occurs if the prefetched buffer scheme is applied. In Table 1 we also show the required buffer size by using equation (1).

$\Delta_{i,j}$ (ms)	$\delta_{i,j}$ (ms)	Number of discontinuities	Total discontinuity time (ms)	No. of discontinuities with prefetched buffer	Buffer size required $(videoframes)$
50	40	3	9.98	0	1
100	75	6	14.92	0	1
150	90	10	36.17	0	3
200	100	10	88.67	0	4

Table 1. The discontinuity caused by delay jitter

4.2 Performance of the Dynamic Period Adjustment Protocol

Here our simulation model consists of three users. Users 1 and 2 send media units to user 3 who is the *chairman*. Both users 1 and 2 suffer clock drift between 1.01 to 0.99. The maximum and minimum end-to-end delay among users are $\Delta_{1,2} = 150ms$, $\delta_{1,2} = 80ms$, $\Delta_{1,3} = 50ms$, $\delta_{1,3} = 40ms$, $\Delta_{2,3} = 200ms$, $\delta_{2,3} = 100ms$. The collection and playback periods of a medium are $\eta_i = \mu_j = 30ms$ for $i = 1, 2$ and $j = 3$. We assume that user 1 and 2 start to send 10000 media units to user 3 at the same time. The local clocks at user 1 and 2 are 1 in the beginning and start drifting after transmission. The drifts[7] distribute uniformly between $-0.0005 \sim 0.0005$ per collection period. Once the clocks reach the boundaries, 0.99 or 1.01, they jump back to 1. Note that the typical local clock drift rate is less than 0.001. In our simulations we use the large clock drift rate to show that our mechanisms can handle the worst cases.

In Table 2, we show the simulation results for the maximum, minimum, and average asynchrony with different adjustment periods. We observe that the average asynchrony increases monotonically when the adjustment period increases.

---

[7] If the drifts are constant, then we need to perform the dynamic period adjustment protocol only once instead of doing it periodically.

Adjustment Period (m.u.)	Maximum Asynchrony (ms)	Minimum Asynchrony (ms)	Average Asynchrony (ms)
1	0.303	-0.363	-0.269
2	0.459	-0.951	-0.252
10	1.611	-3.856	-1.04
50	4.577	-18.076	-3.341
100	13.459	-39.564	-8.747
200	9.628	-46	-16.598

Table 2. The average asynchrony with different adjustment period

The difference between maximum and minimum asynchrony implies that the variance of the asynchrony also increases with the increasing adjustment period. Note that the average asynchrony is $49.9ms$ if the dynamic period adjustment mechanism is not used in the simulation model.

According to the quality of service (QoS) requirements of applications and the level of clock drift, we need to choose appropriate adjustment periods to meet the QoS requirements while keeping the synchronization overhead low. Note that our simulation model, as described, spans in a WAN environment. The dynamic period adjustment mechanism is still able to keep the average asynchrony within a few milliseconds while the period of adjustment action is 50 media units.

4.3 Performance of the Initial Collection Time Synchronization Protocol

Here we simulate a teleconference with 3 users. We assume that user 3 is the chairman and users 1 and 2 have local clock drift between 1.01 and 0.99 and user 3 has local clock with speed 1. The maximum and minimum end-to-end delays $\Delta_{i,j}$ and $\delta_{i,j}$ among three users are assumed to be the same and are used as simulation variables.

Although the global initial collection G_c can be computed for all three users from equation (15), the local clock drift among them will cause the small amount of asynchrony as shown in Table 3. The asynchrony is so small that it is far below human perception.

$\Delta_{i,j}$ (ms)	$\delta_{i,j}$ (ms)	Asynchrony (ms)
50	40	0.0096
100	75	-0.016
150	90	-0.018
200	100	0.01

Table 3. The asynchrony when the initial collection time synchronization scheme applied

4.4 Performance of the Initial Playback Time Synchronization Protocol

The simulation model has the same input parameters as in the previous case. As shown in Table 4 the asynchrony is caused by different speeds of timers ω_j for $j = 1, 2, 3$, which are used to countdown the initial playback time. Note that the asynchronies in Table 3 and 4 are small and almost unchanged while we vary the values of $\Delta_{i,j}$ and $\delta_{i,j}$. In other words, the two synchronization mechanisms for initial collection and playback times perform well in both LANs and WANs.

$\Delta_{i,j}$ (ms)	$\delta_{i,j}$ (ms)	Asynchrony (ms)
50	40	-0.024
100	75	-0.013
150	90	0.015
200	100	0.016

Table 4. The asynchrony when the initial playback time synchronization scheme applied

5 Conclusion

Our work provides synchronization protocols for multimedia traffic in integrated services networks. Using proposed synchronization protocols for initial collection time and initial playback time, the global initial collection time G_c and global initial playback time G_p can be determined in a distributed manner. Then the dynamic period adjustment mechanism takes over the task of resynchronizing local clocks, and maintains the synchronization achieved by synchronizing G_c and G_p. With the cooperation of intermediate nodes in delay estimation, the dynamic period adjustment protocol provides fast estimation of VGT. The frequency of performing the dynamic period adjustment protocol should be added to the QoS requirements of applications to further reduce the overhead caused by the protocol. In addition, our protocols have the following advantages:

- No additional connection is needed for transmission of feedback units.
- Faster and more accurate response to asynchrony.
- In real-time multimedia applications, users are free to enter/leave without affecting the existing connections.
- No a priori knowledge of the distribution of end-to-end delay and of the clock drift rate are required.
- Network messages are significantly reduced by implicit transmission of time information and prefetched buffering.

We also point out that our protocols heavily rely on a reliable broadcast election protocol and precise measurement on delay stamp at intermediate nodes.

Acknowledgement: We would like to thank Hui Zhang, Julio Escobar, Tom Little, and Venkat Rangan for their suggestions and constructive comments.

References

1. A. Campell, G. Coulson, F. Garcia, D. Hutchison, and H. Leopold, "Integrated Quality of Service for Multimedia Communications," *Proc. of the INFOCOM '93 Conference*, pp. 732-740, Apr. 1993.
2. D. C. Verma, H. Zhang and D. Ferrari, "Delay Jitter Control for Real-Time Communication in a Packet Switching Network," *Proc. of the TRICOMM '91 Conference*, pp. 35-43, Dec. 1991.
3. D. Ferrari and D. C. Verma, "A Scheme for Real-Time Channel Establishment in Wide-Area Networks," *IEEE Journal on Selected Areas in Comm.*, Vol. 8, No. 3, pp. 368-379, Apr. 1990.
4. D. L. Mills, "Internet Time Synchronization: The Network Time Protocol," *IEEE Transactions on Communications*, Vol. 39, No. 10, pp. 1482-1493, Oct. 1991.
5. H. Zhang and D. Ferrari, "Rate-Controlled Static-Priority Queueing," *Proc. of the INFOCOM '93 Conference*, pp. 227-236, Apr. 1993.
6. J. Escobar, D. Deutsch, and C. Partridge, "Flow Synchronization Protocol," *IEEE Global Communications Conference*, pp. 1381-1387, Dec. 1992.
7. L. Besse, L. Dairaine, L. Fedaoui, W. Tawbi and K. Thai, "Towards an Architecture for Distributed Multimedia Applications Support," *Proc. of ICMCS'94*, Boston, May 1994.
8. L. Lamont and N. D. Georganas, "Synchronization Architecture and Protocols for a Multimedia News Service Application," *Proc. of ICMCS'94*, Boston, May 1994.
9. P. V. Rangan and H. M. Vin, and S. Ramanathan, "Communication Architectures and Algorithms for Media Mixing in Multimedia Conferences," *IEEE/ACM Transactions on Networking*, Vol. 1, No. 1, pp. 20-30, Feb. 1993.
10. R. Gusella and S. Zatti, "An Election Algorithm for a Distributed Clock Synchronization Program," *IEEE 6th International Conference on Distributed Computing Systems*, pp. 364-371, Boston, May 1986.
11. R. Gusella and S. Zatti, "The Accuracy of the Clock Synchronization Achieved by TEMPO in Berkeley UNIX 4.3BSD," *IEEE Transactions on Software Engineering*, vol. 15, no. 7, pp. 847-853, Jul. 1989.
12. R. Steinmetz, "Synchronization Properties in Multimedia Systems," *IEEE Journal on Selected Areas in Comm.* vol. 8, no. 3, pp. 401-412, Apr. 1990.
13. R. Steinmetz and C. Engler, "Human Perception of Media Synchronization," IBM European Networking Center, Technical Report 43.9310, 1993.
14. S. J. Golestani, "A Framing Strategy for Congestion Management," *IEEE Journal on Selected Areas in Comm.*, Vol. 9, No. 7, pp. 1064-1077, Sep. 1991.
15. S. Ramanathan and P. V. Rangan, "Adaptive Feedback Techniques for Synchronized Multimedia Retrieval over Integrated Networks," *IEEE/ACM Transactions on Networking*, Vol. 1, No. 2, pp. 246-260, Apr. 1993.
16. T. D. C. Little and A. Ghafoor, "Synchronization and Storage Models for Multimedia Objects," *IEEE Journal on Selected Areas in Comm.*, Vol. 8, No. 3, pp. 413-427, Apr. 1990.

17. T. D. C. Little and A. Ghafoor, "Multimedia Synchronization Protocols for Broadband Integrated Services," *IEEE Journal on Selected Areas in Comm.*, Vol. 9, No. 9, pp. 1368-1382, Dec. 1991.
18. W. A. Montgomery, "Techniques for Packet Voice Synchronization," *IEEE Journal on Selected Areas in Comm.*, Vol. SAC-1, No. 6, pp. 1022-1027, Dec. 1983.

Efficient Support for Multiparty Communication

Clemens Szyperski[*][1] and Giorgio Ventre[2]

[1] The Tenet Group, International Computer Science Institute, 1947 Center Street, Suite 600, Berkeley, CA, 94704
[2] Dipartimento di Informatica e Sistemistica, Università di Napoli Federico II, Via Claudio 21, 80125, Napoli, Italy

Abstract. Communication of multimedia data among multiple parties can be costly in terms of use of network resources if guaranteed quality of service is requested. We believe, however, that knowledge of the functional characteristics of real applications can be helpful to devise new mechanisms to support isochronous communication efficiently. In this paper we follow this approach to introduce a new communication abstraction, called the *Half-Duplex Real-Time* channel. The new abstraction can reduce the amount of resources needed by real-time protocols to offer guaranteed performance networking by exploiting the interaction patterns typical of multimedia applications such as tele-conferencing and tele-education.

1 Introduction

Multimedia applications, in particular those based on multi-party interactive multimedia (MIM) [9], show an inherent complexity, stemming from the different media they use and the large number of participants they might involve.

In this paper we propose a new solution towards an efficient network support of MIM applications; the solution is based on a careful investigation of this particular class of applications. Several proposals have been presented in the literature for their classification from the viewpoint of the support they require from the communication subsystem [2, 7, 11, 12, 13, 14, 15]. We present a different approach to such classification based on the identification of a set of basic functional characteristics that this class of multimedia applications exhibits. The first goal of this paper is to show that such characteristics can be effectively exploited to reduce the amount of resources required to provide real-time communication. The second goal is to show that this result can still be accomplished with a guaranteed *quality of service* (QoS) according to the Tenet approach to real-time networking [3].

Indeed, the issue of efficiency is particularly important in the framework of network support for multiparty interactive multimedia, and has to be faced by

---

[*] Currently with Oberon Microsystems Inc., Basel, Switzerland

taking into account two diverging requirements: the request of real-time communication for isochronous media such as video and audio, and the cost of distributing such media within large sets of participants.

The solution we propose is based on a new abstraction, called the *Half-Duplex Real-Time Channel*. We introduce this abstraction as an extension of the unicast-based mechanisms of the original Tenet scheme, in order to cut down the problems linked to the adoption of resource reservation protocols. We show that our proposal does not harm the capability of the scheme to offer always a guaranteed performance for communication of isochronous data. In particular, the introduction of the Half-Duplex Real-Time channel allows to reduce the complexity of the creation of network support for common MIM applications, and to sensibly decrease the amount of resources to be reserved in the network to offer guaranteed QoS.

2 Model of Interaction

Interpersonal communication models show differences in the way data is exchanged among involved participants. In order to offer computer-based multimedia tools implementing these and other models, we need to analyze how such differences influence the kind of communication service these tools require from the network. This can be done by considering, as the prevailing aspect, the *model of interaction* criterion, defined as the model according to which information exchange is allowed in a multiparty application.

We will present our conjectures by considering a computer supported multimedia environment where participants to a multi-party communication are geographically distributed entities (*nodes*), interacting through communication abstractions (*channels*) over a real-time internetwork.

A *dynamic* interaction is required for MIM applications in which the amount of information that can be exchanged among all the participants is almost equivalent and the characteristics of the applications do not allow the determination of a particular subset of participants that will be the main source of data. Since the coexistence of multiple flows of information should be, in these cases, permitted, support for simultaneous communications among the nodes should be provided by the network, in spite of any arbitration or control mechanisms that might be managed by or imposed to the users.

A *static* interaction is required in all MIM applications where the information flows only from a statically determined subset of nodes, or where the amount of data flowing from the other nodes does not justify the creation of a dedicated, real-time communication channel. Typical applications requesting such kind of interaction are those generally referred to as *telepresence* [6]. An example of this kind of application is *teleprogramming* [5], where a user receives a number of different media from remote devices, sensors or computers, typically to control or manipulate a remote apparatus, according to a many-to-one information flow.

A *controlled* interaction is required when the application specifically permits the identification of a dynamically variable subset of participants that contributes to the data flow. The subset may depend on the specific attributes of the participants. For example, in a conference, there exist subsets including the conference speakers and the conference managers respectively. Both sets are expected to contribute to the data flow more than the attendees. The identification of the subset of senders can also depend on application-specific rules and agreements. For example, in a class lecture students are expected to ask for permission to interrupt the professor. In such cases, those rules and agreements dynamically modify the subset of participants authorized to transmit data. In the lesson example, the student who is authorized to intervene will become the speaker and the multimedia data produced by him will be transmitted to the other participants in addition to or in alternation with the data flowing from the professor.

An important example characterized by this model of interaction is the one generally classified as CSCW, or *computer supported collaborative work*. The goal of this service is to support the complex interactions required to allow effective meetings among geographically distributed professionals. In addition to video and audio data, participants in this model of interaction share applications such as editors, spreadsheets and drawing spaces. Information flows according to a highly dynamic, all-to-all pattern. Since the target environment is often that of technical and commercial enterprises, access control has usually to be enforced.

Another example is the so-called Virtual Café. The idea is to allow users at different sites to communicate informally through an electronic common space [1]. However, due to the informal nature of this model, access control is generally not of concern and scheduling is usually limited to the specification of some "opening hours".

Table 1 lists these new applications. The entries are meant to indicate the dominant cases, but should not be taken as discriminative in the mathematical sense. Two examples have been added to this table (i.e. *broadcasting services* and *distributed computing*), even though they do not directly fall into the definition of MIM applications. We included them for their relevance in terms of network support requirements.

3   Problems of Scaling

So far we considered purely functional aspects of MIM applications and no limits of any sort were assumed for the capability of a real-time communication system to provide the services required by these applications.

We believe, however, that the important issue of scaling should be taken into account [12]. We refer to the *size* of a multiparty application as to the bound on the number of expected participants. In particular, we are interested in how support from a real-time communication subsystem has to cope with this aspect.

In fact, for conversational multimedia the maximum number of participants that may be supported depends on when it becomes infeasible either for the human user or for the service provider to accommodate one more participant. For example, from a logical viewpoint, the dynamic model of interaction of some

Type	Description	Model of Interaction	Data Flow	Accessibility	Event Scheduling
CSCW	All participants actively send/receive data	Dynamic	N -> N	Controlled	Planned
Telepresence	Many senders, one receiver	Static	N -> 1	Controlled	Unplanned
Virtual Cafe	All participants actively send/receive data	Dynamic	N -> N	Uncontrolled	Unplanned
Broadcasting Services	One sender many receivers	Static	1 -> N	Controlled	Planned
Distributed Computing	All participants actively send/receive data	Dynamic	N -> N	Controlled	Unplanned

Table 1. Characteristics of Selected MIM Applications.

MIM applications might require a fully interconnected communication structure. This requirement seems to dictate a small bound on the maximum number of participants, since the connection complexity grows quadratically, and the information to be consumed by each participant grows linearly with the number of participants. Therefore, the existence of such a bound is due not only to implementation reasons, as those related to the complexity of the required communication structure, but also to the limits inherent in the human capability of effectively perceive several simultaneous information flows.

We studied how size relates to the various applications introduced in the previous section (Table 1). The result is presented in Figure 1, where the number of participants and the model of interaction have been used as criteria.

From Figure 1 we observe that the upper bound on the number of participants is somehow related to the specific application. It should be noted that this figure is intended to be only indicative. Still, we can distinguish among small, large, and very large scale applications. As a first guess, we expect the bounds for each of the three classes to be around 20, around 100, and practically unbounded, respectively. Of course, more precise numbers require an analysis of individual applications.

4 Principles of a Solution

As an example, we will consider the **Learnet Project**, an experiment ongoing at the University of Napoli "Federico II" for the application of distributed mul-

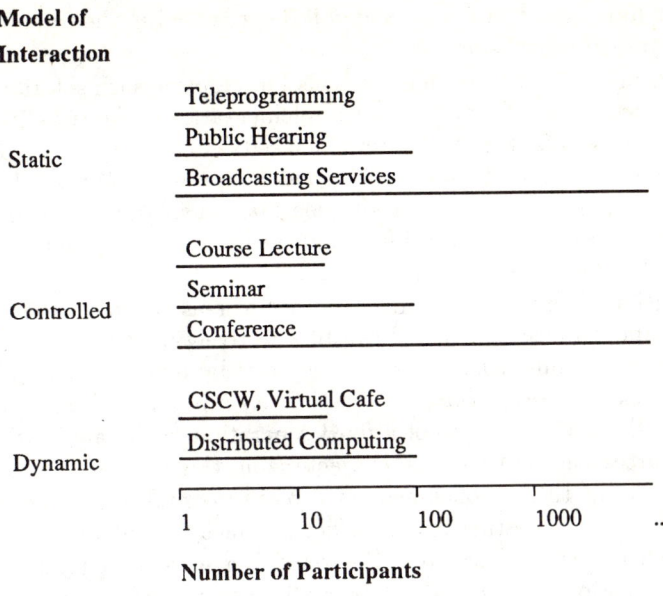

Fig. 1. Size Ranges of MIM Applications.

timedia technology to remote education. The goal of this project [8] is to use an ATM-based metropolitan area network to distribute over a number of remote classrooms the didactical information produced by the **Nettuno** consortium. Nettuno is a consortium among the major italian universities to experiment new technologies for distance learning in the framework of the courses of the first level academic degree in Engineering.

In the experimental scenario, a student can access from a remote classroom the multimedia database of the consortium, to retrieve lessons recorded by professors of the participating universities. To improve the didactical validity of teaching, students have also access to on-line sessions of tele-tutoring, in which tutors made available by the universities answer questions over the topics covered by the lessons.

A tele-tutoring session (in our terms, the *conference*) is typically one-to-many, where the *participants* are the tutor (the *conference manager*) on one side, and the students on the other. To allow a high level of interaction, in addition to the tutor also the students are allowed to be *senders* (e.g. when they answer or ask a question) and *receivers* (e.g. when they simply attend to the session) of multimedia data, such as audio, video, text, images.

For a better didactical impact the switching between these two roles has to be highly dynamic and controlled by the tutor. Consequently, the communication system has to support a two-way interaction characterized by a well defined and adequate QoS.

We propose now the principles for a solution of the problem of communication support for this kind of applications. The first solution we propose is based on the simplex multicast connection, and will be presented in the framework of the Tenet real-time protocol suite [3].

We have individuated two main reasons for adopting such solution. The first one is that mechanisms for an efficient implementation of multicast communication begin to be available in a number of packet-switching networks. The second is that a simplex multicast channel is a good compromise between the need of a mechanism more powerful than the simplex channel solution (on which the first Tenet scheme was originally based [4]), and the need of not to adopt mechanisms too powerful to be efficiently implemented.

An additional, and probably more important reason can be seen in the fact that the simplex multicast channel can be effectively used as a base for the realization of the communication services required by multimedia application. In fact, in the case of a static type of interaction, a simplex multicast channel seems to fit naturally in the scenario of a fixed separation of the application participants in sources and destinations, characterizing this kind of applications. For these applications the establishment of a real-time multicast connection from each source to all the destinations is sufficient, since the only form of data communication from the destinations to the source is generally related to application and communication control, and can be realized by means of non real-time data transmission. In the case where interaction among the sources is also required, it is sufficient to implement the connections by considering the destination set as including also the elements of the source set.

When the level of interaction among the participants increases, then a richer communication structure has to be provided. However, the simplex multicast channel can still provide the basic component for an efficient implementation of such a structure. For example, in the case of a dynamic type of interaction, where all the application's participants are allowed to freely interact with the others, a complete network of real-time multicast channels connecting each participant with the others can be adopted. The same solution, of course, can be applied in the case of applications characterized by a controlled type of interaction among the participants.

Even though the above sketched solution appears to be straightforward and ease to implement, some considerations might be made regarding its efficiency. As we pointed in a previous section, MIM applications can be defined as requiring a controlled interaction when the participants who can actively contribute to the data flow constitute a subset that can vary dynamically. Such subset can be usually identified by considering specific attributes of the application.

In the case of a teleconference, for example, all the speakers could simultaneously appear on the machines of the conference attendees, even though in practice only one among the speakers or a smaller subset of them is actively involved in the conference[3].

---

[3] Of course, some form of feedback from the conference attendees should be allowed, like questions or criticisms. In the case of large scale teleconferences, such feedback could be implemented by means of a non real-time communication channel, while for smaller scale MIM applications such as a teleseminar, a different solution could be feasible or appropriate.

In the case of a computer supported lecture, the floor will be mainly taken by the professor; however, a good interaction between him and the students and also among the students themselves is highly desirable. For this reason, since in theory all the students might intervene during the lesson, it could be argued that a communication structure offering a real-time connection among all the participants is required. However the inherent order regulating the course of the lesson might be exploited to reduce the communication requirements.

Both the examples presented show that a more efficient solution for the implementation of the communication structure can be found, by considering the characteristics of the applications. In the teleconference example, the source set is usually small and only a subset of the speakers actively contributes to the conference at the same time; this characteristic should be exploited during the establishment of the channel, to determine some form of optimization in the reservation of the resources needed to guarantee a certain quality of service.

5 The Half-Duplex Real-Time Channel

We will consider again the lesson example; as we have seen, even though the professor is the only source of data for most of the time, all the students are supposed to intervene for a limited time during the course of the lesson. Again, a first solution to this problem could be to provide every participant with a multicast connection to all the others.

However, due to characteristics of the application, the multicast connections originating from the students would be mostly unused. A solution to this problem could be to implement a dynamic management of connections, such that only the connection from the professor to the students is established at the beginning, while the other connections are created on demand, whenever a communication is requested by the students. This solution appears to be cost-effective, but it shows two major drawbacks: the first is that the time required to establish a new connection could be non-negligible, and, in general, unpredictable. The second is that the resources required for new connections on demand could be not promptly available.

The second solution is again based on an analysis of the application characteristics and on some ideas derived from the original Tenet scheme. In the lesson example, a Tenet-based approach would establish a group of simplex unicast channels with guaranteed QoS from the professor to the students. In an extended scheme with multicast connections capability the communication structure would appear as in Figure 2.a . Let us now suppose that in our application only one participant at a time can transmit data (i.e. is authorized to intervene). Since the sources are mutually exclusive, if the network could offer a guaranteed QoS in both directions of the same multicast connection, we could simply allow communication from the students by implementing, at the application level, control mechanisms realizing the required mutual exclusion. Figure 2.b shows the data flow when one of the students (d_m) starts to transmit data.

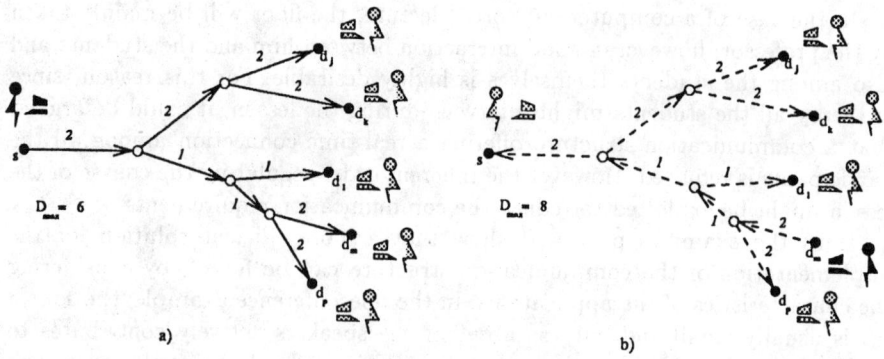

Fig. 2. A half-duplex multicast connection.

At the network layer, this control mechanism needs to be enforced to avoid that an ill-behaving application could jeopardize the performance of other existing applications. In the Tenet suite, this enforcement can be done by means of the already existing mechanisms for rate control. With such mechanisms, whenever packets flowing from an application are exceeding the amount of resources originally allocated to it, thus violating the quality-of-service agreement with the network, they will be discarded with no harm to the other connections. Floor-control mechanisms within the network protocols can thus be avoided (e.g. [16]).

We can now introduce a new communication abstraction, called the *Half-Duplex Real-Time* channel (HD). A HD channel can be defined as a two-way communication structure connecting a network entity, called the *main source*, to a set of other network entities, called *secondary sources*. The structure can offer guaranteed performance only if communication flows from the main source to the secondary ones, or, alternatively, from one of the secondary sources to the other network entities connected by the HD channel.

The major problem of the introduction of this new abstraction in the Tenet real-time communication scheme is that its reservation mechanisms are dependent on the data direction over a channel. In the example of Fig. 2, all the network links have been assigned a delay bound in time units. When the data flows from the professor s to the students, the maximum delay the data will suffer is bounded by 6 units, while in the case where the data flows from d_m to all the other participants, the maximum delay will be 8 units. To avoid a too complex channel establishment mechanism, we believe that the specification of the required quality of service should be allowed only for the main direction of the data flow. In the lecture example, the QoS requirements will be specified by the *main source* corresponding to the professor, while each one of the other possible sources (*secondary sources*), corresponding to the students, will only have from the network a fixed, yet guaranteed QoS.

The interconnection scheme given by a HD channel can offer, in a real-time protocol suite, an efficient and cost-effective alternative to the already available, fully interconnected scheme. Even though the former solution seems to restrict the capability of an application to select a real-time communication perfectly matching its needs, as instead allowed by the latter, we believe that the advantages characterizing our proposal are well worth the limitations that might be associated to it.

First, this solution greatly simplifies the establishment of a communication structure with guaranteed QoS, particularly for MIM applications characterized by a large number of participants, since only the establishment of a multicast channel is required. This is particularly true in all cases in which a sort of asymmetry in the flowing of multimedia data among the participants can be observed; in such cases, the possibility of a reduced, but still guaranteed quality of service for less important connections is usually an acceptable drawback for a faster establishment of the complete communication structure.

Second, this solution provides savings in terms of resources required to establish a real-time connection. Following the traditional approach, the same application would have required the establishment of as many multicast channels as the number of participants, to allow each of them to have real-time interactions with all the others. In the next section we will focus on this last aspect and we will discuss the implementation of the proposed scheme within the Tenet protocol suite.

6 Guaranteed Quality of Service

To analyze the problems related to the implementation of a half-duplex multicast connection with guaranteed QoS, we will consider a simple example. In Fig 3.a we have a network node n_j with three channels passing through it. We now suppose that one of these channels, c_2, has to be half-duplex. In the Tenet approach this would imply to split channel c_2 in two different channels, $c_{2'}$ and $c_{2''}$ with similar characteristics, but flowing through node n_j in opposite directions (Fig. 3.b).

To determine the amount of resources required to accommodate both channels we have to perform first the local tests in all the nodes originally traversed by c_2 and then the final tests at the respective destinations. To understand if the mutual exclusion condition can be exploited to reduce the amount of resources required we will focus our attention to the local tests. The procedure for the establishment of a guaranteed real-time channel is described in detail in [4]. Here we will only remind that in each node three tests have to be performed: the *deterministic test*, the *statistical test*, and the *delay bound test*.

6.1 The Deterministic Test

According to the Tenet scheme, the traffic requirements of a connections are expressed in terms of the minimum inter-arrival times for the transmitted packets, x_{min}, the minimum value of the average packet inter-arrival time, x_{ave}, and the interval I on which these value have been computed. The quality of service required by the connection is expressed in terms of an absolute bound on the

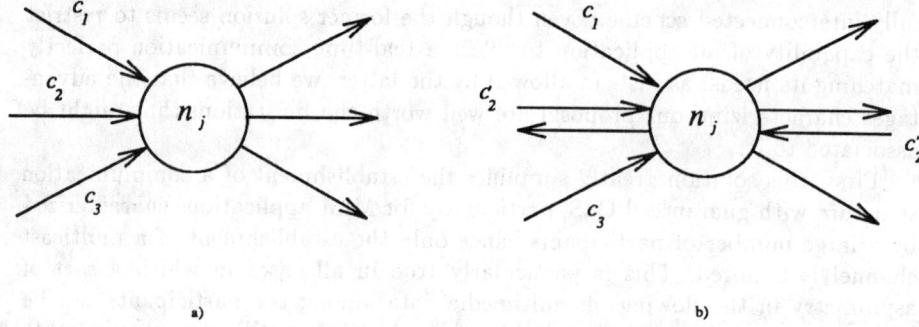

Fig. 3. A possible solution for half-duplex connections.

source-destination transmission delay, D_{max}. In this case the channel is called *deterministic*; if a deterministic channel j is established, in each node n on the channel path the packets will be subjected to a maximum delay $d_{j,n}$ computed at the end of the establishment phase. Alternatively, a requirement on the expected delay can be expressed in terms of a maximum value D_{max} to be satisfied by the packets with a probability greater than or equal to a value Z_{min} (*statistical* channel) [4].

For our simple example the deterministic test is :

$$\frac{t_1}{x_{min,1}} + \frac{t_{2'}}{x_{min,2'}} + \frac{t_{2''}}{x_{min,2''}} + \frac{t_3}{x_{min,3}} \leq 1 \qquad (1)$$

where t_j is the maximum packet processing time for for channel j. If the real-time communication service requested has the same characteristics in both directions, channels $c_{2'}$ and $c_{2''}$ will contribute in the same way to the left term of the deterministic test. However, if the mutual exclusion is enforced they will never contribute at the same time. Indeed, since the deterministic test is based only on x_{min}, to have it passed is sufficient that both the new flows have the same value for this parameter of the original simplex channel. If the deterministic test has been passed by channel c_2, it will be passed also by channel $c_{2'}$ and $c_{2''}$. In the case that the flows in the two directions show different values for x_{min} and x_{ave}, then we have to verify the test for a channel c_2 such that $x_{min,2} = min(x_{min,2'}, x_{min,2''})$.

6.2 The Statistical Test

Given the mutual exclusion hypothesis, the two channels $c_{2'}$ and $c_{2''}$ can never be included in the same *overflow combination*. An overflow combination is a set of channels that, when simultaneously active for a sufficient long time, may cause packets to violate the local delay bounds. Moreover, if we ensure that the

---

[4] For the sake of brevity, a third type of real-time service, called jitter controlled, will not be considered.

number of packets received by node n_j over both channels $c_{2'}$ and $c_{2''}$ is, for each temporal window of duration I, equal to the number of packets that channel c_2 would have transmitted in the worst case in the same interval, we have exactly the same number of overflow combinations as if we had only channel c_2. To ensure this requirement, we have to enforce that the constraints for x_{min} and x_{ave} will be respected even in the worst case.

The constraint over x_{min} is given by that of the deterministic test. Regarding the condition over x_{ave}, we have two possibilities. The first one is to impose that whenever there is a change in the transmission direction (i.e. $c_{2'}$ stops while $c_{2''}$ starts to transmit, or vice-versa), the condition is respected in the worst case. Such case is verified when, immediately before stopping, channel $c_{2'}$ transmits at the maximum possible rate $x_{min,2'}$ for the longest possible interval $I_{min,2'} = I_{2'} \frac{x_{min,2'}}{x_{ave,2'}}$, and immediately after starting, channel $c_{2''}$ transmits at the maximum possible rate $x_{min,2''}$ for the longest possible interval $I_{min,2''} = I_{2''} \frac{x_{min,2''}}{x_{ave,2''}}$.

To avoid this problem, we have to impose a time interval δ between stopping the transmission in one direction and starting to transmit in the opposite direction. In the case of a switch from $c_{2'}$ to $c_{2''}$, we can determine such interval by imposing the following condition:

$$\delta_{2'} + I_{min,2'} \geq max(I_{2'}, I_{2''}) \qquad (2)$$

In the case that $I_{2'} = I_{2''} = I_2$, from the above condition we have that the time interval should be:

$$\delta_{2'} \geq I_2 - I_{min,2'} = I_2(1 - \frac{x_{min,2'}}{x_{ave,2'}}) \qquad (3)$$

Similarly, for a switch from $c_{2''}$ to $c_{2'}$, we have to impose the following time interval:

$$\delta_{2''} \geq I_2 - I_{min,2''} = I_2(1 - \frac{x_{min,2''}}{x_{ave,2''}}) \qquad (4)$$

To avoid to have to impose this time interval before switching from one transmission direction to the other, a second solution is possible. It consists of over-dimensioning the original channel, (in our example c_2) in order to have that it could endure the worst case situation we depicted above. In this case we have to properly choose I_2 and $x_{ave,2}$ such that:

$$\frac{I_2}{x_{ave,2}} = (\frac{I_{min,2'}}{x_{min,2'}} + \frac{I_{min,2''}}{x_{min,2''}}) \qquad (5)$$

and from the definition of I_{min},

$$\frac{I_2}{x_{ave,2}} = (\frac{I_{2'}}{x_{ave,2'}} + \frac{I_{2'}}{x_{ave,2''}}) \qquad (6)$$

In the hypothesis that $I_{2'} = I_{2''} = I_2$, we have that:

$$x_{ave,2} = \frac{x_{ave,2'} \cdot x_{ave,2''}}{x_{ave,2'} + x_{ave,2''}} \qquad (7)$$

And indeed, if we suppose that $x_{ave,2'} = x_{ave,2''}$ we have exactly the result we would expect:

$$x_{ave,2} = \frac{x_{ave,2'}}{2} \tag{8}$$

The advantage of this last solution is that in this case there is no need for a transition interval during the direction switch over the channel. Since the imposition of this interval should be done at the application level, the last solution seems to be simpler to implement, for it does not require any additional information flow from the protocols to the application. The drawback, of course, is that we now have to reserve more resources than by adopting the former solution.

6.3 The Delay Bound Test

The delay bound test determines the minimum delay bound to be assigned to the channel being established so that scheduler saturation can be avoided in the node. To do so, the test takes into account the service time of packets flowing through the node. This depends on the local processing time plus the data link processing delays. The former can be shared for $c_{2'}$ and $c_{2''}$ under the mutual exclusion assumption, while the latter is based on two different outgoing links that individually require protection in the worst case.

Hence, the delay bound test requires checking for the sum of the data link processing delays. In simple words, the delay bound test ends up in doubling the resources allocated in terms of overall use of bandwidth. This is to be expected since the two alternate flow directions present at each edge of a half-duplex tree typically map to different links, both requiring reservations.

To conclude, we can say that the costs of establishing a half-duplex multicast tree are at most twice that of a simplex multicast tree, independently of the number of participants. In the case where the processing nodes are the bottleneck (i.e. bandwidth allocation is not expected to be a limiting factor), and where the optimizations described above are applied, the costs for a half-duplex tree can even approach that of a simplex tree.

7 Conclusions

MIM applications are very demanding in terms of network support. We have shown that some characteristics of such applications can be effectively exploited to reduce the amount of resources required to provide quality-of-service guarantees in communication.

We have introduced a new communication abstraction, called the half-duplex real-time channel. We have shown how this abstraction can be efficiently implemented over the existing support for simplex multicast communication characterizing recent proposals for packet-switching networks. The proposed solution reduces the complexity of the establishment of proper network support for applications with certain communications patterns, and greatly decreases the amount of resources to be allocated. These results have been accomplished in the framework of the Tenet approach to real-time communication, i.e. still providing guaranteed QoS. We believe that this result can be of interest for multiparty interactive multimedia such as tele-conferencing and tele-education, especially when involving large sets of participants.

Several problems remain to be solved. Among them, the improvement of the algorithm for the establishment of a half-duplex real-time channel. In particular, we are interested in devising mechanisms such that, within certain bounds, the specification of quality-of-service requirements can be allowed to all the nodes interconnected by a half-duplex channel. We are also interested in the extension of the mechanism towards a higher level of dynamicity in the management of connections. This would allow a secondary source to temporarily become the primary one and to obtain the quality of service best suited to its own requirements. In this direction we are currently investigating the application to the Tenet scheme of mechanisms for dynamic management of real-time connections [10].

References

1. S.A. Bly, S.R. Harrison, and S. Irwin. Media Spaces: Bringing People Together in a Video, Audio and Computing Environment. *Communications of ACM*, 36(1), January 1993.
2. Yee-Hsiang Chang. Remote Conferencing: Statement of Desired Functions. *Internet Engineering Task Force, Remote Conferencing BOF Draft*, October 1992.
3. D. Ferrari, A. Banerjea, and H. Zhang. Network Support for Multimedia: a Discussion of the Tenet Approach. Technical Report TR-92-072, International Computer Science Institute, November 1992.
4. D. Ferrari and D. Verma. A Scheme for Real-Time Channel Establishment in Wide-Area Networks. *IEEE Journal on Selected Areas in Communications*, 8(3), April 1990.
5. J. Funda. Teleprogramming: Towards Delay-Invariant Remote Manipulation. Technical Report MS-CIS-91-40, University of Pennsylvania, Dept. of Computer and Information Science, May 1991.
6. D.R. Hofstadter and D.C. Dennett. *The Mind's I: Fantasies and Reflections on Self and Soul*. Basic Books, New York, 1981.
7. T. D. C. Little and A. Ghafoor. Network Considerations for Distributed Multimedia Object Composition and Communication. *IEEE Network Magazine*, November 1990.
8. M. Longo and G. Ventre. The Learnet Project. Technical Report 93/15, Centro di Ricerca sul Calcolo Parallelo e Supercalcolo, Università di Napoli Federico II, Napoli, in Italian, 1993.
9. M. Moran and R. Gusella. System Support for Efficient Dynamically-Configurable Multi-Party Interactive Multimedia Applications. In *Proceedings of the Third International Workshop on Network and Operating System Support for Digital Audio and Video*, San Diego, November 1992.
10. C. Parris, G. Ventre, and H. Zhang. Graceful Adaptation of Guaranteed Performance Service Connections. In *Proceedings of IEEE GLOBECOM '93*, Houston, December 1993.

11. J.C. Pasquale et al. The Multimedia Multicast Channel. In *Proceedings of the Third International Workshop on Network and Operating System Support for Digital Audio and Video*, San Diego, November 1992.
12. E.M. Schooler. The Impact of Scaling on a Multimedia Communication Architecture. In *Proceedings of the Third International Workshop on Network and Operating System Support for Digital Audio and Video*, San Diego, November 1992.
13. H. Schulzrinne. A Transport Protocol for Audio and Video Conferences and other Multiparticipant Real-Time Applications. *Internet Engineering Task Force, Internet-Draft*, October 1992.
14. C. Topolcic, editor. *Experimental Internet Stream Protocol, Version 2 (ST-II)*. Request for Comments. Networking Working Group, N. 1190, October 1990.
15. R. Yavatkar. Issues of Coordination and Temporal Synchronization in Multimedia Communication. In *Proceedings of IEEE Multimedia '92*, Monterey, 1992.
16. R. Yavatkar. MCP: A Protocol for Coordination and Temporal Synchronization in Multimedia Collaborative Applications. In *Proceedings of 12th IEEE International Conference on Distributed Computer Systems*, June 1992.

QoS Negotiation for Multicast Communications[1]

Laurent Mathy and Olivier Bonaventure

University of Liège

Institut d'Electricité Montefiore B28, B-4000 Liège, Belgium
{mathy, bonavent}@montefiore.ulg.ac.be

Abstract. This paper deals with the Quality of Service (QoS) negotiation for multicast connections. First, we show that in the multicast case, the QoS parameters may be separated into two classes, namely the parameters whose scope is the whole multicast connection and those whose scope is limited to each receiver separately. Then, after a brief presentation of the enhanced QoS defined in the OSI95 transport service, we examine how the QoS negotiation schemes used in the peer-to-peer ISO and OSI95 transport services can be extended to multicast connections. Finally, a practical issue about QoS negotiation for multicast connections is presented.

1. Introduction

The emerging multimedia applications impose several requirements on the underlying communication architecture. The main requirements are the support of Quality of Service (QoS) parameters and multipeer communications (communications where several can send and several receive). Several solutions have been proposed to fulfil these two requirements. However, the proposed solutions are often strong on QoS and weak on multipeer communications [CCH94][DBL92][DBL94][FeV90] or the opposite [DCS93][Hen94][PPA92]. Surprisingly, the negotiation of the QoS parameters on point-to-multipoint (multicast) connections, which are a simple case of multipeer communications, has only received a very limited interest.

Only a few works try to cover adequately these two requirements. Among those works, one can cite ST-II [Top90], the proposed signalling protocol for the B-ISDN [ATM93] and the half-duplex real-time channel from the Tenet Group [SzV93].

ST-II has been designed to support point-to-multipoint connections in the network layer. ST-II is a network level protocol that also supports QoS parameters [Par92]. The negotiation scheme included in the ST-II specification is limited and not completely specified from a service point of view.

The current version of the signalling protocol for ATM, as defined in the ATM Forum UNI Specification version 3.0, includes some support for one-way point-to-multipoint connections. However, no negotiation scheme is defined in Q.93B, and the QoS parameters of the ATM connection are only selected by the calling user.

---

[1]This work was partially supported by the Commission of the European Communities under the RACE 2060 project CIO: "Coordination, Implementation and Operation of Multimedia Services".

The Tenet Group has devised solutions to support real-time communications with guaranteed QoS [FBZ92][FeV90] and studied the application of those solutions in the multicast case [SzV93]. However, those real-time communications are always based on a source traffic description and do not allow a QoS negotiation between the source and the receivers.

In this paper, we will concentrate on the enhanced QoS [DBL92][DBL94], developed in our department for the OSI95 peer-to-peer transport service [Bag93], and study how the QoS parameters can be negotiated on (1→N) multicast connections (or point-to-multipoint connections), that is communications where one sends and several receive.

This paper is structured as follows. In section 2, we will present a classification of the QoS parameters for multicast communications. After a brief presentation, in section 3, of the enhanced QoS defined in the OSI95 Transport Service, we will then study, in section 4, how the negotiation schemes for both those enhanced QoS and the QoS defined in the OSI Transport Service can be extended to cover (1→N) multicast connections. We will then present a formalisation of the negotiation scheme in section 5 and a practical issue concerning negotiation in section 6.

2. A Classification of the QoS Parameters

In the peer-to-peer case, one set of QoS values (minimum value, peak value, average value, ...) is associated with each parameter on each direction of data transfer. On a (1→N) multicast connection, a new dimension is added to the problem due to the presence of several receivers.

Indeed, in the multicast case, one can distinguish between two classes of QoS parameters:
- The QoS parameters whose scope is the whole connection (and thus affect the sender and all the receivers). These QoS parameters will be called connection-wide QoS parameters.
- The QoS parameters whose scope is limited to one receiver. These QoS parameters will be called receiver-selected QoS parameters.

In the following paragraphs, we will examine how the most common QoS parameters can be classified.

2.1. The Throughput

2.1.1. Generalities

While there are several definitions for the throughput (e.g. instantaneous throughput, average throughput, ...), the idea behind this QoS parameter is always the same. The throughput is a measure of the rate at which SDUs are exchanged. The throughput is usually one of the major QoS parameters for a connection-oriented data transfer.

To determine how to classify the throughput, we have to examine what happens when the sending service user issues SDUs. These SDUs have to be conveyed by the service provider and delivered to the receiving service users.

Obviously, if the throughput measured by one receiver is lower than the throughput measured at the sender's side, SDUs must be stored by the service provider. If this situation is lasting, SDUs will accumulate in the service provider, and this will eventually cause the release of the connection by the service provider, due to lack of resources. Of course, to avoid these problems, the service provider may use several mechanisms, such as interface flow control, to force the sender to slow down to a value of the throughput acceptable for the receiver.

Now, suppose that the throughput, measured by one receiver, is higher than the throughput measured at the sender's side. Clearly, such a situation cannot last indefinitely. Indeed, the service provider is going to eventually run out of SDUs and be forced to slow down the delivery of SDUs to the receiver.

Therefore, we see that the throughput at the receivers' sides and the throughput at the sender's side are not independent. Consequently, there is no point in selecting different sets of values for the throughput parameter at the different endpoints of a (1→N) multicast connection and the throughput can then be classified as a connection-wide QoS parameter.

2.1.2. Influence of Filtering on the Classification of the Throughput

In order to cope more easily with heterogeneity among receivers due, for instance, to different underlying networks or end-system capabilities, the idea of filters has been introduced [PPA92][ZDE93].

A filter is a function used to deliver to a receiver a data flow whose data rate is different (generally lower) than the data rate measured at the sending side. The definition and use of filters are discussed in [PPA92, 93][HSF93][ZDE93].

Whatever the filter be, it needs, to operate correctly, additional information on the data flow. Indeed, it is rather difficult to ask a filter to reduce the data rate of a given flow if that filter has no clue on which data units to discard in order to achieve the required reduction in the data rate. A random discarding would generally have disastrous consequences.

To avoid such a problem, a sender has several ways to provide the receivers (and subsequently the filters) with information on the data flow.

For instance, the data flow may be "layered", each level (called a sub-flow) being carried out on different (1→N) connections [PPA92, 93][HSF93]. In that case, the data flow is conveyed on several connections, and a receiver can ask a filter to reduce the data rate of that flow by simply "blocking" some well chosen connections.

Another way to define sub-flows is to differentiate the data units of a (1→N) multicast connection thanks to timestamps or tags. In such a case, the filtering function would be based on a relation involving the values of the timestamps or tags. However, those techniques using timestamps or tags are nothing but a way to define and multiplex several logical connections, each carrying a sub-flow, on a real one.

The use of filters is illustrated on figure 2.1.

Whatever the technique used to create sub-flows, the throughput on each (real or logical) connection carrying a sub-flow is still connection-wide.

Therefore, it should be clear that filtering allows a receiver to choose among the (real or logical) connections composing a data flow on a receiver-selected basis, but filtering can in no way be considered as modifying the connection-wide nature of the throughput on a connection.

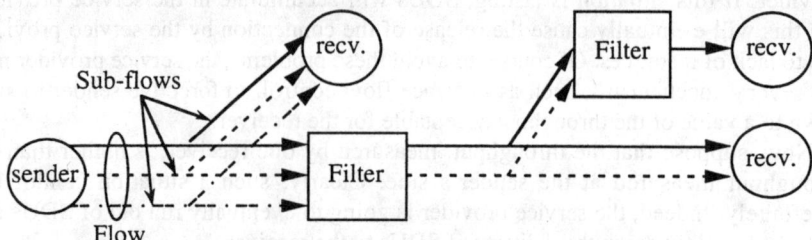

Fig. 2.1. Effect of Filters on a Data Flow.

2.2. The Transit Delay

The transit delay is usually defined as the elapsed time between the submission of a SDU by the sending service user to the service provider and the delivery of this SDU by the service provider to a receiving user. An interesting property of the transit delay parameter is that it is mainly relevant for the receivers.

As the receivers may be at different locations on the network and due to different network capabilities, local network congestion or different processing capabilities in end-systems, the measured values of the transit delay may be different from one receiver to another.

However, unlike what happens with the throughput parameter, permanent differences in the measured transit delay values for some receivers cause no problem to the service provider. Moreover, these differences do not have a significant influence on the behaviour of the sender.

As those differences in the measured transit delay cause no problem and may be permanent, differences in the selected values will not either. Consequently, the transit delay is a receiver-selected QoS parameter.

2.3. The Transit Delay Jitter

The transit delay jitter is usually defined as the maximal variation of the transit delay. It permits to specify the acceptable distortion of the SDU delivery pattern at a receiver compared to the SDU submission pattern at the sender's side. As the transit delay jitter is defined in term of transit delays, it is not surprising that it has also a meaning for the receivers only.

One way for the service provider to control the transit delay jitter is to buffer data before delivery, in order to minimise the differences between the transit delays experienced by the SDUs. Therefore, the more buffer space is available for the service provider to serve a receiver, the less is the transit delay jitter measured for that receiver.

Thus, the measured values of the transit delay jitter will only depend on local conditions (for instance, the amount of buffers available in a receiver's host). That is why, as it has already been the case for the transit delay, no problem is caused by permanent differences in the transit delay jitter measured for the different receivers. The transit delay jitter is thus a receiver-selected parameter.

2.4. The Error Control

The error control can be expressed as a combination of several error rate parameters (e.g. corruption rate, loss rate).

These two parameters are usually defined for peer-to-peer connections. However, in the multicast case, a new dimension to the problem of error control has to be taken into account, due to the presence of several simultaneous receivers. Indeed, the semantics of a reliable data transfer must now take into account the number of receivers that have to receive an error free SDU, at each invocation of the data transfer facility. That is why the concept of degree of reliability [MLB94] is used to specify the minimum number of receivers to which the data are to be delivered in a fully reliable way. A list of mandatory receivers may also be associated with the degree of reliability.

Obviously, as it relates to the whole (1→N) multicast connection, the degree of reliability is a connection-wide parameter.

When the degree of reliability specifies a minimum number of receivers that is lower than the total number of receivers on the multicast connection, some receivers may experience errors. In this case, error rate values have to be specified for those receivers. Since those error rates are measured independently for each receiver, it is natural to allow that each receiver selects its own error rate values independently. The error rate parameters are thus receiver-selected QoS parameters.

A classification of the common QoS parameters can be sketched as in table 2.1.

Table 2.1. Classification of the QoS parameters.

QoS Parameter	Classification
Throughput	Connection-wide
Transit Delay	Receiver-selected
Transit Delay Jitter	Receiver-selected
Error Rates	Receiver-selected
Reliability	Connection-wide

3. The Enhanced QoS

The enhanced QoS has been defined in the OSI95 Transport Service [DBL94] [DBL92]. In the OSI95 Connection-oriented Transport Service a QoS parameter is seen as a structure of three values, respectively called "compulsory", "threshold" and "maximal quality". Each value has its own well-defined meaning and is the result of a contract between the service users and the service provider.

The main idea behind the introduction of the enhanced QoS is that the service provider is subject to some well-defined duties, known by each side. In other words, the rules of the game are clear.

In the following sections, we will present the QoS defined in the OSI95 peer-to-peer Transport Service and we will see how to extend the definitions in order they can be supported on a (1→N) multicast connection.

3.1. The Compulsory QoS value

The idea behind the introduction of a "compulsory" QoS value is the following one: *when a compulsory value has been selected for a QoS parameter of a service facility, the service provider will monitor this parameter and abort the service facility when it notices that it cannot achieve the requested service.*

No obligation of results is linked to the idea of compulsory value. The service provider tries to provide the requested service facility and, by monitoring its execution, it will:
- either execute it completely without violating the selected compulsory value;
- or abort it if the selected compulsory value is not fulfilled.

In the case of a multicast connection, the concept of aborting a service facility needs to be clarified. Indeed, here, the service facility will only be aborted for the concerned service user, which means that the concerned service user will be excluded from the (1→N) multicast connection. Of course, if an Active Group Integrity (AGI) condition, that specifies conditions on the membership of the set of the service users that have to participate in the data transfer phase of the (1→N) multicast connection [MLB94], has been defined and is violated due to the exclusion of a service user, the (1→N) multicast connection is then released.

3.2. The Threshold QoS value

Some service users may find a little too radical the solution of aborting the requested service facility when one of the compulsory QoS values is not reached. They may prefer to get information about the degradation of the QoS value.

To achieve that "threshold" QoS value has been introduced with the following semantics: *when a threshold value has been selected for a QoS parameter of a service facility, the service provider will monitor this parameter and indicate to the service user(s) when it notices that it cannot achieve the selected value.*

On a (1→N) multicast connection, when the violation is detected at the sender's side, an indication may be delivered to the sender and the receivers. On the other hand, if the violation concerns a receiver, only that receiver may be provided with an indication.

This threshold QoS value may be used with or without an associated compulsory value.

3.3. The Maximal Quality QoS value

In most cases, if the service provider is able to offer a "stronger" value of the QoS parameter than the threshold, the service user will not complain about it. But it could happen, for reasons of cost or limited resources, that the service user wants to put a limit to a "richer" service facility.

To achieve that a "maximal quality" QoS value has been introduced with the following semantics: *when a maximal quality value has been selected for a QoS parameter of a service facility, the service provider will monitor this parameter and avoid occurrence of interactions with the service users that would give rise to a violation of the selected value.*

It is possible to associate, with the same QoS parameter, a maximal quality value, a threshold value and a compulsory QoS value with, of course, the maximal quality "stronger" than the threshold value, itself "stronger" than the compulsory value.

3.4. The Negotiation in the Peer-to-Peer Case

For each type of QoS value, precise rules are defined for each participant in the negotiation. We will first present the negotiation rules for each type of value separately and then a complex negotiation involving both a compulsory and a maximal quality QoS values.

In this paper, we will use the terms "weakening" and "strengthening" a parameter to indicate the trend of the modification. Weakening a throughput means reducing its value, whereas weakening a transit delay means increasing its value.

3.4.1. Negotiation of a Compulsory QoS value

The negotiation scheme for a compulsory QoS value has been defined in accordance with the semantics of this QoS value. The main idea of this negotiation scheme is the following:

If a service user introduces a compulsory value for a QoS parameter to be negotiated, the only possible modification is the strengthening of this compulsory value. In particular, it is absolutely excluded for the service provider to modify this value in order to relax the requirement.

However, as the calling user may not accept an unlimited strengthening of the proposed compulsory QoS value, it introduces a bound indicating to what extent the proposed compulsory QoS value may be strengthened.

When the service provider analyses the request of the calling service user, it has to decide whether it rejects it or not. If it chooses to accept the request, it has to examine the bound of strengthening. This bound may be made weaker (brought closer to the compulsory value) by the service provider. However, the service provider cannot modify the proposed compulsory value (which is application dependent exclusively).

After receiving the indication primitive, the called service user may accept or reject the request. If it accepts it, it may strengthen the compulsory QoS value up to the value of the bound and return it in its response (figure 3.1, where "s" stands for "stronger" and "w" for "weaker").

In the peer-to-peer case, if the negotiation is successful, the bound is of no interest anymore and the selected compulsory QoS value reflects now the final request to the service provider from both service users.

3.4.2. Negotiation of a Threshold QoS value

The negotiation for a threshold value is similar to the negotiation for a compulsory value. Here also the only possible modification is the strengthening of the threshold value. Here also the calling service user introduces, in the request primitive, a bound indicating to what extent the proposed threshold QoS value may be strengthened.

If a compulsory and a threshold values are associated with the same QoS parameter, there exist a set of order relationship between the compulsory, the threshold and their bounds values which must be verified in the request primitive and maintained during the negotiation.

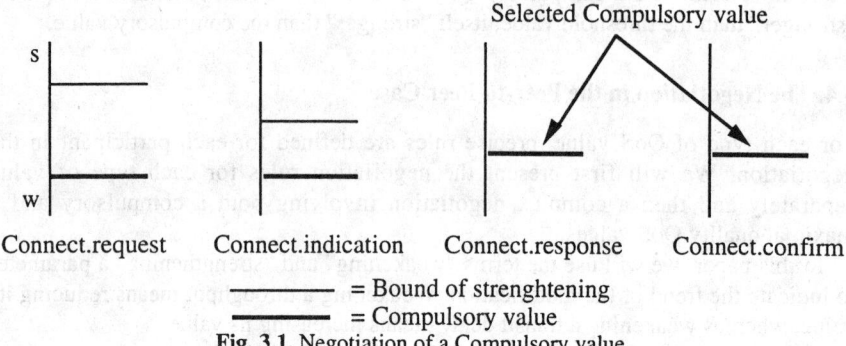

───── = Bound of strenghtening
───── = Compulsory value

Fig. 3.1. Negotiation of a Compulsory value.

3.4.3. Negotiation of a Maximal Quality QoS value

The negotiation scheme for a maximum quality QoS value has been defined in accordance with the semantics of this QoS value. The main idea of this negotiation scheme is the following:

If a service user introduces a maximal quality QoS value for a performance parameter, the only possible modification is the weakening of this maximal quality QoS value. This value can be weakened during the negotiation by the service provider that indicates by this way the limit of the service it can provide, and by the called service user (figure 3.2).

If the maximal quality value and a compulsory value or/and a threshold value are associated with the same QoS parameter, there exist an order relationship between the maximal quality value and the bound value on the threshold (or the bound value on the compulsory value if no threshold value is specified) which must be verified in the request primitive and preserved during the negotiation.

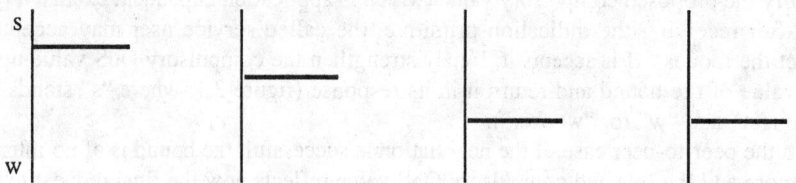

Fig. 3.2. Negotiation of a Maximal Quality QoS value.

As an example, a negotiation involving both a compulsory value and a maximal quality value is presented in figure 3.3. More complex negotiation schemes involving the compulsory, the threshold and the maximal quality values are presented in [Bag93].

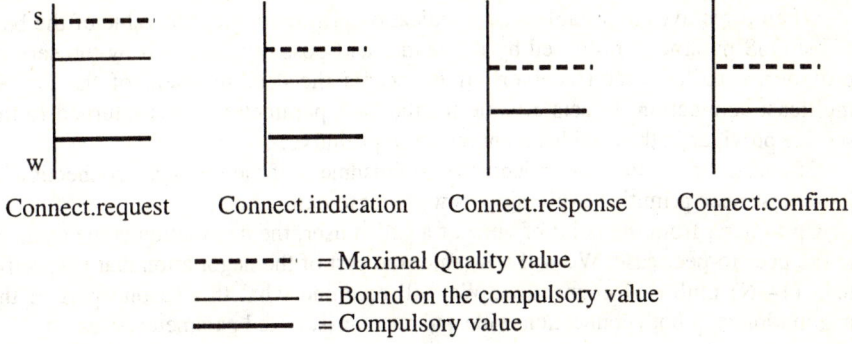

Fig. 3.3. Negotiation of both Compulsory and Maximal Quality values.

4. Negotiation on a (1→N) Multicast Connection

In this section, we consider the negotiation of the QoS parameter values on a (1→N) multicast connection. To simplify the presentation, we consider that the connection is established by the sender and that there are two receivers. The negotiation schemes proposed hereafter may be extended directly to more general (1→N) multicast connection (i.e. multicast connections where there are more than two receivers and/or which are not established by the sender).

Before examining how the negotiation of the compulsory, threshold and maximum values, described in the previous section for the peer-to-peer case, can be extended to (1→N) multicast connections, we will first examine how the ISO QoS negotiation scheme used in the ISO transport service [ISO 8072] can be extended to the multicast case.

4.1. Negotiation of a Best Effort QoS Value

For both connection-wide and receiver-selected parameters, the first part of the negotiation of a best effort QoS parameter value is the same as in the peer-to-peer case.

The calling service user (sender) supplies a proposed value for the parameter to the service provider in the establishment (connect) request primitive.

Then, the service provider tries to provide each called user (receiver) with an establishment indication service primitive. In those indication primitives, the proposed value may have been weakened by the service provider. Here, it is worth noting two things. Firstly, the proposed value may be different from one indication primitive to another. This is because the capabilities of the service provider may be different from one called user to another. This may be due, for instance, to different underlying networks, different protocol implementations, and so on. Secondly, for the same reasons, it may happen that the service provider will not be able to offer the service to some receivers whatever the value of the QoS parameter. In such a case, the service provider does not issue an establishment indication primitive to the concerned called users, which are "ignored".

When it receives an establishment indication primitive with the value of the best effort QoS parameter proposed by the sender and possibly modified by the service provider, a called user (receiver), if it accepts the establishment of the (1→N) multicast connection, selects a value for the QoS parameter that is returned to the service provider in the establishment response primitive.

Of course, if a called user rejects the establishment of the multicast connection, it issues a service primitive expressing its wish not to participate.

Up to here, from the point of view of a called user, the negotiation is the same as in the peer-to-peer case. We will now enter the part of the negotiation that is specific to a (1→N) multicast connection. We will examine what this second part of the negotiation is in both connection-wide and receiver-selected parameter cases.

4.1.1. Connection-wide parameters

During the first part of the negotiation, the service provider collects the QoS parameter values selected by the called users which agree to participate in the (1→N) multicast connection. The service provider must still select among the collected values for the connection-wide parameters, the ones that will apply to the whole connection and then issue an establishment confirm primitive with those connection-wide values to the sender and the receivers (that have agreed on the establishment of the connection).

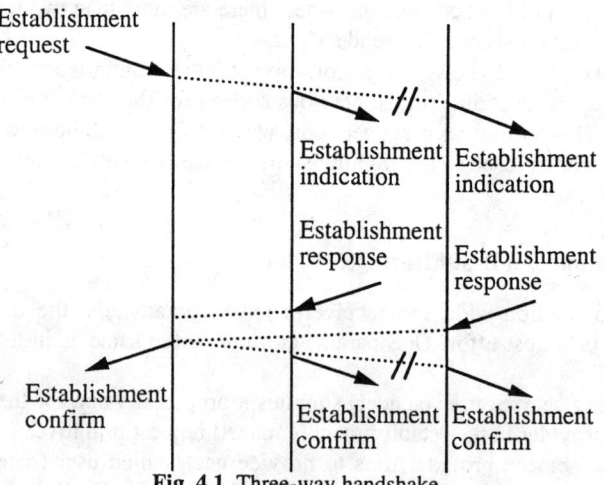

Fig. 4.1. Three-way handshake.

From the description above, it is clear that the negotiation of a connection-wide QoS parameter will require a three-way handshake. A three-way handshake between a sender and two receivers is illustrated in figure 4.1.

Let us now examine what the selected values are for the different connection-wide parameters.

4.1.1.1. Throughput

In the ISO transport service, the result of the negotiation of the throughput is the lowest value that is acceptable by either the calling user, the called user or the service provider. In the multicast case, this rule can be easily extended and the selected value will be the lowest value proposed by the called users.

4.1.1.2. Degree of Reliability

The degree of reliability represents a wish of the sender of the (1→N) multicast connection. If it was included in the multicast extension of ISO 8072 [ISO 8072], its value would then be imposed by the sender and not negotiated.

4.1.2. Receiver-selected parameters

As a receiver-selected parameter has a local meaning for each receiver separately, there is no need for the service provider to select single common values for such parameters. Therefore, from the point of view of the service provider, the negotiation of a receiver-selected value appears like several independent peer-to-peer negotiations but with the same initial value proposed by the sender.

However, the service provider may take advantage of the third handshake of the connection-wide parameter value negotiation to inform the sender and the receivers of the range of the receiver-selected values on the (1→N) multicast connection.

4.2. Negotiation of a compulsory QoS Value

For both the connection-wide and the receiver-selected parameters, the first part of the negotiation of a compulsory QoS parameter is the same as in the peer-to-peer case.

The calling user (sender) issues an establishment request primitive with a proposed value for the compulsory QoS parameter and a bound indicating to what extent the proposed compulsory value may be strengthened (see section 3.4.1).

Then, the service provider tries to provide each called user (receiver) with an establishment indication primitive. In those indication primitives, the bound may have been made poorer (brought closer to the compulsory value) by the service provider. Again, the bound may be different from one indication primitive to another. As regards the compulsory value in the indication primitives, it is the same as the value initially proposed by the sender, since the service provider may not alter this value.

If, for some receivers, the service provider deems that it will not be able to maintain performance above those expressed by the compulsory value, it "ignores" those receivers. In other words, the service provider does not issue indication primitives to those receivers.

After having received the indication primitive, a called user (receiver) may accept or reject the establishment of the (1→N) multicast connection. If it accepts, it may strengthen the compulsory QoS value up to the value of the bound and return it in its response primitive. The first part of the negotiation is illustrated on figure 4.2, where "s" stands for "stronger" and "w" for "weaker".

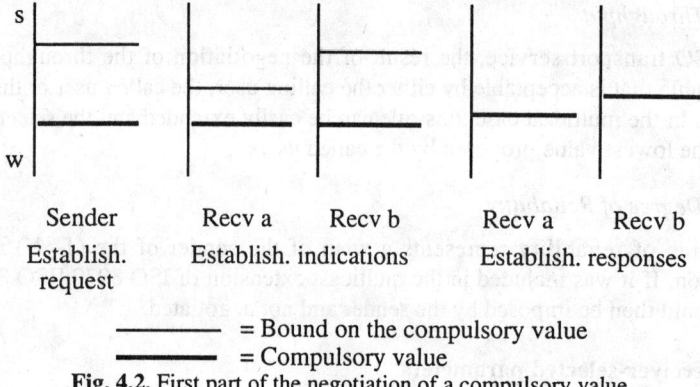

──────── = Bound on the compulsory value
━━━━━━━━ = Compulsory value

Fig. 4.2. First part of the negotiation of a compulsory value.

Again, from the point of view of a called user, the negotiation is the same as in the peer-to-peer case.

We will now examine what are the specific parts of the negotiation on a (1→N) multicast connection, for both connection-wide and receiver-selected QoS parameters.

4.2.1. Connection-wide parameters

Now that the service provider has been collecting the compulsory values proposed by the receivers, it has still to select among them the connection-wide compulsory values that will apply to the connection and inform the sender and receivers of those values thanks to a confirm primitive. Again, the negotiation is realised through a three-way handshake.

In the next sections, we examine what the selected values are for the different connection-wide parameter.

4.2.1.1. Throughput

The compulsory value of the throughput parameter represents a minimal performance. It is thus referred to as the minimum compulsory throughput value.

According to the semantics of a minimum compulsory value and in order to meet the requirements of the receivers, it seems natural to select as the connection-wide minimum compulsory value the highest value returned to the service provider by the receivers.

Unfortunately, things are not that easy. Indeed, recall that the service provider makes use of a bound (in this case called the upper minimum compulsory throughput) to indicate to a called user to which extent it can provide a minimum compulsory throughput. Those upper minimum compulsory values must now be taken into account by the service provider in order to avoid that the connection-wide minimum compulsory throughput value be higher than the upper minimum compulsory value imposed to some receivers. This would not only violate the rules of negotiation of the minimum compulsory value [Bag93][DBL94], but this would also induce a non-sense. Indeed, it is clear that the service provider would be unable to maintain the level of performance for those receivers whose upper minimum compulsory value was lower than the selected connection-wide minimum compulsory value.

We then see that in the case of a connection-wide compulsory value on a $(1 \rightarrow N)$ multicast connection, the bounds are now of great interest throughout the lifetime of that connection.

Therefore, we see that a condition that must be verified by the selected connection-wide minimum compulsory throughput value in order to cause no problem is that it must be lower than the lowest upper minimum compulsory value imposed to the receivers. We may call this latter value the connection-wide upper minimum compulsory value. If the connection-wide minimum compulsory is lower than the connection-wide upper minimum compulsory, we say that they are compatible.

Thus, the task of the service provider is in fact to find a subset of receivers which have compatible connection-wide minimum compulsory and connection-wide upper minimum compulsory throughput values (computed from the values selected by the receivers in the selected subset). The receivers included in the subset are said to have compatible selected QoS values. The other receivers, which are not included in the selected subset, are not taken into account any longer. Of course, several of such subsets of receivers may be found, the selected one depending on the algorithm used by the service provider. The description of such algorithms is outside the scope of this paper.

The figure 4.3 illustrates the complete negotiation for the throughput. The figure 4.3 (a) shows a case of compatible values selected by the receivers, whereas figure 4.3 (b) shows a case of incompatible selected values. In such a case, the service provider has to choose one of the receiver and no longer take the other one into account.

4.2.1.2. Degree of Reliability

The degree of reliability, which represents a minimum value, is imposed by the sender and not negotiated.

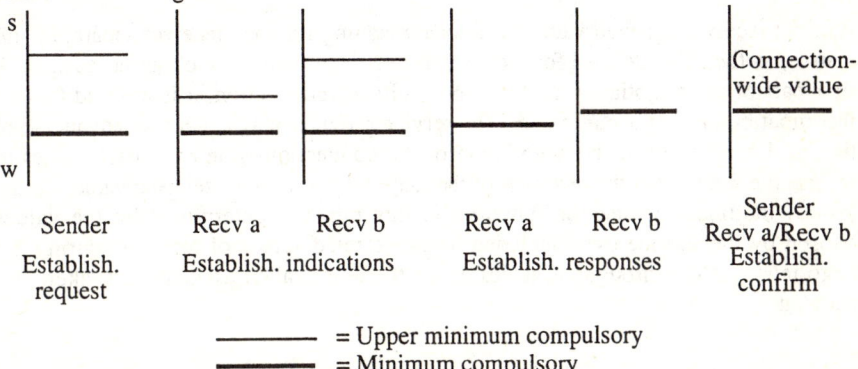

Fig. 4.3 (a). Compatible selected values.

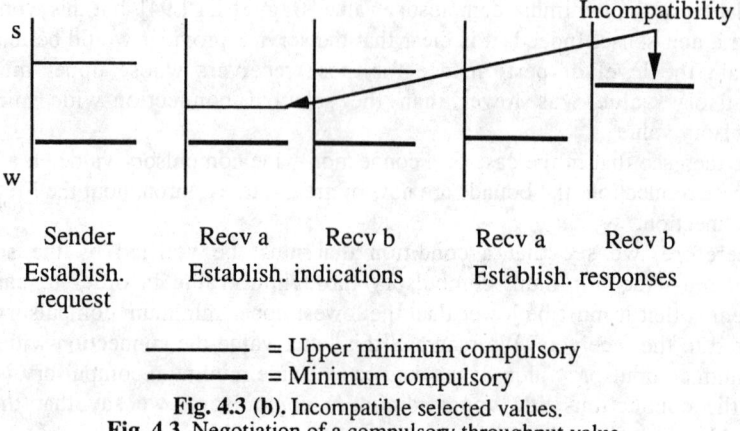

Sender   Recv a   Recv b   Recv a   Recv b
Establish.   Establish. indications   Establish. responses
request

─────── = Upper minimum compulsory
─────── = Minimum compulsory
Fig. 4.3 (b). Incompatible selected values.
Fig. 4.3. Negotiation of a compulsory throughput value.

4.2.2. Receiver-selected parameters

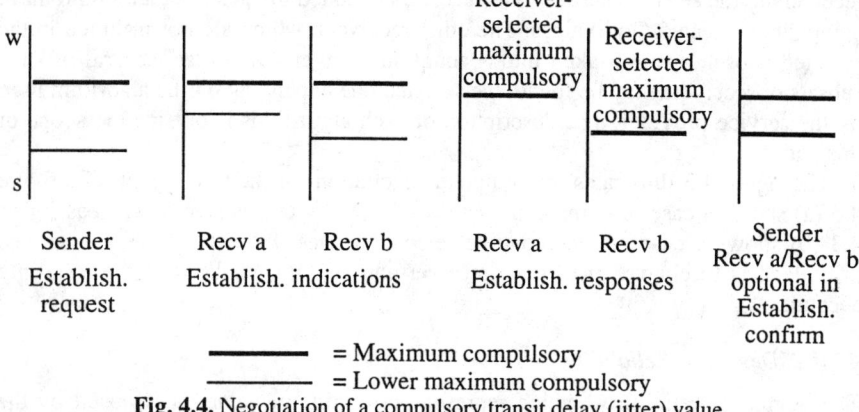

Sender   Recv a   Recv b   Recv a   Recv b   Sender
Establish.   Establish. indications   Establish. responses   Recv a/Recv b
request   optional in
                                                                             Establish.
                                                                             confirm

─────── = Maximum compulsory
─────── = Lower maximum compulsory
Fig. 4.4. Negotiation of a compulsory transit delay (jitter) value.

Again a receiver-selected value has a local meaning for each receiver separately. Of course, as the rules defined for the peer-to-peer negotiation are observed during the first part of the negotiation on a (1→N) multicast connection, it is ensured that no incompatibility may occur locally. The service provider may again take advantage of the third handshake of the negotiation of the connection-wide parameter values to inform the sender and the receivers of the range of the receiver-selected values on the (1→N) multicast connection. Obviously, this range is determined by the values chosen by the service users included in the selected subset of receivers during the negotiation of the throughput values, since the other called users are not taken into account.

4.2.2.1. Transit Delay / Transit Delay Jitter

A compulsory value for a QoS parameter represents a minimal performance for that parameter. As far as the transit delay is concerned, a minimal performance corresponds to a maximal value of the transit delay. That is why a compulsory value for the transit delay is called a maximum compulsory value and the bound on that value is called the lower maximum compulsory value. The complete negotiation of the transit delay (or transit delay jitter) is illustrated on figure 4.4.

Those designations also apply to the transit delay jitter and the error rates.

4.2.2.2. Error Rates

The service provider does not intervene in the negotiation of the error rate values [Bag93], which means the values proposed to the called users (receivers) in the indication primitives are the ones proposed by the sender (compulsory and bound).

4.3. Negotiation of a Threshold QoS Value

The rules for the negotiation of a threshold value are similar to the rules for the negotiation of a compulsory value [DBL94].

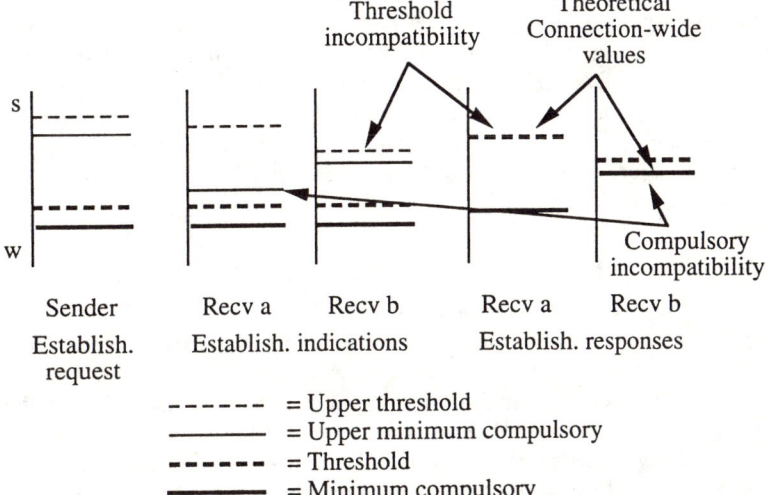

Fig. 4.5. Problem during simultaneous compulsory and threshold value negotiations.

However, the simultaneous use of a compulsory and a threshold value for a same connection-wide QoS parameter may cause problems and requires further study. Such a situation is illustrated on figure 4.5. On this figure, we see that, if we apply the rules of negotiation described in section 4.2, the connection-wide compulsory value is the compulsory value proposed by receiver$_a$ while the compulsory value proposed by receiver$_b$ is incompatible. On the other hand, we see that the connection-wide threshold value is the threshold value proposed by receiver$_b$ while the threshold value proposed by receiver$_a$ is incompatible. Since one receiver rules out the other and vice versa, the question that arises is which of the receiver to choose to be included in the subset of selected receivers?

Thus, when a compulsory value and a threshold value are used simultaneously for a same connection-wide QoS parameter, they cannot be negotiated separately, which introduces many difficulties in the negotiation scheme.

In the case of a receiver-selected QoS parameter, no problem is caused by the simultaneous use of a compulsory value and a threshold value since the rules of the peer-to-peer negotiation are always verified locally.

4.4. Negotiation of a Maximum Quality QoS Value

For both connection-wide and receiver-selected parameters, the negotiation scheme of the maximum quality value is the same as the negotiation scheme of a best effort QoS parameter value, except there exists a little difference when a compulsory (or threshold) value and a maximum quality value are used simultaneously for a same QoS parameter.

Indeed, in the case of a connection-wide parameter, it is natural to select as the connection-wide maximum value the most restrictive one, that is the weakest one. However, if the maximum quality value is associated with a compulsory (threshold) value for a same QoS parameter, there exists an order relationship between the maximal quality value and the bound value of the compulsory or threshold (the maximum quality value has to be "stronger" than the bound on the compulsory (threshold) value). This relation has to be preserved in both the request and the indication primitives. However, the maximum quality value selected by a receiver in its response can be weaker than the bound on the compulsory (threshold) value, provided the selected maximum quality value is still stronger than the selected compulsory (threshold) value itself.

4.4.1. Connection-wide parameter

The negotiation of a connection-wide maximum quality value requires a three-way handshake. If the maximum quality value is the only one associated with a QoS parameter, the negotiation is exactly the same as in the best effort case.

However, if a compulsory (threshold) value is also associated with the QoS parameter, the service provider must take care that the selected connection-wide maximum quality value is not weaker than the selected connection-wide compulsory (threshold) value, which would result in a non-sense. Thus, there is another condition that must be verified by the connection-wide compulsory (threshold) value, when it is associated with a maximum quality value. This condition is that the connection-wide compulsory (threshold) value must be weaker than the connection-wide maximum quality value.

4.4.1.1. Throughput

From the discussion above, it should be clear that the receivers included in the subset of receivers during the negotiation of the throughput values, must not only have compatible minimum compulsory (threshold) and upper minimum compulsory (threshold) values but also compatible minimum compulsory (threshold) and maximum quality values.

The figure 4.6 illustrates the simultaneous negotiation of a compulsory value and a maximum value.

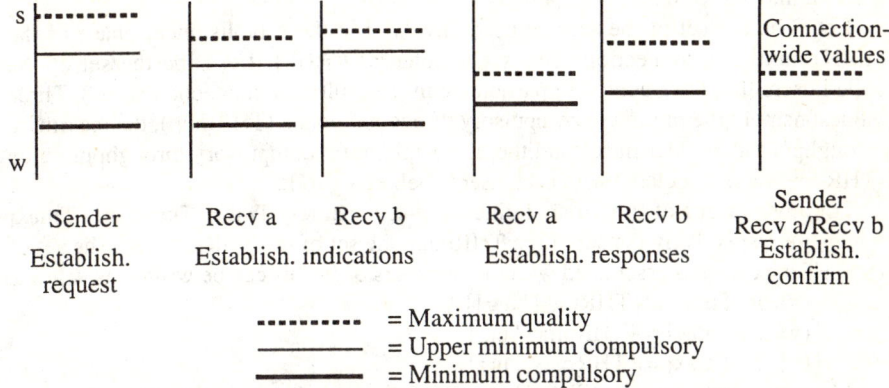

```
----------    = Maximum quality
_____    = Upper minimum compulsory
██████████    = Minimum compulsory
```

Fig. 4.6. Simultaneous negotiation of compulsory and maximum quality values.

4.4.2. Receiver-selected parameters

Again a receiver-selected value has a local meaning for each receiver separately and there is no need for the service provider to unify those values. The negotiation of a receiver-selected maximum value follows the same scheme as in the best-effort case. Of course, when a compulsory (threshold) value is associated with the maximum value, it is ensured that no incompatibility may occur locally, as the rules defined for the peer-to-peer negotiation are observed during the first part of the negotiation on a $(1 \rightarrow N)$ multicast connection. The service provider may again take advantage of the third handshake of the negotiation of the connection-wide parameter values to inform the sender and the receivers of the range of the receiver-selected values on the $(1 \rightarrow N)$ multicast connection.

Before concluding this section on the negotiation of the QoS parameter values on a $(1 \rightarrow N)$ multicast connection, we would like to highlight the fact that only the receivers with compatible values for the throughput will take part in the data transfer phase of the multicast connection. Moreover, although the choice of a subset of such receivers will depend on the algorithm used by the service provider, it seems fair to require that if the whole set of the called users that accept the establishment of the connection verifies the conditions (see section 4.4.1.1), then this set has to be the chosen one.

Finally, another condition that has not been taken into account yet, but that has to be verified by the selected subset of receivers, is the AGI condition of the $(1 \rightarrow N)$ multicast connection that specifies conditions on the membership of the set of service users that has to participate in the data transfer phase of the connection [MLB94].

In the next section, we will give a more formal description of the negotiation of simultaneous compulsory and maximum values of the throughput parameter.

5. Formalisation of the Negotiation

The negotiation scheme for the throughput parameter described in the previous section, may be formalised as follows.

Let C be the set of the service users involved in the establishment phase of the (1→N) multicast connection. This set is noted as $C=\{u_i\}$. Let P be the set of the called users that have agreed to participate in the multicast connection ($P \subseteq C$). Three values, namely the minimum compulsory throughput value ($THRmin_i$), the maximum throughput value ($THRmax_i$) and the upper minimum compulsory throughput value ($THRumc_i$) are associated with each user u_i belonging to P.

For any subset I of P, we can define three sets, namely $Tmin_I$, $Tumc_I$ and $Tmax_I$ which are respectively the set of the $THRmin_i$, the set of the $THRumc_i$ and the set of the $THRmax_i$ values associated with the members of I. This can be written as:

($\forall I \subseteq P$) $Tmin_I = \{THRmin_i \mid u_i \in I\}$
($\forall I \subseteq P$) $Tumc_I = \{THRumc_i \mid u_i \in I\}$
($\forall I \subseteq P$) $Tmax_I = \{THRmax_i \mid u_i \in I\}$

Moreover, we note $\phi_{AGI}(S)$ the logical function that is true if the set S of service users verifies the AGI condition, and false otherwise.

We can now define the function IsTHRvalid(I) that is true if the members of I have selected compatible throughput values and satisfy the AGI condition, and false otherwise:

IsTHRvalid(I) = ($THRmin_I \leq THRumc_I$) $\wedge$ ($THRmin_I \leq THRmax_I$) $\wedge$ $\phi_{AGI}(I)$

where $THRmin_I = \max(Tmin_I)$, and
$THRumc_I = \min(Tumc_I)$ and
$THRmax_I = \min(Tmax_I)$.

The negotiation of the throughput parameter values may then be described by the following procedure:

<u>If</u> IsTHRvalid(P) <u>then</u> P
<u>else</u> any $I \subseteq P$ such that IsTHRvalid(I)

6. A Practical Issue

In the previous sections, we have shown what the connection-wide values for the throughput parameter are for a given subset of receivers and we have also shown what relations must be verified between those connection-wide values in order to avoid problems for either the service provider or any of the receivers in the subset.

Using the same notation as in the previous section, the set of selected receivers is called I, the highest minimum compulsory value associated with I is called $THRmin_I$, the lowest upper minimum compulsory and maximum values associated with I are called respectively $THRumc_I$ and $THRmax_I$. The relations that must exist between the connection-wide values, in order that those values be compatible with the values that were proposed by each receiver of I, are:

$THRmin_I \leq THRumc_I$ (1) and
$THRmin_I \leq THRmax_I$ (2).

We now examine the practical problem to find the conditions that must be verified by the throughput values proposed by a service user u_x so that this service user can be included in an existing set I without violating the compatibility of the values of the throughput parameter associated with the receivers in I. Such conditions on the throughput values are useful when, for instance, a service user joins a $(1{\rightarrow}N)$ multicast connection. They may also be used by the algorithms searching for a set I during the negotiation of the QoS parameters.

Let us call $THRmin_x$, $THRmax_x$ and $THRumc_x$ respectively the minimum compulsory, the maximum and the upper minimum compulsory values associated with the user u_x. Those values are compatible with the ones associated with the service users in the set I if:

$THRmin_x \leq THRumc_I$ (3), and
$THRmin_x \leq THRmax_I$ (4), and
$THRumc_x \geq THRmin_I$ (5) and
$THRmax_x \geq THRmin_I$ (6).

This means that if these four conditions are fulfilled, u_x may be included in I and that, if we call the resulting set I', the following conditions:

$THRmin_{I'} \leq THRumc_{I'}$
$THRmin_{I'} \leq THRmax_{I'}$

where $THRmin_{I'} = max(THRmin_x, THRmin_I)$ (7), and
             $THRumc_{I'} = min(THRumc_x, THRumc_I)$ (8) and
             $THRmax_{I'} = min(THRmax_x, THRmax_I)$ (9)

are still verified.

The demonstration of those properties is obvious.

Remind that the rules of the peer-to-peer negotiation [Bag93][DBL94] ensure that:

$THRmin_x \leq THRumc_x$ (10) and
$THRmin_x \leq THRmax_x$ (11).

7. Conclusion

In this paper, we have examined how the QoS parameters defined in OSI95 can be negotiated on point-to-multipoint connections. We have also shown that in the multicast case, the QoS parameters can be separated into two classes, namely the connection-wide and the receiver-selected parameters. The properties of each class of parameters have been discussed. The negotiation scheme presented can be applied directly to several transport services (e.g. [Hen94], [DCS93]).

Finally, when many-to-many connections are considered as a set of several point-to-multipoint connections [Hen94][MLB94], this negotiation scheme may also be used for each point-to-multipoint connection.

Acknowledgements

We are particularly grateful to two of our colleagues, Guy Leduc and Yves Baguette, for their comments and suggestions on this work.

References

[ATM93] ATM Forum: ATM User-Network Interface Specification, Version 3.0, September 1993.
[Bag93] Y. Baguette: Enhanced Transport Service Definition (Informal specification in English - Version 1), ISO/IEC JTC1/SC6/WG4/N822, 19 Apr. 1993, 85 p.
[CCH94] A. Campbell, G. Coulson, D. Hutchison: A Quality of Service Architecture, *ACM SIGCOMM Computer Communication Review*, April 1994, Vol.24, No.2, pp 6-27.
[DBL92] A. Danthine, Y. Baguette, G. Leduc, L. Léonard: The OSI 95 Connection-mode Transport Service - The Enhanced QoS, *High Performance Networking, IV*, IFIP Transactions C-14, A. Danthine, O. Spaniol, eds., Elsevier (North-Holland), pp 235-252.
[DBL94] A. Danthine, O. Bonaventure, G. Leduc: The QoS Enhancements in OSI95, *The OSI95 Transport Service with Multimedia Support*, A. Danthine, ed., Springer-Verlag, pp 124-149.
[DCS93] C. Diot, P. Cocquet, D. Stunault: Specification of ETS, the Enhanced Transport Service, ISO/IEC JTC1/SC6/N7883, January 1993
[FBZ92] D. Ferrari, A. Banerjea, H. Zhang: Network Support for Multimedia - A Discussion of the Tenet Approach, Technical Report TR-92-072, International Computer Science Institute, Berkeley, November 1992.
[FeV90] D. Ferrari, D. Verma: A Scheme for Real-Time Channel Establishment in Wide-Area Networks, *IEEE Journal on Selected Areas in Communications*, April 1990, Vol.8, No.3, pp 368-379.
[Hen94] L. Henckel: Multipeer Transport Services for Multimedia Applications, Participant's *Proc. 5th IFIP Conference on High Performance Networking*, Grenoble, June 27-July 1, 1994, S. Fdida, ed., pp 165-183.
[HSF93] D. Hoffman, M. Speer, G. Fernando: Network Support for Dynamically Scaled Multimedia Data Streams, *Proc. 4th International Workshop on Network and Operating Systems Support for Digital Audio and Video*, Lancaster, November 3-5, 1993, pp 251-262.
[ISO 8072] ISO/IEC JTC1: Information technology - Transport service definition for Open Systems Interconnection, ISO/IEC JTC1/SC6 N7734 (ISO DIS 8072 - DIS ballot terminates on April 1993).
[MLB94] L. Mathy, G. Leduc, O. Bonaventure, A. Danthine: A Group Communication Framework, *3rd International Broadband Islands Conference*, Hamburg, June 7-9, 1994, O. Spaniol, W. Bauerfeld and F. Williams, eds., Elsevier Science Publishers (North-Holland), 1994, pp 167-178.
[Par92] C. Partridge: A Proposed Flow Specification, Internet RFC 1363, 1992
[PPA92] J. Pasquale, G. Polyzos, E. Anderson, V. Kompella: The Multimedia Multicast Channel, *Proc. 3rd International Workshop on Network and Operating Systems Support for Digital Audio and Video*, San Diego, November 12-13, 1992, P. Venkat Rangan, Ed., Springer-Verlag, pp 197-208.
[PPA93] J. Pasquale, G. Polyzos, E. Anderson, V. Kompella: Filter Propagation in Dissemination Trees: Trading Off Bandwidth and Processing in Continuous Media Networks, *Proc. 4th International Workshop on Network and Operating Systems Support for Digital Audio and Video*, Lancaster, November 3-5, 1993, pp 269-278.
[SzV93] C. Szyperski, G. Ventre: Efficient Group Communication with Guaranteed Quality of Service, *4th IEEE Workshop on the Future Trends in Distributed Computing Systems*, Lisboa, September 1993.
[Top90] C. Topolcic: Experimental Internet STream Protocol, Version 2 (ST-II), Internet RFC 1190, 1990.
[ZDE93] L. Zhang, S. Deering, D. Estrin, S. Shenker, D. Zappala: RSVP: A New Resource ReSerVation Protocol, *IEEE Network*, September 1993, Vol.7, No.5, pp 8-18.

Support for High-Performance Multipoint Multimedia Services

Georg Carle, Jochen Schiller, Claudia Schmidt
Institute of Telematics, University of Karlsruhe,
Zirkel 2, 76128 Karlsruhe, Germany
Phone: +49 721 608-[4027,4021,3414], Fax: +49 721 388097
Email: [carle,schiller,schmidt]@telematik.informatik.uni-karlsruhe.de

Abstract. Existing and upcoming distributed multimedia applications require highly diverse services to satisfy their communication needs. Service integrated communication systems should be capable of providing high-performance real-time multipoint communication service with guaranteed quality of service (QoS). Existing communication systems and known strategies for resource reservation face increasing difficulties in fulfilling these requirements, in particular in high-speed wide area networks. Therefore, new concepts are required to support the variety of emerging applications in a heterogeneous internetworking environment. In this paper, a framework for real-time multipeer services is presented. It is based on the separation of service requirements into network bearer and transfer service requirements. The transfer service enhances the network bearer service in order to meet the service requirements of the applications. Applications and transport systems interact using an enhanced service interface, which offers several QoS parameters. The transfer service is supported by transfer components (layer 2b-4 protocol functions in end and intermediate systems), resource management functions and the integration of specialized VLSI modules for time-critical processing tasks. The hardware components can be parametrized and selected individually dependent on the required service. The transfer system guarantees specified service qualities by assigning processing resources to specific connections. Selection of the appropriate combination of components and their parametrization enhances the bearer service of the underlying network in order to provide the required multimedia service at the transport service interface. Guaranteed services are realized by the reservation of resources both in the network (for guaranteeing network bearer service) and in transfer components (for guaranteeing transfer service). The paper describes a general approach towards high performance multipeer services, and dedicated parts in more detail. Preliminary performance results of VLSI components are presented and compared with measurements of typical software implementations.

1 Introduction

In the recent years, several new distributed applications with diverse service requirements have been developed. Frequently, these emerging applications require both high performance as well as support of a wide variety of real-time communication services. High-speed networks, (e.g., ATM-based networks) are able to fulfill bearer service requirements by providing data rates exceeding a gigabit per second and supporting different kinds of services. However, the service required by applications usually differs from the bearer service. Protocol processing in the end systems is needed to enhance a basic bearer service in order to meet application requirements. Yet, current communication systems (including higher layer protocols) are even facing problems in delivering the available network performance to the applications and demand for enhancements. Different approaches [1], [2], [3], [4], [5], [6] on implementing high performance communication subsystems have been undertaken during the last few years: software optimisation, parallel processing, hardware support and dedicated VLSI components. Some of the approaches deal with efficient implementations of standard protocols such as OSI TP4 or TCP. Others developed protocols especially suited for advanced implementation environments. The use of dedicated VLSI components is mostly limited to very simple communication protocols (e.g., [7], [8]).

Another issue in the evolution is the increasing importance of group communication scenarios. Upcoming applications, for example in the areas of computer-supported co-operative work (CSCW), distributed systems, and virtual shared memory systems require point-to-multipoint (Multicast, 1:N) as well as multipoint-to-multipoint (Multipeer, M:N) communication [9]. For a growing number of applications such as multimedia collaboration systems, the provision of a multicast service associated with a specific quality of service (QoS) is necessary. If multipoint communication is not supported by the network or by the end-to-end protocols, multiple point-to-point connections must be used for the distribution of identical information to the members of a group. The use of multicasting is beneficial in various ways: It saves bandwidth, reduces processing effort for the end systems and the mean delay for the receivers, and furthermore, simplifies addressing and connection management. Various issues need to be addressed in order to provide group communication services in high-speed networks [10], [11]. Intermediate systems need to incorporate a copy function to support 1:N connections. Communication protocols must be capable of managing multipoint connections, and group management functions need to be provided for the administration of members joining and leaving a group. Another key problem that must be solved for a reliable multipoint service is the recovery from packet losses due to congestion in the network nodes and end systems. In this paper, we present VLSI support that is targeted towards complex multicast protocols. Especially, time-critical and processing intensive functions, e.g., selective retransmission are provided by dedicated VLSI components.

This paper is organised as follows: Section 2 gives an overview of an integrated approach for multimedia communication in high-speed networks and presents the conceptual framework for integrating VLSI components for multicast functions into end systems and group communication servers. The framework is not limited to a certain protocol and allows the use of high-speed protocols with fixed size packet headers. Section 3 presents a performance evaluation of different scheduling mechanisms and of different error control strategies in multicast scenarios. Section 4 discusses the functionality of a dedicated connection processor for managing multicast retransmission and presents some preliminary complexity and performance results of the discussed component. Section 5 summarises the paper and points out some future directions.

2 Multipoint Multimedia Communication Systems

Advanced multimedia applications are sensitive to the quality of service parameters throughput, delay, jitter, and reliability. Furthermore, they need to concurrently process different data streams, e.g., audio, video, and conventional data communication. Each of these streams requires a different QoS combination. As not only point-to-point but multipoint-to-multipoint communication models with several senders and receivers are needed, specific QoS parameters are associated with all members of a group or only with a subset of receivers.

There is a need to accommodate traditional communication systems to these new application requirements. Therefore, several new components have to be introduced, and existing components of the communication system have to be modified. In the following, the distinct components of an enhanced communication subsystem for multimedia multipoint services are described.

2.1 Enhanced Service Interface

Traditionally, a limited set of QoS parameters was used to describe the requested service, and often no enforcement of these parameters was provided. Moreover, most communication protocols targeted towards pure data transfer, like TCP/IP, offer only a reliable point-to-point service (with the single QoS parameter 'reliability'). The service interface of these protocols no longer reflects the requirements of upcoming multimedia applications. These applications have strong throughput and delay requirements, while they often tolerate a reduced reliability for some communication channels. In particular, for networks under heavy load, certain data losses may be better tolerated by humans than additional delay introduced by retransmissions.

Enhanced service models and interfaces are required in order to serve emerging applications. Applications and the communication system need a common language, so a source is able to specify the traffic characteristics of the flow and, in turn, the communication system guarantees a specific service. The QoS parameters throughput, delay, jitter, and reliability can be specified by a minimum, maximum, and average value. Applications prefer to specify their requirements in an application related syntax and semantics, which usually differs from the one used in the communication system. In this case, a mapping of the parameters used at the service interface valid for TSDUs (e.g., frames of a video source) to parameters inside the communication system valid for TPDUs, is necessary. Additionally, the requested service needs to be described by different service classes, which specify how the service parameters should be guaranteed. Service classes can be classified as deterministic, statistic, and best effort services. Deterministic services guarantee QoS parameters even in the worst case, statistic services guarantee QoS parameters with a given probability, and best effort services are targeted towards applications which have no specific requirements. Real time services based on statistical multiplexing achieve either statistical or best effort reliability. Fully reliable services need additional mechanisms for error correction and are therefore limited to statistical or best effort delay. The complete service is described in a *service contract*. In this contract, the application agrees to limit the traffic passed to the service interface, while the service provider is obliged to reserve the resources required to maintain the specified service quality.

2.2 Resource Management

To deliver a requested service to a particular connection, it is usually necessary to reserve certain resources in all involved communication systems for that connection [12]. Such resources are the bandwidth of a link, processing power, and buffer space. Resource management is an important feature, not yet included in most existing communication systems, that describes the ability to create and maintain resource reservations. Resource management consists of three different parts:

- *admission control* and

- *reservations* during the connection establishment, and
- *assignment of resources* during the data transfer phase.

Because the resources of communication systems are finite, not all connection requests can be accepted. In order to meet all service commitments, a communication system must contain an admission control algorithm that determines if a connection request with specific QoS requirements can be accepted without violating already accepted requests or if it must be denied. If all admission control tests are positively passed, reservations are made.

During the data transfer phase the communication system must provide mechanisms to assign the resources dependent on the reservations to the connections. Resources can be classified as active and passive resources [13], where active resources execute a process and passive resources hold data of the process. Dependent on the selected service class, the passive resource buffer space can be used exclusively by one connection or can be shared among several connections. In this case it is important to determine which packets have to be dropped if there is no more buffer available. All active resources are exclusive resources and are assigned by a *scheduler*. A scheduler decides which packet is processed next. There exist many proposed scheduling algorithms with different capabilities.

Scheduling Algorithms. Recently, several scheduling disciplines have been proposed [14], [15], [16] and compared [17]. The algorithms differ in the ability to provide guarantees for the QoS parameters throughput, delay, and jitter. In particular for the delay parameter, considerable research has been devoted towards the development of methods for providing a provable analytic upper bound, which will be experienced by all data packets. However, in many cases a very high percentage of packets will observe a delay far lower than the computable worst case bound.

In [17] a classification of scheduling disciplines in *Sorted Priority Queue Mechanisms* and *Framing Strategies* is proposed. Sorted Priority Queue Mechanisms use a state variable for each connection. Upon the arrival of a packet from this connection, the variable is updated and the packet is stamped with the new value. Packets are served in order of increasing stamps. In Framing Strategies, the time axis is divided into periods of some constant length, each called a frame. Bandwidth is allocated to each connection as a certain fraction of the frame time. Sorted Priority Queues are flexible in reserving different delays and bandwidth combinations to connections, while framing strategies couple the allocation of delay and bandwidth. The advantages of framing strategies are an easier implementation and no need of a test for schedulability at connection establishment time. However, for several disciplines of both classes it has been shown that they provide throughput as well as worst case delay guarantees. Although these disciplines guarantee a delay bound, these delays are often very long. Experiments [18] have shown that traffic scheduled by a simple FIFO discipline provides significantly lower delays even for heavy traffic load. If the implementation overhead of complex scheduling schemes is taken into account, FIFO seem to be a good trade-off for high performance networks.

In this paper we want to present preliminary simulation results of three different scheduling disciplines: FIFO, Virtual Clock [14], and a packet-by-packet version of Round Robin [16]. Virtual Clock is a Sorted Priority Queue Mechanism, which aims to emulate Time Division Multiplexing. Each packet is associated with a virtual transmission time, which is the time at which the packet had been transmitted under Time Division Multiplexing. Virtual Clock is able to guarantee a specified throughput. The packet-by-packet round robin server uses separate queues for each connection and serves in each round the specified bandwidth of a connection. Statistical multiplexing using FIFO scheduling may provide a satisfying mean delay, while allowing a simple implementation. However, its firm delay bounds may often be far longer than applications wish to accept [19]. More sophisticated scheduling algorithms allow to reduce the firm delay bounds, while leading to more complex implementations.

Protocols for Resource Reservation. To reserve resources, the service requirements have to be distributed to all involved end and intermediate nodes. Several special reservation protocols have been developed [12], [20], [21] during the last years. They are characterised by a connection oriented simplex concept. To establish a connection with service commitments, the reservation protocols have to distribute a data structure, the so-called *flowspec*, which describes both, the characteristics of the traffic sent by the source and the service requirements. The protocol entities in the involved communication systems activate the resource management entities, which decide to accept or deny the connection. Furthermore, reservation protocols define rules to modify service parameters or mechanisms to detect errors. To support emerging multimedia applications efficiently, the protocols are not only required to provide mechanisms for guaranteeing a requested service in point-to-point communications, but also to accommodate receivers of a multicast group. Specific problems occur if each of these receivers has different service requirements.

2.3 Reliable Services

Packet Loss due to Congestion. The dominant factor which causes high speed networks to discard packets is buffer overflow due to congestion. In packet switched networks, statistical multiplexing allows a high degree of resource sharing. Short periods of congestion may occur due to statistical correlations among variable bit rate traffic sources, resulting in buffer overflow. The probability for packet loss may vary over a wide range, depending on the applied strategy for congestion control. Packet losses due to buffer overflow are caused by superposition of traffic bursts. Therefore, they do not occur randomly distributed, but in bursts and show a highly correlated characteristic. If a reliable service has to be provided, mechanisms are required which are able to handle this type of error efficiently. For multicast connections, the problem of packet losses is even more crucial than for unicast connections. Collisions of the multicast connections with independent unicast connections may occur independently at every output port of a node. Therefore, the packet loss probability in multicast connections will be higher than in unicast connections.

Error Control Mechanisms. For applications that cannot tolerate packet losses of the network, error control mechanisms are required. Error control consists of two basic steps: error detection and error recovery. For error recovery, two mechanisms are available: Automatic Repeat ReQuest (ARQ) and Forward Error Correction (FEC). Error control is difficult in networks that offer high bandwidth over long distances. High data rates in combination with a long propagation delay result in

high bandwidth-delay products. A large amount of data may be in transit. For example, at a distance of 5000 km and a data rate of 622 Mbit/s, more than 2 MByte may be stored by the link. This causes problems for the following reasons:
- End-to-end control actions require a minimum of one round-trip-delay, and retransmissions require large buffers and may introduce high delays;
- Efficient error control with timer-based loss detection is difficult, because delay variations do not allow very accurate timer setting, causing deterioration of the service quality;
- Processing of error control needs to be performed at very high speeds, if no bottle-neck is to be introduced.

ARQ Methods: ARQ (Automatic Repeat ReQuest) mechanisms are widely used in current data link and transport protocols. In every retransmission based scheme, the transmitter needs to store messages upon acknowledgement. At least the data of one round-trip delay needs to be stored. For go-back-N protocols, implementation of transmitter and receiver may be very simple, and no buffering is required for the receiver. For selective repeat protocols, implementation of transmitter and receiver is more complex, and a large buffer is required by the receiver. Processing overhead of ARQ methods is proportional to the number of data and acknowledgement packets that are processed. For point-to-point communication, ARQ mechanisms are well understood, and a number of protocols for data link layer or transport layer, employing these mechanisms, are known. For multicast communication, there are still many open questions concerning acknowledgement and retransmission strategy, achievable performance and implementation. Large groups require that the transmitter stores and manages a large amount of status information of the receivers. The number of retransmissions is growing for larger group sizes, decreasing the achievable performance. Additionally, the transmitter must be capable of processing a large number of control information. If reliable communication is required to every multicast receiver, a substantial part of the transmitter complexity is growing proportionally to the group size. In addition, individual receivers may limit the service quality of the whole group. To overcome these problems, a scheme that provides reliable delivery of messages to K out of N receivers may be applied (K-reliable service).

FEC Methods: Forward error correction (FEC) methods promise a number of advantages [22]. All FEC methods have in common that redundant information is transmitted to the receiver, together with the original information, permitting detection and correction of errors. The delay for error recovery is independent of the distance, and large bandwidth-delay products do not lead to high buffer requirements. Therefore, FEC is a promising approach in high-speed networks. In contrast to ARQ mechanisms, FEC is not affected by the number of receivers. However, FEC has three main disadvantages when applied for error correction in high speed networks. It is computationally demanding, leading to complex VLSI components. It requires constantly additional bandwidth, limiting the achievable efficiency and increasing packet loss during periods of congestion. FEC achieves best performance for random errors. However, packet losses frequently occur in bursts [23]. If FEC is used for packets of variable length, parity packets will be calculated for data units of fixed size, leading to inefficient use of parity information.

The question when to apply FEC for real-time applications in high-speed networks requires extended assessments of various trade-offs. While investigations of FEC are subject of our ongoing work, this paper concentrates on multicast error control by ARQ protocols.

2.4 Conceptual Framework for Support of Real-Time Multicast Services

A conceptual framework was developed that allows to select the mechanisms best suited to provide a specific service. Depending on the required combination of delay and reliability for a specific scenario with given distances, and depending on the targets for utilisation of network resources, different strategies are appropriate to achieve the most appropriate trade-off. These strategies differ in the mechanisms for scheduling and resource reservation, and in the error control mechanisms that are applied.

While the integration of specialised multicast components into the end systems represents an important step towards a high performance reliable multicast service, further improvements of performance and efficiency may be achieved by the integration of dedicated servers in the network that provide support for group communication. In many cases of multicasting, the achievable throughput degrades fast for growing group size. A significant advantage can be achieved if a hierarchical approach is chosen for multicast error control.

Group Communication Servers. Figure 1 presents a network scenario with multicast mechanisms in the transport component of end systems and in dedicated servers. The proposed Group Communication Server (GCS) integrates a range of mechanisms that can be grouped into three main tasks:
- Provision of a high-quality multipoint service with efficient use of network resources;
- Provision of processing support for multicast transmitters;
- Support of heterogeneous hierarchical multicasting.

For the first task, performing error control in the server permits to increase network efficiency and to reduce delays introduced by retransmissions. Allowing retransmissions originating from the server avoids unnecessary retransmissions over common branches of a multicast tree. For the second task, the GCS releases the burden of a transmitter that deals with a large number of receivers, providing scalability. Instead of communicating with all receivers of a group simultaneously, it is possible for a sender to communicate with a single GCS or with a small number of GCSs. The servers may be used hierarchically, where every server provides reliable delivery to a subset of the receivers. Integrating support for reliable high performance multipoint communication in a server allows better use of such dedicated resources. The servers will exploit possible multicast capabilities of the bearer service. For bearer services that do not offer multicasting, it is preferable to integrate a copy function into a server instead of integrating it into the transmitter. If multiple transmitters are connected to a server and a single multicast connection is used originating from the server, the transmitters will receive a copy of their own transmissions. To avoid this, multiple multicast connections originating from the server must be used. For the third task, a GCS may use the potential of diversifying outgoing data streams, allowing conversion of different error control schemes and support of different service qualities for individual servers or subgroups. The group communication server may offer a large range of error control mechanisms. For end systems, it is not required to have VLSI components for multicast error control. It will be sufficient to have access to a local GCS for participation in a high performance multipoint communication over long distances. Then, the error control mechanisms of individual end systems have only negligible influence on the overall performance, as simple error control mechanisms are sufficient for communication with a local GCS. If a priority field is used in the frame format, the server is able to distinguish packets of different importance. One example application would be hierarchically coded video. For information with different importance, specific packets may be suppressed for certain outgoing links. This may be used to reduce unnecessary traffic in case some receivers in a multicast group do not need the complete information. As an example, colour information may be eliminated for receivers that do not use it. Additionally, it may be used to protect packets with higher priority against loss due to buffer overflow.

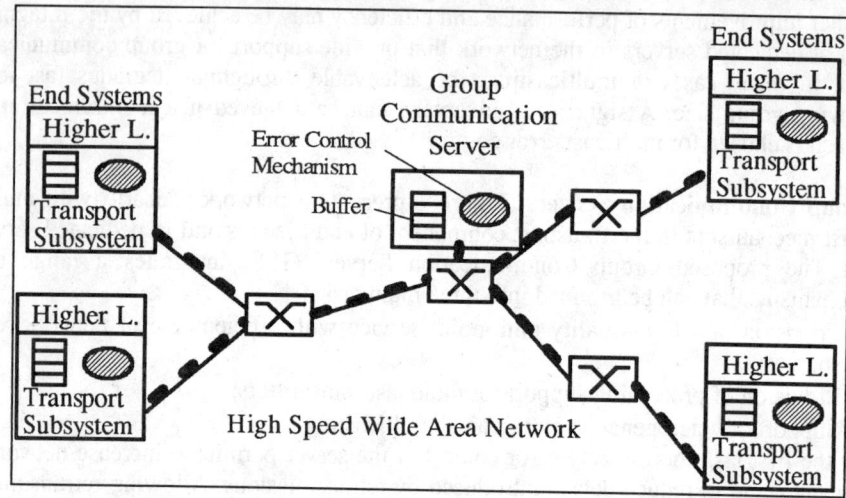

Figure 1. Real-time Multicast Support in Server and End Systems

Configuration and Reservation. While existing protocols for resource reservation are limited to exchange information required for the provision of a bearer service, our framework plans to extend this information by the choice of error control mechanisms that may be applied in intermediate and end systems. The following strategy is proposed for reservation of resources and selection of appropriate error control mechanisms: After a connection setup request is passed from the application across the service interface, the transmitter evaluates its resources that are required in worst case in order to fulfill the required service quality. It preallocates these maximum resources and passes the connection setup request together with the list of preallocated resources and supported error control mechanisms to intermediate and end systems. This allows the receivers to evaluate the combination of bearer service and transfer service best suited for the required service, based on the selection of the appropriate error control mechanism in combination with an economical reservation of resources. The result of the evaluation is passed back to the transmitter, resulting in a two-way handshake for minimum information exchange. Resources that were preallocated in excess are subsequently released. If the application requested to guarantee the QoS only to a subset of the receivers, an additional information transfer from transmitter towards the receivers may be performed, resulting in a three-way handshake for minimum resource allocation for QoS guarantees to K out of N receivers.

3 Performance Evaluation

3.1 Simulation of Scheduling Algorithms

It is of high importance to identify a scheduling algorithm that allows to guarantee a specific service quality, while efficiently using network resources. The following simulations allow to determine differences in the achievable performance. However, the final selection of a suitable scheduling mechanism will not only be determined by a performance evaluation based on simplifying assumptions. Implementation complexity leading to additional processing delay and to high system costs also needs to be considered.

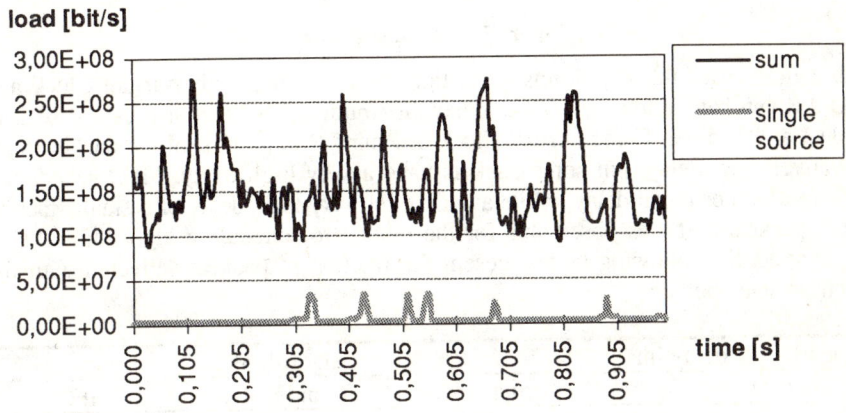

Figure 2. Simulated load (sum = 22 single sources)

A detailed comparison of the following three scheduling algorithms was performed: FIFO, Round Robin, and Virtual Clock. Simulations were performed to investigate the influence of the three scheduling algorithms on mean and maximum delay for real-time services. The sources for the simulation are modeled using Markov modulated Bernoulli processes with three states [24]. The sources represent a typical data stream produced by video applications. The minimum data rate of a source is 3.0 Mbit/s, the average data rate 6.3 Mbit/s, and the peak rate 30.0 Mbit/s. In the first step, the simulation model consists of a single node with several input and output links and only homogenous sources. An almost saturated output link was assumed. 19 or 22 sources were multiplexed for a load of 86% and 98% (155Mbit/s link), respectively. In the simulation, ideal processing delay of the scheduling algorithms was assumed, while the only time consuming process in the node was the sending of data.

The delay distribution of Virtual Clock, Round Robin, and FIFO shown in Figure 3 gives an example of the different behavior of the algorithms under the load shown in Figure 2. They show different end-to-end delay variance and end-to-end delay maxima.

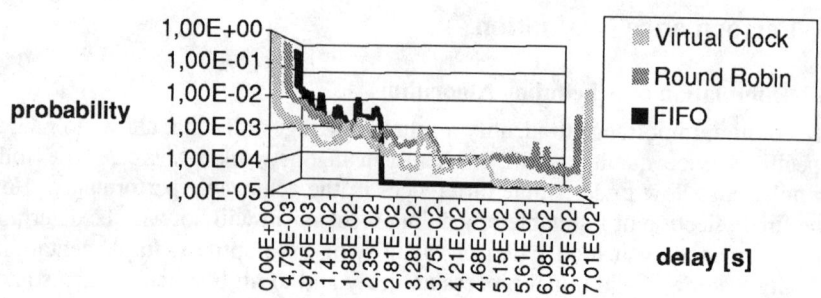

Figure 3. Delay distribution

Although none of the algorithms gives tight delay bounds, only Virtual Clock and Round Robin have a long tail in the delay distribution. This fact also can be seen in Table 1 and Table 2. Using Virtual Clock or Round Robin, most of the packets have a relatively low delay, but some packets have a very high delay. FIFO scheduling leads to a higher mean delay, while avoiding the very high delay of a small fraction of the packets that was observable for the other two scheduling algorithms. For a single node, the following tables present the fraction of packets delivered with the specified delay bound:

load	delay [ms]	algorithms		
		Virtual Clock	Round Robin	FIFO
86.40%	< 0.122	0.854	0.639	0.497
	< 1.0	0.868	0.701	0.574
	> 70.117	0.017	0.003	0
98.16%	< 0.122	0.730	0.379	0.146
	< 1.0	0.745	0.458	0.213
	> 70.117	0.030	0.011	0

Table 1. Delay distribution for different load

The Virtual Clock algorithm is based on packets with time stamps. A large maximum delay is the result of the sometimes fast growing gap between the virtual and the real clock. In these cases, timestamps of some packets are very large.

The simulations show that under certain conditions even very simple algorithms may have a satisfying behavior. The implementation complexity of a simple algorithm, such as Round Robin or, in particular, FIFO, is much lower than, for exam ple, Virtual Clock. For the latter, complex search and insert operations for the packet queue have to be implemented.

delay [ms]	load	algorithms		
		Virtual Clock	Round Robin	FIFO
minimum	86.40%	0.0054	0.0054	0.0054
	98.16%	0.0054	0.0054	0.0054
average	86.40%	3.4585	3.4585	3.4585
	98.16%	8.1802	8.9743	9.1624
maximum	86.40%	134.5072	80.7198	22.4315
	98.16%	182.1402	113.5866	31.2514

Table 2. Absolute delay of packets in a single node

Preliminary hardware designs showed that for all three algorithms, implementations for more than 1 million packets/s are feasible based on standard semi-custom VLSI technology. A detailed hardware design of a more complex type of round robin, the hierarchical round robin algorithm, was developed using the hardware description language VHDL. The implementation allows for a maximum of 1024 different connections, 16 levels of hierarchy, 64 slots per frame, up to 15 packets per connection queued in the node, and a maximum of 1024 packets over all. Gate level synthesis and simulation has shown that the implementation can easily handle more than 2.5 million packets/s for 1.0 µm CMOS technology, the area is 12696 gates for the control logic and 66784 bits of memory for control registers. Processing effort of the scheduling algorithms is per packet. Increasing the packet size will increase the achievable throughput, if the component for processing of the scheduling algorithm happens to be the bottleneck.

3.2 Delay Distribution for Retransmissions

Simulations were performed and analytical methods were applied in order to evaluate the influence of the proposed framework on delay and throughput for the envisaged multicast scenarios. The deployment of GCSs permits a significant delay reduction in case of packet losses and allows to increase efficiency substantially.

Figure 4 shows how a GCS may save one round-trip-time from transmitter to GCS for errors that occur between GCS and receiver. For a GCS serving many groups, extremely large retransmission buffers might be necessary to fulfill all retransmission requests. For GCSs with limited retransmission buffers, negative acknowledgements may be passed to the transmitter if requested packets are not available in the buffer of the GCS. Simulations were performed for a variable bitrate video source, a GCS with limited retransmission buffer, and a network element under congestion. For the simulation scenario, cf. Figure 5. Figure 6 shows the packet delay times observed by the receiver for a distance of 1000 km between transmitter and GCS, and a distance of 100 km between GCS and receiver. In the simulation, between 2000 and 2500 packets were transmitted for every encoded image. The majority of packets was delivered without retransmissions, while some packets retransmitted by the GCS observed a small delay, and other packets retransmitted by the source were significantly delayed.

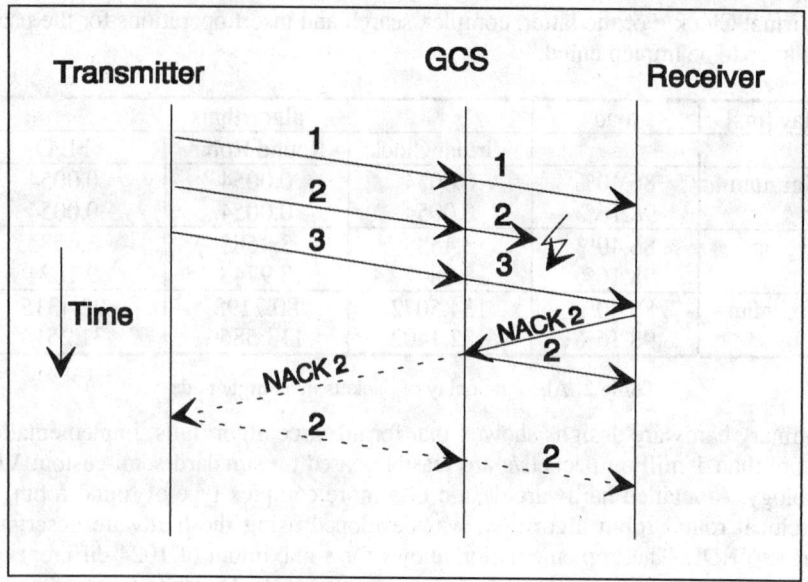

Figure 4. Delay Reduction by Retransmissions from GCS

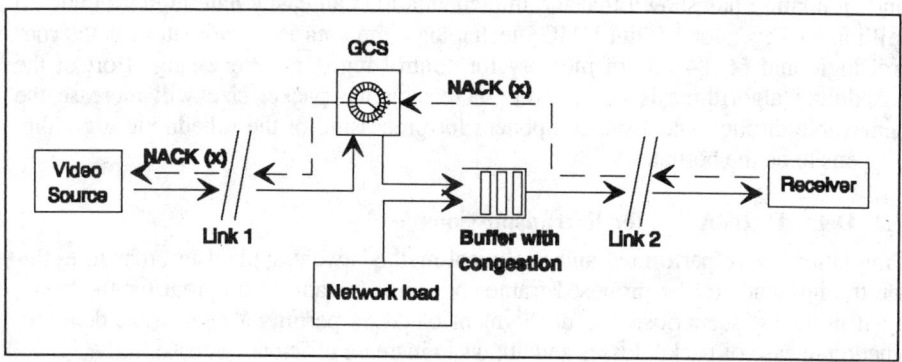

Figure 5. Simulation Scenario

3.3 Efficiency Analysis of Error Control Scenarios

Using Weldon's approximation and G/G/1 queuing models, the achievable performance of RMC (Reliable MultiCast)-AAL in selective repeat and go-back-N modes and the potential gain by deployment of GCSs was evaluated. The analysis is based on the following assumptions: protocol processing times may be neglected and acknowledgements are transmitted over a reliable connection. In correspondance with the results of [25], the throughput efficiency of a memoryless multicast go-back-N protocol may be expressed to

$$\eta_{1:N,GBN} = \frac{1}{\overline{m}} = \frac{1-Q}{1+sQ}$$

In this formula, $\overline{m}$ denotes the mean of the number of transmission attempts for successful transmission of a packet, Q denotes the loss probability of a packet, and s denotes the number of packets in transmission. In correspondance to [26] and [27], the efficiency of a selective-repeat protocol for a receiver buffer size of one path capacity may be expressed to

$$\eta_{1:N,SR} = \frac{1}{\overline{m}} = \frac{1-Q}{1+sQ^2}$$

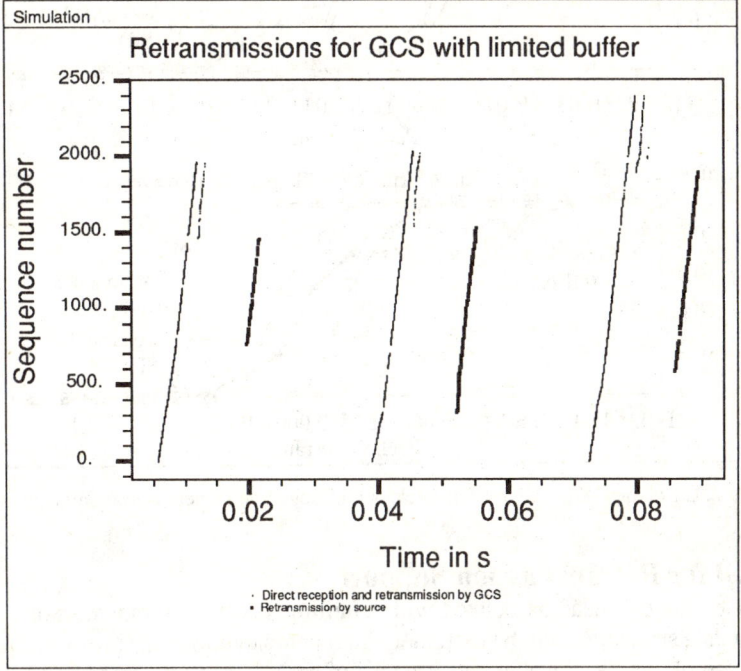

Figure 6. Reception with Retransmissions from GCS and Source

Figure 7 shows the efficiency of the two retransmission modes in three different scenarios. Scenario 1 represents a basic 1:N multicast without GCS. Scenario 2 represents 1:N multicasting with a GCS that performs retransmissions as multicast. In scenario 3, the GCS uses individual VCs for retransmission. The analysis is based on the following assumptions: protocol processing times may be neglected, acknowledgements are transmitted over a reliable connection, and buffers are sufficiently large. A group of 100 receivers and a data rate of 622 Mbit/s are assumed. Two cases are distinguished. The upper diagram of Figure 7 shows the efficiency for an overall distance of 1000 km (distance of 500 km from GCS to the receivers), and the lower diagram shows an overall distance of 505 km (distance of 5 km from GCS to the receivers). The analysis shows that in all cases, the efficiency is increased significantly by the GCS. Highest efficiency may be achieved for scenario 3 and selective repeat. Scenario 2 improves significantly for a shorter distance between GCS and the receivers. Go-back-N retransmissions show acceptable performance only for moderate bandwidth-delay products. Regarding efficiency, scenario 3 and selective repeat should be selected. However, this solution requires the highest implementation complexity for end systems and GCS.

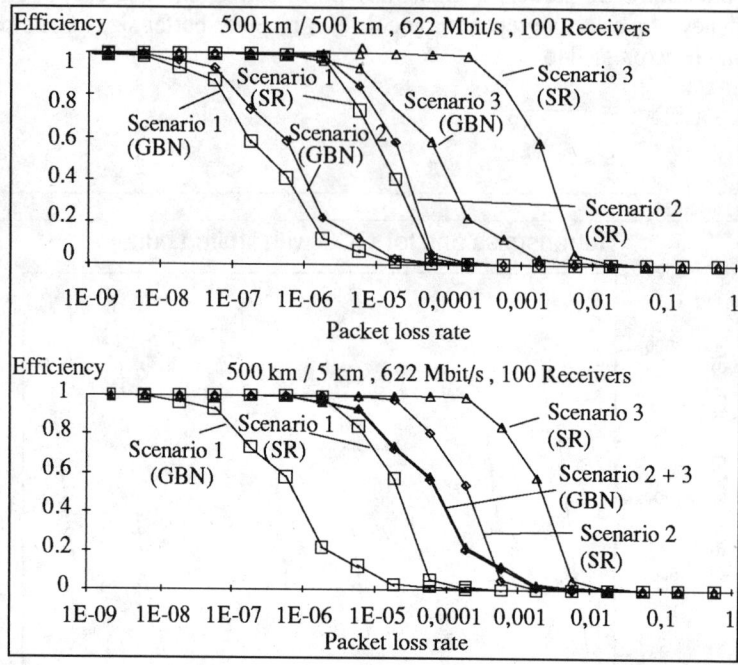

Figure 7. Efficiency Analysis for Go-back-N and Selective Repeat Retransmission Modes

4 VLSI for Retransmission Support

The processing overhead associated with handling selective retransmissions and the required data structures may be extremely high compared to other protocol functions. As an example, the achievable performance of an XTP [2] implementation on Digital Alpha Computers (150 MHz, 6.66 ns cycle time) may be regarded. In best case, the function to insert a new gap in a list needs 824 4-byte commands and takes, therefore, approximately 5.4 µs. The best case occurs if the new entry can be inserted at the beginning of the list. If the new entry has to be inserted after the first 10 entries, it needs 4054 commands or approximately 27.03 µs due to the search operations in the list. These calculations assume that the processor is not interrupted during execution of this function and all data is stored in the fast processor cache. XTP was implemented using C without special inline assembly code.

If data is sent at a rate of 1 Gbit/s, this results in more than 122,000 1024-byte packets per second. If the retransmission of each packet has to be controlled, this results in a new entry in the list in less than 10 µs.

Clearly, retransmission support forms a time critical task especially in a multicast environment. Therefore, we are implementing dedicated VLSI support for this task. The retransmission support presented in the following can handle negative selective, positive selective, and positive cumulative acknowledgements. It can be used for gaps managed by the receiver to support the acknowledgement function or for gaps managed by the transmitter to support the retransmission mechanism. The ALU has a set of commands to set, delete, insert, and read gaps for unicast connections and to manage multicast groups.

4.1 Logical Representation of Data

A dynamic linked list stores gaps of transmitted data in the following representation: *[seq_no_1, seq_no_2]* with *seq_no_1* and *seq_no_2* representing the beginning and ending of a gap. These gaps are connected via linked lists (cf. Figure 8). For every multicast group (MC) the pointers to the connections participating in that group are stored. A connection can be a member in different groups at the same time. For every connection the ALU stores a pointer to the appropriate list of gaps. Additionally, the ALU manages special lists for every multicast connection.

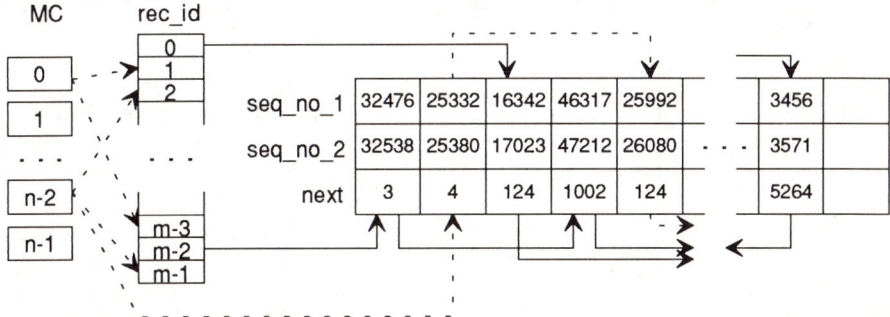

Figure 8. Logical Structure of Linked Lists for Retransmission Support

A simple example is given in Figure 9. First, the gap lists of the receivers R1, R2, and R3 are shown. The receiver R1, for example, has two gaps in transmitted data,

1) Lists of gaps for unicast connections

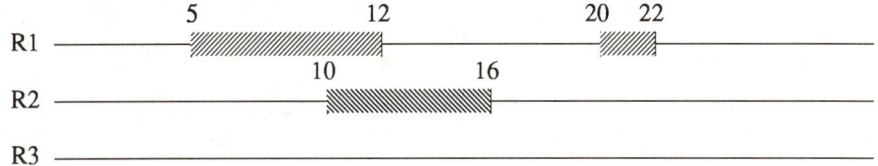

2) List of gaps to support k-reliability

Figure 9. Example of unicast and multicast lists of gaps

one from sequence numbers 5 to 12 and one from 20 to 22. The second type of lists stores the gaps needed to support full reliability or *k*-reliability of a multicast connection. In the example the gap (10,12) is stored, because less than 2 receivers have received this data.

Depending of the implemented protocol, retransmission of data can be performed by a multicast to the group or by individual retransmissions to the appropriate receivers.

4.2 Operations of the Retransmission ALU

The following table shows the operations supported by the specialised ALU.

operation	input parameters	output parameters	comment
`init_list`	rec_id, seq_no		initializes a new list for the connection *rec_id* with the initial sequence number *seq_no*, sets the error flag if *rec_id* is already in use
`close_list`	rec_id		closes the list for connection *rec_id*, sets the error flag if the list does not exist
`init_mcg`	mc_con_id, rel		initializes a new multicast group with the identification *mc_con_id* and the reliability *rel* (*rel* ≥ number of connections denotes full reliability)
`close_mcg`	mc_con_id		closes a mulitcast group and deletes all linked lists
`add_mcg`	mc_con_id, rec_id		adds a new connection *rec_id* to an existing multicast group *mc_con_id*
`del_mcg`	mc_con_id, rec_id		deletes an existing connection from a multicast group
`set_rel`	mc_con_id, k		sets the value *k* for the reliability of the multicast group *mc_con_id*
`set_high_ack`	rec_id, seq_no		sets the *high_ack* register to the value of *seq_no*; sequence numbers less than *high_ack* have been already acknowledged
`shift_high_ack`	rec_id, length		shifts the *high_ack* register to *high_ack* + *length*
`set_high_seq`	rec_id, seq_no		sets the *high_seq* register to the value of *seq_no*; *high_seq* represents the highest se-

			quence number in use
shift_high_seq	rec_id, length		shifts the *high_seq* register to *high_seq* + *length*
set_gap_1	rec_id, seq_no, length		inserts new entry (*seq_no*, *seq_no* + *length*); overlapping entries are automatically joined or deleted, respectively
set_gap_2	rec_id, seq_no_1, seq_no_2		analogous to *set_gap_1*, but the new entry is of the form (*seq_no_1*, *seq_no_2*)
del_gap_1	rec_id, seq_no, length		deletes an existing entry, a part of an existing entry, or several existing entries, the deleted part is of the form (*seq_no*, *seq_no* + *length*); if necessary an entry is divided into two new entries
del_gap_2	rec_id, seq_no_1, seq_no_2		analogous to *delete_gap_1*, but the deleted part is of the form (*seq_no_1*, *seq_no_2*)
read_reg	rec_id, reg_id	cont	reads the contents *cont* of the register *reg_id* (e.g. high_ack, high_seq, number_of_gaps)
read_mc_reg	mc_con_id, reg_id	cont	reads the contents *cont* of the multicast register *reg_id* (e.g. number_of_gaps)
get_gap_1	rec_id, ptr	seq_no, length, next	reads the entry *ptr* points to; if *ptr* = 0, the first gap is read out, if *next* = 0 the entry represented by (*seq_no*, *length*) is the last one, otherwise *next* point always to the next entry of the list
get_gap_2	rec_id, ptr	seq_no_1, seq_no_2, next	analogous to *get_gap_1*, but the entry is represented by (*seq_no_1*, *seq_no_2*)
get_mc_gap_1	mc_con_id, ptr	seq_no, length, next	analogous to *get_gap_1*, but now the entries of the multicast group *mc_is* are read out

get_mc_gap_2	mc_con_id, ptr	seq_no_1, seq_no_2, next	analogous to *get_gap_2*, but now the entries of the multicast group *mc_is* are read out

Dimensioning of the component: *rec_id, k* ∈ [0, 255]; *mc_con_id* ∈ [0, 63]; *seq_no, seq_no_1, seq_no_2, length, cont* ∈ [0, 2^{32}-1]; *reg_id* ∈ [0, 15]; *ptr, next* ∈ [0, 2^{16}-1]

Table 3. Operations of the Retransmission ALU

Every operation sets the error flag if it failed due to memory overflow or violation of several conditions, such as *high_ack* ≤ *seq_no* ≤ *high_seq* and other range checking.

4.3 Implementation Architecture for Retransmission Support

Figure 10 shows an overview of the internal structure of the ALU. The retransmission ALU consists of 5 memory banks (A through E) that store sequence numbers representing gaps (memory A and B), pointers of linked lists (memory C), and state information, such as connection and multicast identification, register number of an anchor element, and other flags indicating the state of a connection (memory D and E). The I/O-bus connects the input/output-port (32 bit) of the retransmission ALU with the 5 register banks (Ai through Ei, 0≤i≤3). From these registers data can be transferred to the memory.

Two simple ALUs (*ALU A*, 32 bit and *ALU D*, 8 bit) perform operations like OR, XOR, AND, ADD, SUB, and NEG. Two specialized 32 bit modulo 2^{32} comparators (*comp A* and *comp B*) perform fast comparisons needed for list operations. The ALUs and the comparators can work concurrently if no data dependencies exist.

For the command set_gap_2, for example, first of all the command itself and the connection identification are read from the I/O-bus into the registers E and D, respectively. In the next two cycles the central control unit reads the sequence numbers (seq_no_1, seq_no_2) into the registers A and B, respectively. After reading the complete command and several range checking operations the loop for searching the right position to insert the new entry in the list starts. Therefore, the first entry of the appropriate list is loaded into the registers A and B, respectively, and compared with the new entry. The loop terminates if the new entry fits, otherwise, the next entry is loaded and compared.

4.4 Microcode Examples of the Retransmission ALU

To provide a maximum of flexibility all functions of the *Retransmission ALU* are translated into a sequence of microcode operations. These operations are specially adapted to the implementation architecture shown in Figure 10. The *central control unit* controls the microprogram via a special micro sequencer.

Example microcode operations of the *Retransmission ALU* are listed in Table 4. In addition, there are the microcode operations of the *ALU D* and complex comparison operations of the two comparators *comp A* and *comp B*. The operations of the ALUs and the comparators are always executed in parallel in one clock cycle.

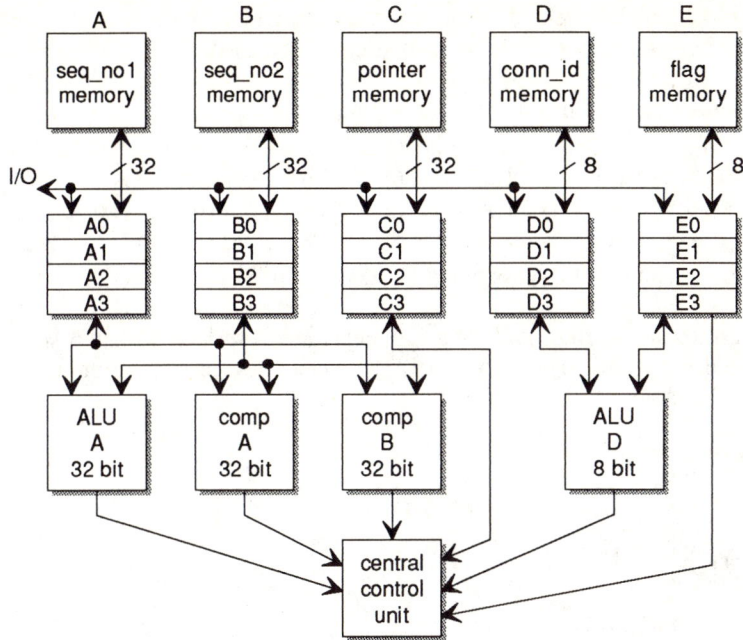

Figure 10. Structure of the Retransmission ALU

operations	comment
RMOVE S, D	move a complete row of entries from the registers or RAM into the registers or RAM. S, D ∈ {Ri, RAM; $0 \leq i \leq 3$}, Rn = (An, Bn, Cn, Dn, En), S ≠ D
ANOP	no operation, ALU A
AMOVE I/O, D	move data from the I/O-bus into the register D; D ∈ {Ai, Bi; $0 \leq i \leq 3$}
ACLR D	clear register D; D ∈ {Ai, Bi; $0 \leq i \leq 3$}
AINC S, D	S + 1 -> D; S, D ∈ {Ai, Bi; $0 \leq i \leq 3$}
ADEC S, D	S - 1 -> D; S, D ∈ {Ai, Bi; $0 \leq i \leq 3$}
AMOVE S, D	S -> D; S, D ∈ {Ai, Bi; $0 \leq i \leq 3$}
AADD S, D	S + D -> D; S, D ∈ {Ai, Bi; $0 \leq i \leq 3$}
ASUB S, D	S - D -> D; S, D ∈ {Ai, Bi; $0 \leq i \leq 3$}

Table 4. Microcode Examples of the Retransmission CPU

4.5 Implementation Detail: A 32 bit Modulo 2^{32} Comparator

Due to the complexity of the complete design, only a selected part is presented in detail. This section shows the design of a 32 bit modulo 2^{32} comparator (cf. *comp A* and *comp B* in Figure 10). Such a comparator is needed to compare three 32 bit values at the same time. This comparison is done while searching for the right position in the retransmission list to insert a new entry or to delete an existing entry. Due to the modulo 2^{32} arithmetic and the need for fast search operations in the dynamic lists common comparators with only two inputs cannot be used.

Four functions are calculated at the same time:
$$\text{le\_le}(a, b, c) = \text{true} \Leftrightarrow ((a \leq c) \land (a \leq b \land b \leq c)) \lor ((a > c) \land (a \leq b \lor b \leq c))$$
$$\text{le\_lt}(a, b, c) = \text{true} \Leftrightarrow ((a < c) \land (a \leq b \land b < c)) \lor ((a > c) \land (a \leq b \lor b < c))$$
$$\text{lt\_le}(a, b, c) = \text{true} \Leftrightarrow ((a < c) \land (a < b \land b \leq c)) \lor ((a > c) \land (a < b \lor b \leq c))$$
$$\text{lt\_lt}(a, b, c) = \text{true} \Leftrightarrow ((a < c) \land (a < b \land b < c)) \lor ((a > c) \land (a < b \lor b < c))$$
$a, b, c \in \mathbb{N} \bmod 2^{32}$; le: less or equal; lt: less than

The complete architecture and its components are described with the standardized hardware description language VHDL[28]. This allows for simulation and synthesis based on the same language.

```
library IEEE;
  use IEEE.std_logic_1164.all;
  use IEEE.std_logic_arith.all;

entity MOD_KOMPA is
  port ( A, B, C : IN std_logic_vector (31 downto 0);
         le_le, le_lt, lt_le, lt_lt: OUT boolean);
end MOD_KOMPA;

architecture BEHAVIORAL of MOD_KOMPA is
begin
  process (A,B,C)
  variable h1, h2, h3, h4, h5, h6 : boolean;
  begin
    h1 := A < B; h2 := A <= B;
    h3 := A < C; h4 := A >  C;
    h5 := B < C; h6 := B <= C;
    le_le <= (NOT(h4) AND (h2 AND h6)) OR (h4 AND (h6 OR h2));
    le_lt <= (h3       AND (h2 AND h5)) OR (h4 AND (h5 OR h2));
    lt_le <= (h3       AND (h1 AND h6)) OR (h4 AND (h6 OR h1));
    lt_lt <= (h3       AND (h1 AND h5)) OR (h4 AND (h5 OR h1));
  end process;
end BEHAVIORAL;
```

The above listed VHDL description was synthesized into a gate level description using a high level synthesis tool. The area of the design is 1084 gates and the estimation of the critical path 9.6 ns. The implementation of only the *le_le* function on an Alpha processor needs 20 4-byte commands which results with 6.6 ns cycle time (150 MHz) in a duration of more than 132 ns.

5 Summary and Future Work

Within this paper, a framework for the provision of high performance real-time multicast services has been presented which has the potential to fulfill the requirements of upcoming distributed multimedia applications. It is based on VLSI components dedicated to specific processing tasks that are to be integrated in end systems, special network elements called group communication servers, and the deployment of resource reservation in intermediate and end systems. Simulation results on the suitability of selected scheduling algorithms for services with guaranteed delay are presented. A performance evaluation is given which shows the potential benefits of selective retransmissions in multipoint connections, and potential improvement of efficiency if GCSs are integrated into the network. Implementation details of multicast retransmission support have been discussed. It was shown how certain system and support functions may be implemented efficiently by the use of dedicated hardware.

Not only high performance, efficient use of network resources and resource reservation, but specifically service integration will be a major requirement for forthcoming communication subsystems. System components that may be selected and parametrized based on the requested application service will be important for providing a high degree of flexibility. The presented framework thus may be viewed as a hardware implementation of the function-based communication subsystem F-CSS presented in [29]. Currently, the implementation of additional components for FEC and memory management are under development. A more detailed evaluation of the achievable performance is also subject of ongoing work, including investigation of the influence of processing times and of limited buffers.

Acknowledgement. The authors would like to thank Martina Zitterbart, Torsten Braun, and Burkhard Stiller for valuable discussions. The support by the Graduiertenkolleg „Controllability of Complex Systems" (DFG Vo287/5-2) is also gratefully acknowledged.

6 References

[1] Ito, M.; Takeuchi, L.; Neufeld, G.; *Evaluation of a Multiprocessing Approach for OSI Protocol Processing;* Proceedings of the First International Conference on Computer Communications and Networks, San Diego, CA, USA, June 8-10, 1992

[2] Strayer, W.T.; Dempsey, B.J.; Weaver, A.C.; *XTP: The Xpress Transfer Protocol;* Addison-Wesley Publishing Company, 1992

[3] Feldmeier, D.C.; *An Overview of the TP++ Transport Protocol;* in: Tantawy A.N. (ed.): High Performance Communication, Kluwer Academic Publishers, 1994

[4] Sterbenz, J.P.G.; Parulkar, G.M.; *AXON Host-Network Interface Architecture for Gigabit Communications;* in: Johnson, M. J. (ed.): Protocols for High-Speed Networks, II, North-Holland, 1991, pp. 211-236

[5] Braun, T.; *A Parallel Transport Subsystem for Cell-Based High-Speed Networks;* Ph.D. Thesis (in German), University of Karlsruhe, Germany, VDI-Verlag, Düsseldorf, 1993

[6] Braun, T.; Zitterbart, M.; *Parallel Transport System Design;* in: Danthine, A.; Spaniol, O. (eds.): High Performance Networking, IV, IFIP, North-Holland, 1993, pp. 397-412

[7] Krishnakumar, A.S.; Kneuer, J.G.; Shaw, A.J.; *HIPOD: An Architecture for High-Speed Protocol Implementations*; in: Danthine, A.; Spaniol, O. (eds.): High Performance Networking, IV, IFIP, North-Holland, 1993, pp. 383-396

[8] Balraj, T.; Yemini, Y.; *Putting the Transport Layer on VLSI - the PROMPT Protocol Chip;* in: Pehrson, B.; Gunningberg, P.; Pink, S. (eds.): Protocols for High-Speed Networks, III, 1992, North-Holland, pp. 19-34

[9] Heinrichs, B.; Jakobs, K.; Carone, A.; *High performance transfer services to support multimedia group communications*; Computer Communications, Volume 16, Number 9, September 1993

[10] Waters, A. G.; *Multicast Provision for High Speed Networks*; 4th IFIP Conference on High Performance Networking HPN'92, Liège, Belgium, December 1992
[11] Bubenik, R.; Gaddis, M.; DeHart, J.; *Communicating with virtual paths and virtual channels*; Proceedings of the Eleventh Annual Joint Conference of the IEEE Computer and Communications Societies INFOCOM'92, pp. 1035 - 1042, Florence, Italy, May 1992
[12] Ferrari D., Banerjea A., Zhang H.; *Network Support for Multimedia - A Discussion of the Tenet Approach;* Technical Report TR-92-072, International Computer Science Insitute, Berkeley, California, November 1992
[13] Herrtwich R.G.; *An introduction to real-time scheduling*; Tech. Rept. TR-90-035, International Computer Science Institute, Berkeley, July 1990
[14] Zhang L.; *Virtual Clock: A New Traffic Control Algorithm for Packet Switching Networks*; Proceedings of ACM SIGCOMM'90, Philadelphia, September 1990
[15] Parekh A.K.; *A Generalized Processor Sharing Aproach to Flow Control in Integrated Services Networks*; PhD Thesis, Department of Electrical Engineering and Computer Science, MIT, February 1992
[16] Kalmanek, C.R.; Kanakia, H.; Keshav, S.; *Rate controlled servers for very high-speed networks;* in proceedings of GLOBCOM '90, San Diego, December 1990
[17] Zhang H., Keshav S.; *Comparison of Rate-Based Service Disciplines;* Proceedings of SIGCOMM '91, Communications Architecture and Protocols, Zurich, Switzerland, September 1991
[18] Kurose, J. F.; *Open Issues and Challenges in Proving Quality of Service Guarantees in High-Speed Networks;* ACM Computer Communication Review, Vol. 23, No. 1, January 1993, pp. 6-15
[19] Partridge, C.; *Gigabit Networking;* Addison-Wesley, 1994
[20] Topolcic, C. (Editor).; *Experimental Internet Stream Protocol, Version 2 (ST-II);* Internet Request for Comments RFC 1190, October 1990
[21] Zhang L., Deering S., Estrin D., Shenker S., Zappala D.; *RSVP: A New ReSource ReserVation Protocol*; IEEE Network Magazine, Vol. 9, No. 5, September 1993
[22] McAuley, A.; *Reliable Broadband Communication Using a Burst Erasure Correcting Code*; Presented at ACM SIGCOMM '90, Philadelphia, PA, U.S.A., September 1990
[23] Biersack, E. W.; *Performance Evaluation of Forward Error Correction in an ATM Environment*; IEEE Journal on Selected Areas in Communication, Volume 11, Number 4, pp. 631-640, May 1993
[24] Kleinewillinghöfer-Kopp, R.; Lehnert, R.; *ATM Reference Traffic Sources and Traffic Mixes;* RACE 1022 BLNT Workshop, Munich 1990
[25] Gopal, I.; Jaffe, J.; *Point-to-Multipoint Communication Over Broadcast Links*; IEEE Trans. Commun., Vol. Com-32, No. 9, pp. 1034-1044, September 1984
[26] Sabnani, K.; *Multidestination Protocols for Satellite Broadcast Channels*; Ph. D. Thesis, Columbia University, NY, U.S.A., 1982
[27] Wang, J.; Silvester, J.; *Performance optimisation of the go-back-N ARQ protocols over broadcast channels*; Computer Communications Vol. 14, No. 7, pp. 393-402, September 1991
[28] IEEE; *Standard VHDL Language Reference Manual;* IEEE Std 1076-1987
[29] Zitterbart, M.; Stiller, B.; Tantawy, A.; *A Model for Flexible High-Performance Communication Subsystems;* IEEE-JSAC, May 1993, pp. 507-518

Providing Support for Data Transfer in a New Networking Environment

Rainer Schatzmayr & Radu Popescu-Zeletin

TU - Berlin, Franklinstr. 28/29, FR5-14, 10587 Berlin, F.R.Germany
E.mail: schatzmayr@fokus.gmd.de
Phone: +49 30 254 99 288, fax: +49 30 254 99 202

Abstract. Future networking environments will have different characteristics from actual ones. The network service is expected to provide more bandwidth and be more reliable, and applications data will be generated from different type of sources making different requirements from data transfer protocols. Rather than trying to adapt an existing transport protocol to this new environment, we decided to develop a new one called RAPID. Two important features of RAPID are its unidirectional data transfer capability and its flexibility to adapt to different Quality of Service requirements. To provide an unidirectional data transfer service RAPID makes use of forward error correction codes to enhance the network's reliability degree. This work presents a description of this new environment and the protocol architecture, it also contains an analysis about the use of forward error correction codes and some issues concerning implementation aspects.

1 Introduction

The development of physical mediums for data transfer with large bandwidth and low noise and the advent of high speed switches is changing the environment for data transfer protocols [1]. Along with this evolution in the network technology data transfer applications are changing too. The emerge of multimedia applications whose data transfer requirements can pose different demands from the transfer service is imposing changes at the transport protocol.

Transport protocols have as its main function to adequate the services offered by the network to the user's needs. In [2], [3] and [4], the authors describe the idea of adapting existing transport protocols to this new networking environment. We choose a different way. Considering that from one side the bandwidth from the network service will be a large available resource and a very low bit error rate (BER) can be expected, and from the other side that the major part of data will be generated by multimedia applications, we designed the RAPID transport protocol.

The diversity of multimedia data characteristics requires that the transport protocol must be able to adapt itself to different data transfer requirements. As some multimedia data are expected to present timing constraints, an important issue to be resolved by the transport protocol is how the network's reliability degree can be enhanced to achieve a desired error rate. RAPID makes use of forward error correction (FEC) codes to enhance the network's reliability degree, and so avoid the timing implications of retransmissions requests.

The idea of using FEC codes in high-speed networks has been presented in [5]. The difference in our approach to that in [5] is that we don't restrict our scheme for an ATM network and do not require the protocol to handle individual cells.

Considerations about the networking environment are presented in the second section, a description of the RAPID transport protocol is made in section three, FEC codes are described in section four and some considerations about the protocol implementation are made in section five. Finally section six presents the conclusion obtained from this work.

2 The Networking Environment

Traditionally, end-to-end reliability is assured by means of a retransmission scheme. Initially a copy of data to be transferred is made to a local buffer, the data is sent to the peer entity, and the sender keeps waiting for an acknowledge to free its buffer and transmit new data. If a positive acknowledge arrives the buffered copy is cleared and the sender returns to the initial situation. If a retransmission request arrives, or a time-out occurs, a copy of data kept inside the sender's buffer is made and sent to the receiving entity, and the sender returns to the waiting state.

The use of retransmission mechanisms inside high-speed networks can impose severe performance penalties on the data transfer throughput. One restriction comes from the propagation delay and the time data will be delivered to the receiving transport protocol user. Another problem arises from the buffering of data waiting for retransmissions.

The troubles concerning propagation delay become more evident for applications, whose data transfer has timing constrains. If retransmissions are involved, the resulting time needed to transfer data correctly from sender to receiver might be too long. This means that the transport protocol can not be kept waiting for acknowledges or time-out and then retransmitted lost data. It must be guaranteed that data will be received with an acceptable probability at first transmission attempt.

The buffering trouble arises from the necessity of holding a copy of all data transmitted that hasn't been positive acknowledged. The size of the buffer defines how much data can be sent. Future networks can be described as being a very large bandwidth delay transmission medium, meaning that millions and even billions of bits can be inside the network in a single round trip. If the sender does not have enough buffer capacity, it must stop transmitting and keep waiting for an acknowledge to free some buffer space. The buffer's size imposes a restriction on the connection's throughput. Again if the data is successfully transmitted with an acceptable probability at the first attempt such buffering would become superfluous.

From the paragraphs above we can conclude that an open loop mechanism must be used to enhance the network's reliability degree. RAPID uses forward error correction (FEC) codes to enhance the probability of being successful at the first attempt to transfer data from a sender to a receiver. FEC codes require the sender to transmit additional data that can be used by the receiver to correct erroneous data.

2.1 ATM network impact

An important cornerstone on the emergence of this new networking environment is the development of the asynchronous transfer mode (ATM) technology. In this work we won't concentrate on all the impacts that the ATM technology will have on the transport protocol, restricting ourselves on how errors at the ATM layer will influence the reliability degree of a connection.

If an individual ATM cell handling occurs, that means, if cells are treated one independent from each other, errors can be limited to cell sizes. Such situation occurs inside AAL type 3/4 [6][7], where a bit error detection mechanisms (CRC) is applied to individual cells, and cells have an individual sequence number. In this case, bit errors will result in a fail at the CRC checking, and the cell can be discarded. A cell lost can be recognized by a gap in the sequence number of arriving cells.

On the other side, if ATM cells are treated in blocks bit errors won't be limited to a cell, affecting the whole block. Such situation occurs inside AAL type 5, where a CRC is calculated over a AAL Convergence Sublayer protocol data unit (CSPDU). If a bit error is detected, the whole PDU is affected. Cell losses will result in errors at the CRC checking and in a error on the CSPDUs length. Both kind of errors will result in a whole PDU discard, for the transport protocol a TPDU lost in the network service.

For RAPID we considered the network connection as being an ATM AAL type 5 connection, and that the protocol will use the message mode. To avoid segmentations at the AAL CS sublayer, the network service data unit (NSDU) will be smaller than the maximum AAL CS PDU size. For RAPID, following the analysis presented above a whole NSDU will be lost in case of bit errors or cell loss. Taking this approach, a bit error or a cell lost means a whole NSDU lost. Thus, all NSDUs delivered to the transport protocol have a very low probability of being corrupted.

As the ATM network is expected to discard all NSDUs with error and considering that each NSDU corresponds to a single transport protocol data unit (TPDU), the FEC algorithm is applied on a TSDU basis as shown in Figure 1 In this case, the transport service data unit (TSDU) is segmented in TPDUs, redundant TPDUs (in the figure: FEC) are generated and both (TPDU and FEC) are transferred over the network service. An eventual TPDU lost can be corrected by using the FEC TPDUs.

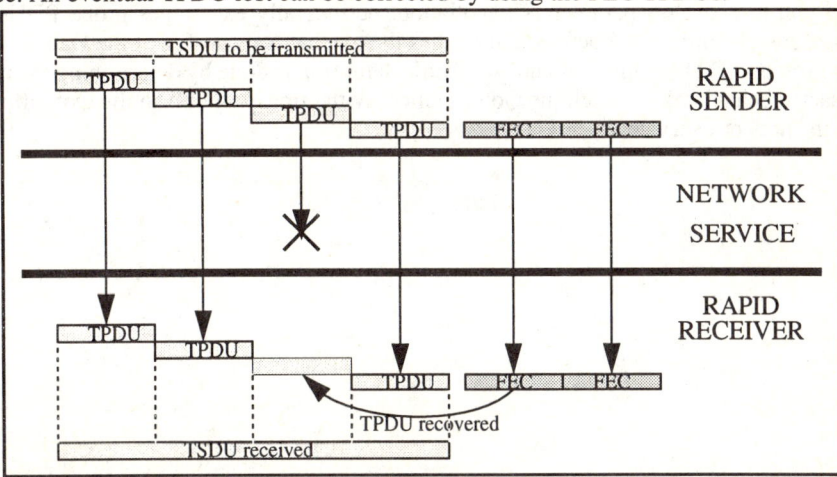

Fig.1. FEC usage

3 The Protocol Architecture

RAPID can be divided in two distinct protocols as shown in Figure 2. One protocol called Connections Control Protocol (CCP) is responsible for signalling functions as connection setup, release, end-to-end Quality of Service (QoS) parameters negotiation and for connection liveliness control. The other protocol is responsible for transferring data from sender to receiver, called Data Transfer Protocol (DTP).

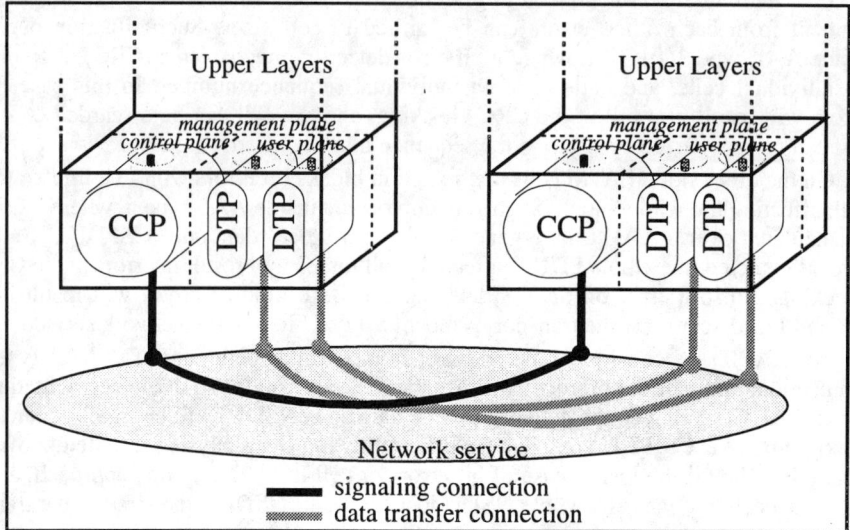

Fig.2. Protocol architecture

3.1 Connection Control Protocol

The Connection Control Protocol (CCP) is responsible for all functions related to connection management. It establishes and releases data transfer connections, makes local and end-to-end QoS parameters negotiation, periodically exchanges probe PDUs to test if remote entities are active, and defines the mechanisms profile of the Data Transfer Protocol (DTP). This mechanisms profile definition is done by determining the necessary mechanisms to match the QoS required by the transport user to the QoS offered by the network service.

The QoS parameters needed for the definition of the DTP profile, are given in Table 1. "QoS parameters".

TABLE 1. QoS parameters

Provider	Parameter	Description
Network Service	bit error rate	residual bit error rate
	loss rate	residual loss error rate
	size	maximal size of a NSDU
	delay	delay to transfer a NSDU
	delay jitter	variation on the transfer delay
	interval	time between two consecutive NSDUs
	burst	size of a NSDU burst
Transport Service	error rate	bit and loss errors
	size	maximal size of TSDU
	delay	delay to transfer a TSDU
	delay jitter	variation on the transfer delay
	interval	time between two consecutive TSDUs
	commitment	quality of service guarantees

3.2 Data Transfer Protocol

The Data Transfer Protocol (DTP) makes use of several mechanisms to transfer data from sender to receiver. The segmentation / reassembling mechanism is necessary to divide transport service data units (TSDUs) in transport protocol data units (TPDUs). Byte padding is used to achieve a determined TPDU length. Sequence numbers are used to detect TPDU losses, and the TPDU position in the TSDU. A rate control defines how TPDUs are delivered to the network service.

RAPID also offers an isochronous data transfer service, being able to compensate delay variations (jitter) that can occur in the network service. Due to the use of a more elaborated addressing scheme than those applied by single point-to-point connections, it is able to provide point-to-multipoint data transfer. Forward error correction (FEC) codes can be used to enhance the network's reliability degree.

3.2.1 Protocols mechanisms profile for DTP

In the following figures (Figure 3 and Figure 4) are the mechanisms profiles for the Data Transfer Protocol. The figures show where mechanisms are located, their relation and execution sequence, and where data buffering takes place.

Figure 3 presents the mechanisms profile for RAPID's sender. Depending on the transport user requirements one of the two kind of interfaces can be present at the transport user interface. At the left side, all data to be transferred is directly sent to the transport protocol (message mode), while at the right side all data is written into a shared buffer (stream mode). In this last mode, RAPID periodically reads the data found inside the shared buffer.

In both cases, after the data has been received by the transport protocol, the TPDU or TPDUs are generated. TPDU generation means segmenting the TSDU and building the protocol header. TPDUs ready to be sent are inserted into a buffer from which they are periodically removed and sent to the network sender through the rate control. This rate control defines when TPDUs are delivered to the network.

If the network's reliability degree must be enhanced, FEC mechanisms are introduced in the DTP mechanisms profile. The FEC functions are executed after TPDUs have been assembled and are waiting in the buffer. The FEC encoding is divided in header encoding (FEC TPDU header is generated) and data encoding (where the FEC TPDU data is generated). The advantage of such division is that FEC TPDU data can be encoded while TPDUs are waiting in the sender's queue.

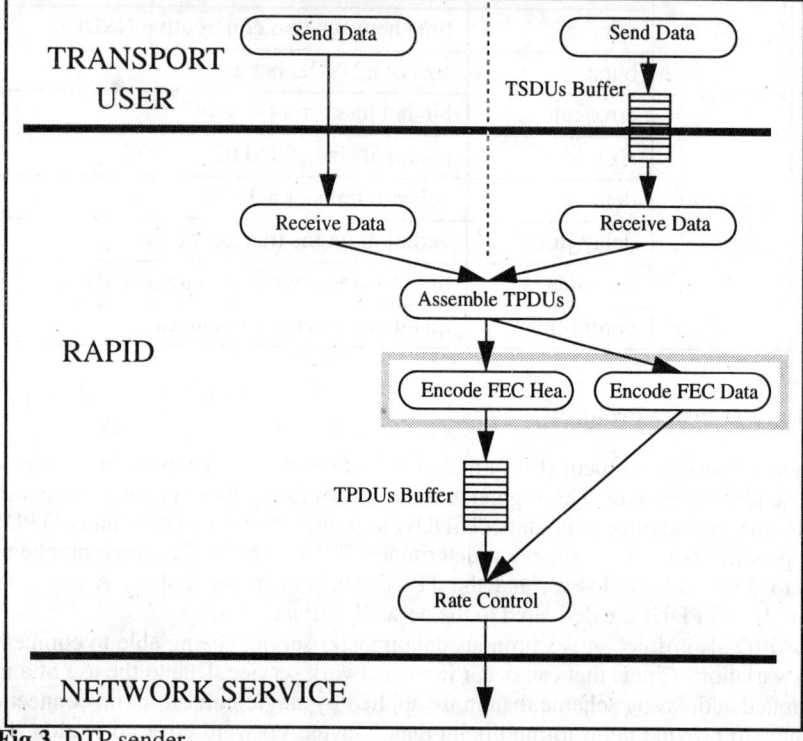

Fig.3. DTP sender

Figure 4 shows the protocol mechanisms profile for the receiver.

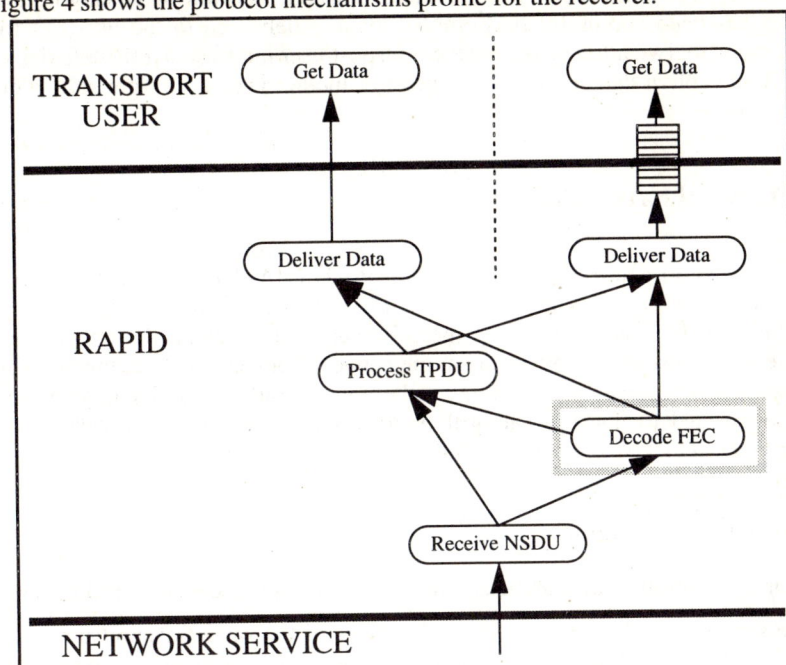

Fig.4. DTP receiver

All information received from the network service at the *Receive NSDU* mechanism are delivered to the FEC decoder. The FEC decoder tests if all informations necessary for the TPDU processing have been correctly received, if so, the TPDU processing can start, even if FEC decoding hasn't finished yet (Figure 5). If the reliability degree of the network is acceptable the decode FEC mechanism can be excluded from the DTP mechanisms profile.

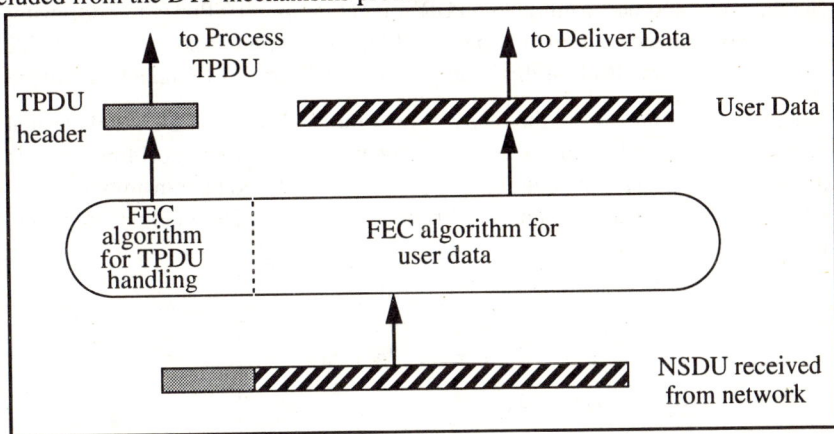

Fig.5. FEC data handling

After the TPDU has been decoded, the FEC decoder has finished its algorithm, and a TSDU has been successful reassembled, data is delivered to the user. As for the sender, there are two alternatives for the transport-user interface. At the left side of the Figure 4 an explicit deliver is shown (message mode), and at the right side, a deliver with help from a shared buffer (stream mode).

4 Forward Error Correction

Forward error correction (FEC) codes will be used in RAPID to enhance the network's reliability degree. There is more than one FEC algorithm, and the choose of which algorithm will be used depends on the following three factors: the network's error characteristic considering how errors at the network connection affect the NSDUs, the enhancement that must be achieved on the connection's reliability degree, and the resources needed for the FEC algorithm. As resources we mean: processing power, memory requirements, and network bandwidth.

4.1 Algorithm description

There are 4 algorithms that can be applied to generate and recover the redundant data.

- The simplest of all is to send multiple copies of the same transport protocol data unit (TPDU). Sending a TPDU more than once makes it more probable to be correctly received. The advantages of this scheme are its simple implementation and low processing requirement. The use of a memory management scheme at the sender allows copies to be done logically without requiring real physical copy of data. At the receiver, a simple discard of all multiple correct received copies must be done (as it is already done by transport protocols). The bandwidth requirement of the network connection increases considerably when the multiple copies algorithm is used.

- A less consuming bandwidth scheme is to send a single parity FEC TPDU. A parity can be calculated performing an 'XOR' operation on all TPDUs to be transferred. The receiver makes use of this parity TPDU to recover from an eventual TPDU lost. The main drawback of this scheme is that only one TPDU can be recovered.

- The third algorithm allows the recovering of more than one TPDU, without making such high bandwidth requirements as in the multiple copies algorithm. The idea of this algorithm consists in a multiplication of all TPDUs to be transmitted by a well-known matrix whose columns are linear independent. The method works as follows. Considering

Data to be sent $C = [c_1 \ c_2 \ c_3 \ c_4 \ c_5 \ ... \ c_n]$

$$\text{Matrix A (columns LI)} = \begin{bmatrix} a_{1,1} & a_{2,1} & a_{3,1} & \cdots & a_{n,1} \\ a_{1,2} & a_{2,2} & a_{3,2} & \cdots & a_{n,2} \\ a_{1,s} & a_{2,s} & a_{3,s} & \cdots & a_{n,s} \end{bmatrix}^T$$

multiplying C*A one gets B = [b_1 b_2 b_3 b_4 ... b_S]
where: n = number of TPDUs to be transmitted
S = number of FEC TPDUs.

The sender transmits both the original data (expressed by C) and the result of the multiplication (B).

The receiver, knowing matrix A and at least $n - S$ elements c_i and b_i is able to rebuild the original data. The recovering consist in solving a linear system. For example, suppose that elements c_2 and c_4 have been lost. At the receiver following information would be available:

Information received = [c_1 c_3 c_5 c_6 ... c_n b_1 b_2 ... b_S], and matrix A.

The recovery means solving following linear system:

$$\begin{bmatrix} c_2 & c_4 \end{bmatrix} \begin{bmatrix} a_{2,1} & a_{2,2} \\ a_{4,1} & a_{4,2} \end{bmatrix} = \begin{bmatrix} b_1 - c_1 a_{1,1} - c_3 a_{3,1} - \ldots - c_n a_{n,1} \\ b_2 - c_1 a_{1,2} - c_3 a_{3,2} - \ldots - c_n a_{n,2} \end{bmatrix}^T$$

- The fourth algorithm is to encode the FEC TPDUs to be transmitted through a Reed-Solomon encoding algorithm. Done in software it works in a similar way as the method described above, although presenting some advantages if done in hardware as encoding can be done with a single shift register. In Reed Solomon codes the previous matrix A is expressed by a polynomial:

$$g(x) = (x - a^1)(x - a^2) \ldots (x - a^S)$$

where a is a primitive element. The advantage is that due to polynomial arithmetical characteristics, the multiplication can be simplified to a cyclical division similar to a CRC calculation.

The decoding at the receiver is almost the same as shown for the matrix encoding, the only changes are at elements [b_1 b_2 ... b_S] which must also be multiplied by a constant at the receiver, as for the Reed - Solomon code matrix A is as large as $n + S$.

An important feature of the Reed Solomon code, is that it allows errors in unknown positions to be corrected, what is impossible in all other methods described above. This feature won't be used here since errors (in fact erasures) are in known positions due to the fact that TPDUs have a sequence number that allows the detection of losses.

4.1.1 FEC time overhead

Two factors play an important rule on the definition of the time overhead introduced by the FEC mechanism. The time overhead is the delay introduced by the algorithm used. The first factor comes from the network, as additional data will take more time to be transmitted. The second factor comes from processing overhead as each algorithm has different processing requirements.

The multiple copies time overhead comes basically from the network. As data will be sent more than once, it will take more time to transmit it. The processing factor will be relative low, since processing at the receiver means redundant copies discard. On the other side, for the other algorithms, the main time overhead is expected to come from the processing factor. As a decoding algorithm must be processed to recover from lost data, the complexity of the algorithm and the receiver's processing power will determine the time overhead.

4.2 Residual error rates

Two kind of errors can be found inside a network connection. One is a bit error, as a bit might get its value changed when it is transferred from one entity to another. The other kind of error are lost errors meaning whole data loss during this transfer. Traditionally such errors are surmounted by the same mechanism.

The residual bit error rate can be reduced with the use of a bit error detection mechanism that can detect, with a high probability, if bits have had their values changed during data transfer. An alternative to enhance the residual loss error rate, without retransmissions, is to use FEC codes since such codes can reconstruct data that has been lost.

4.2.1 Residual bit errors rates

Transport protocols such as TCP, OSI-TP4, XTP, etc. were designed to work over networks with an unacceptable bit error rate. Therefore it was necessary, at the transport layer, to implement a bit error detection mechanisms to detect bit errors inside the TPDU. The use of optical fibre as transmission medium, and the use of CRC bit error detection mechanisms inside the lower levels, should present at the transport network interface an acceptable residual bit error rate. In such scenario, the transport protocol doesn't has to enhance the residual bit error rate with bit error detection mechanisms. For RAPID we considered such scenario.

4.2.2 Residual loss error rate

The residual loss error rate defines the probability that data won't arrive at its destination. Losses inside the network service come not only from information drops inside network's cross connecting points, but also from bit errors if the network protocol discards erroneous data. In RAPID FEC codes are used to avoid retransmissions. FEC codes require a bandwidth overhead as more data has to be transferred. The bandwidth overhead that such code require is analyzed in the next item.

4.3 FEC analysis

An important point on the analysis of the FEC algorithms is the amount of bandwidth necessary to achieve a desired error rate. For the analysis it has been considered that a TSDU will be segmented in one or more than one TPDUs, and that each TPDU will be transmitted in a single NSDU. Therefore error characteristic for NSDUs will be same as for TPDUs. It is also considered that NSDU transfer will be made over a AAL type

5 connection in message mode, so the loss rate will be influenced not only by the ATM cell loss rate, but also by the bit error rate.

Considering the use of an ATM network two kind of errors can be defined:

- byte error, in fact a bit error, with probability of occurring = p_b,
- cell loss, with probability of occurring = p_{cell}.

If one of this errors occur a whole TSDU will be lost, so the TPDU loss error probability is given by:

$$P_{TPDU} = \left(1 - (1-p_b)^{sizeofTPDU}\right) + \left(1 - (1-p_{cell})^{\frac{sizeofTPDU}{cellpayload}}\right)$$

4.3.1 Multiple copies

The multiple copies scheme (Figure 6) increases the reliability by sending TPDUs more than once.

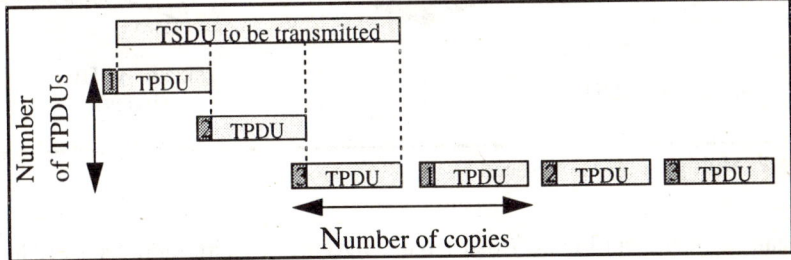

Fig.6. Multiple copies

A TSDU will be erroneous if one of its TPDUs fails to be successfully transferred. Given that each TPDU is sent more than once, the resulting TSDU error probability is given by:

TSDU error probability =

$$1 - (1 - (PTPDU)^{NumberofCopies})^{NumberofTPDUs}$$

Figure 7 shows the bandwidth overhead introduced by the adoption of a multiple copies scheme. It has been considered that the byte error rate is 10^{-8}, that the probability that a cell is lost is 10^{-6}, and that the desired probability that the TSDU has an error is lower than 10^{-12}. Overhead equal to 100% means twice as much bandwidth, equal to 200% three times initial bandwidth, and so on. The number of TPDUs defines the number of TPDUs in which the TSDU is segmented.

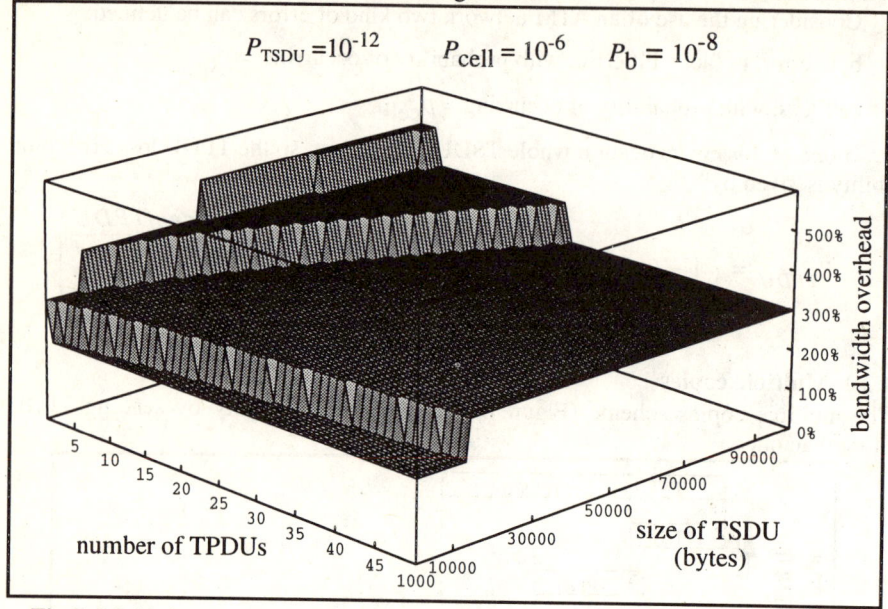

Fig.7. Multiple copies overhead

It can be seen from Figure 7 that large TSDU will require more bandwidth overhead than small TSDUs. The most important point is the influence that segmentation has on the bandwidth overhead. Large TPDUs are more probable to be affected by errors. This explains why segmenting a TSDU in 5 TPDUs may require more bandwidth than segmenting the same TSDU in 20 TPDUs. In the first case, as the TPDUs will be larger, they are more probable to have errors so more copies must be sent. If the same TSDU is segmented in smaller TPDUs, the TPDU error probability will be smaller, and so less copies are needed.

4.3.2 Redundancies

The same analysis done for the multiple copies algorithm can be done for the algorithms that send redundant TPDUs. In Figure 8 the parameter S (redundant TPDUs) defines how many FEC TPDUs will be sent.

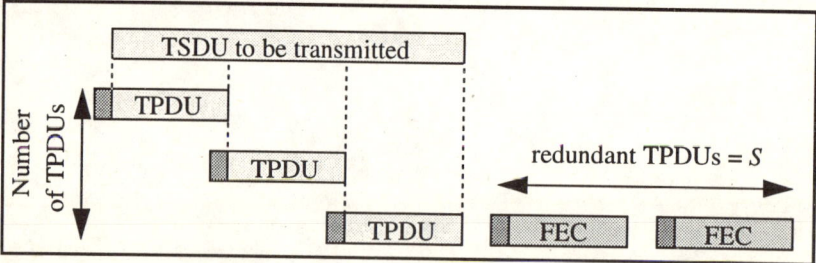

Fig.8. FEC redundancies

A TSDU will be erroneous if more than S TPDUs are lost as S defines the recovering capacity of the algorithm. The TSDU error probability will be:

TSDU error probability =
$$1 - \sum_{i=0}^{S} \left(\left(1 - P_{TPDU}\right)^{(NumofTPDUs + S - i)} P_{TPDU}^{i} \binom{NumofTPDUs + S}{i} \right)$$

Figure 9 shows the bandwidth overhead needed to achieve a desired TSDU error probability of 10^{-12}, for a byte error rate of 10^{-8} and a cell loss error rate of 10^{-6}. As for the multiple copies algorithm, bandwidth overhead equal to 100% means twice as much bandwidth, equal to 10% a 10% increase on the original bandwidth, and so on.

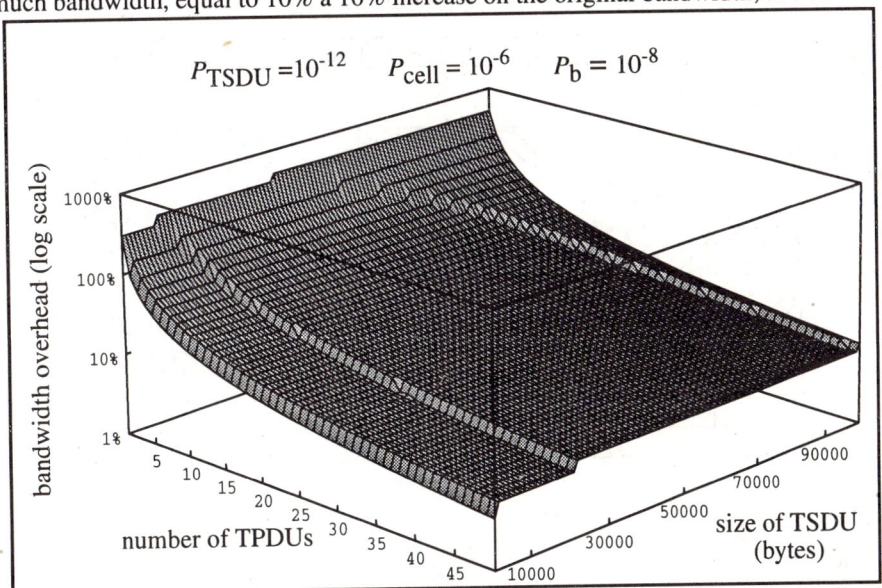

Fig.9. FEC TPDUs overhead

From the graphic in Figure 9 it can be seen that large TSDUs will require a larger bandwidth overhead to achieve a desired error rate (overhead is in logarithmic scale). The number of TPDUs in which a TSDU is segmented influences considerably the bandwidth overhead. This comes from the fact that as larger the number of TPDUs in which a TSDU is segmented as lower will be its loss probability, since each TPDU will be smaller. Small TPDUs also means smaller FEC TPDUs, and so a lower bandwidth overhead.

5 Implementation Issues

In this section are two important issues concerning implementation aspects of the RAPID transport protocol. The first one is a description about how the flexibility needed for the Data Transfer Protocol (DTP) can be implemented. The second point is the influence of the runtime environment on the execution of RAPID.

5.1 Flexibility

Data Transport Protocol (DTP) flexibility is implemented based on a configurator. This configurator provides the communication between the Connection Control Protocol (CCP) and the DTP. It also allows the CCP to make changes on the mechanisms parameters of the DTP. Figure 10 shows the relation among some protocol elements.

The QoS matcher defines the mechanisms profile, based on the Quality of Service (QoS) parameters request from the transport protocol user and the QoS parameters from the network service. During data transfer, changes at the QoS parameters may induce the QoS Matcher to request some changes on the mechanisms profile of the DTP. Such changes are carried out by the configurator.

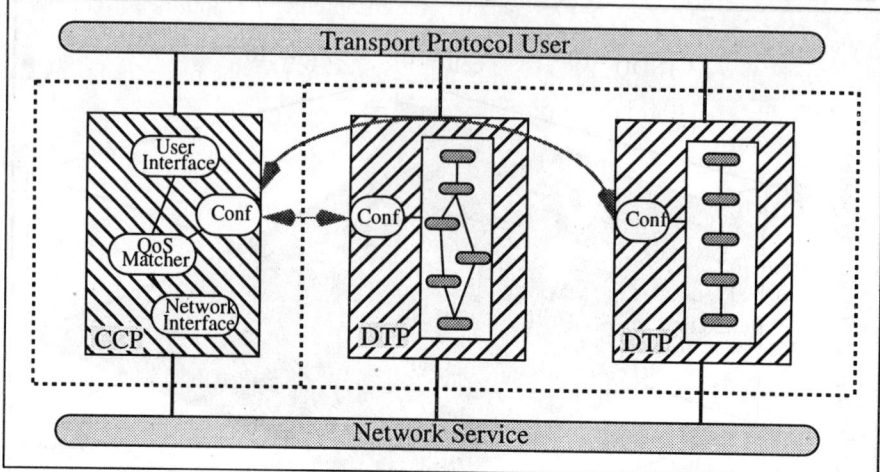

Fig.10. Implementation aspects

5.2 Runtime environment aspects

Transport protocols request the intervention of the operating system in which they are executed to be able to transfer continuous media data. Continuous media data are data generated from audio and video sources. For the Connection Control Protocol (CCP), operating system requirements consist basically of timer management.

There are two types of mechanisms in the DTP protocol. Time dependent mechanisms, which have a specific time to be executed, and the non-time dependent mechanisms, which are usually executed as a consequence of another mechanism. Time dependent mechanisms are mechanisms that involve timing procedures as implicit sending and reading of TSDUs, timestamps values assignment, and rate control.

Another point where the runtime environment plays an important role is on the FEC mechanism implementation. The key idea behind the integration of the FEC algorithm inside the DTP is to allow the FEC algorithms to be processed in parallel with other DTP mechanisms. As the DTP and the FEC algorithm can be run independently, it is possible to execute the protocol in one processor and the FEC algorithm on another processor or on a dedicated hardware component.

6 Conclusion

In the present paper we described the main ideas, design decisions, and implementation architecture for the RAPID transport protocol. RAPID is a transport protocol that takes advantages of new features found inside high-speed networks as low error rates and a high bandwidth. Some peculiarities of RAPID are its unidirectional data transfer capability and the ability to adapt its mechanisms profile to different transport user needs and runtime environment facilities.

To enhance a connections reliability degree, RAPID makes use of FEC algorithms. Initial results from the use of FEC codes show the feasibility of our approach. The mechanisms considered are at present time being implemented, and measurements in comparison to traditional high-speed transport protocols are being performed.

Some of the ideas being developed in this project are also the basis for the extension of the Tempo/TIP [8] protocol architecture.

7 References

[1] L. Kleinrock, The Latency/Bandwidth trade-off in Gigabit Networks, IEEE Communications Magazine, pp 36-40, apr. 1992.

[2] The BERKOM - II MultiMedia Transport System (MMT), Version 3.0, 31 august 1993.

[3] D. Hehmann, R.G. Herrtwich, W. Schulz, T. Schütt, R. Steinmetz, Implementing HeiTS: Architecture and Implementation Aspects of the Heidelberg High-Speed Transport System, Proc. 2 International Workshop on Network and Operating System Support for Digital Audio and Video", pp31-44, Heidelberg, Germany, nov. 1991.

[4] Ø. Kure and I. Sorteberg, XTP over ATM, Computer Network and ISDN Systems, no. 26, pp 253-262, 1993.

[5] E.Biersack, Performance Evaluation of Forward Error Correction in ATM Networks, Proc. SIGCOMM 92, Baltimore, August 1992.

[6] ITU-TS, I.362 Recommendation, B-ISDN ATM Adaptation Layer (AAL) functional description.

[7] ITU-TS, I.363 Draft Recommendation, B-ISDN ATM Adaptation Layer (AAL) specification, jan. 1993.

[8] S. Boecking, The TIP Project: Supporting Advanced Communications on High-Speed Networks, Proc. IEEE GLOBECOM, nov. 1993.

Congestion Avoidance for Video over IP Networks

Toru Sakatani

NTT Human Interface Labs.
Take 1-2356 Yokosuka-Shi Kanagawa, 238-03, Japan
sakatani@nttvdt.ntt.jp

Abstract. Video/Audio packet transport over IP (ver. 4) networks is not guaranteed for real-time communication if network congestion occurs due to excessive traffic load. Congestion avoidance control is thus necessary for real-time video/audio communication over IP networks. An adaptive video bit rate control scheme that detects the onset of congestion is designed to achieve real-time visual communication and efficient data transport. Congestion over a CSMA/CD LAN is detected from the number of retransmissions caused by collisions. The packet delay is used to detect congestion over connected LANs. The effectiveness of the adaptive video bit rate control scheme is verified in experiments that use a number of video/audio communication terminals and data terminals on IP networks. The interaction of the adaptive video bit rate control and TCP congestion control schemes are investigated and suitable bit rate control scheme is discussed. The adaptive bit rate control scheme expands the possibility of visual communication over IP networks.

1. Introduction

Effective communication networks and economical terminals are important to ensure the rapid spread of visual communication services. Visual communication is more cost effective if existing networks can be used because no additional network is needed and service management is not increased. This will foster the spread of personal visual communication services.

IP networks consisting of LANs are widely accepted in offices and factories. LANs are used in local areas, and considering the current use of audio communication within the same building, visual communication services are also needed in local areas. Visual communication over LANs can be used for surveillance, TV broadcasting, and education. Long distance communication can be achieved by accessing ISDN through IP routers or by using frame relay networks. Moreover, the current interest in groupware applications on IP networks indicates that visual communication services will spread rapidly.

The bandwidth utilization of IP (ver. 4) networks is shared by data link access control of the networks, window flow control of the transport protocols, and so on. However, fixed bandwidth allocation to communication sessions and specific transport delay values can not be guaranteed. When video traffic is added to IP networks, some bandwidth is occupied and the possibility of congestion increases. Congestion needs to be avoided to achieve real-time visual communication over IP networks.

Fig. 1 shows the transition in average network utilization (averaging period is 5 s) during one hour on a 10 Mbps CSMA/CD LAN connecting over 100 terminals in

NTT Labs. The highest network utilization rate within a 500 second period is indicated by the gray zone in Fig. 1 and is shown in more detail in Fig. 2. The obvious characteristic is that traffic is very bursty. Network utilization changed rapidly and high network utilization rates were short lived. This result is consistent with other investigations [1-3], so the characteristic is considered to be general. This characteristic suggests that the network does not need to continuously allocate large bandwidth for data traffic. Additional bandwidth is reserved only when the data traffic demands large bandwidth.

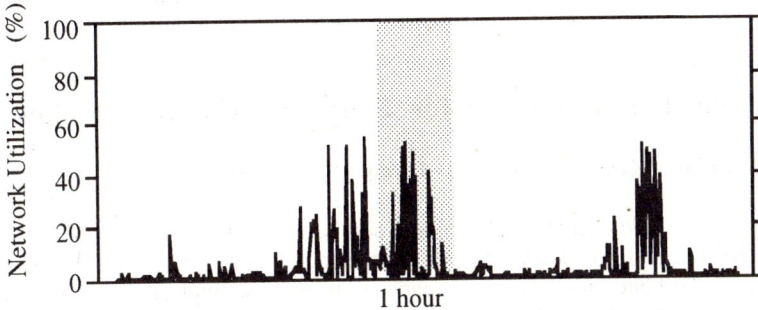

Fig. 1. The transition in average network utilization.

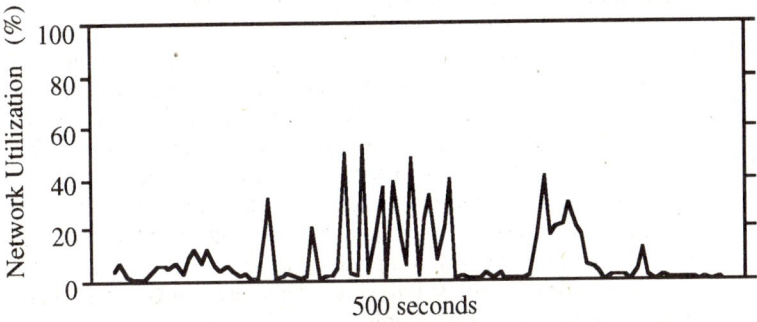

Fig. 2. The transition in average network utilization.

The bandwidth of the video traffic could be allocated based on the peak demands expected from the data traffic to avoid congestion, but the bandwidth of video traffic remains limited. The video coding bit rate may need to be reduced as required to support data traffic. We anticipate that real-time visual communication will be mostly used for meetings and conferences, so users may permit the video quality degradation that would accompany a momentary decrease in the video bit rate.

In previous papers [13, 14], I proposed the adaptive bit rate control schemes by detecting network congestion. Congestion was detected by packet discard, source quench messages, or an increase in the packet round trip time. Experiments using a pair of video/audio communication terminals and traffic generators over 10 Mbps CSMA/CD LANs showed that the increase in the packet round trip time was the best way to measure congestion.

In this paper, the feasibility of the adaptive video bit rate control is verified in experiments using a number of visual communication terminals. Congestion over a

CSMA/CD LAN is detected from the retransmission numbers caused by collisions. The increase in the packet delay is used to detect congestion over connected LANs. The delay of video packets and the data throughput are measured to estimate the achievement of real-time communication and the efficiency of data transport. The transition of the video bit rate is investigated by adding constant rate packets, or TCP data packets over connected LANs to show the interaction between the adaptive video bit rate control and the TCP congestion control.

The rest of the paper is organized as follows. In Section 2, congestion avoidance for real time communication is discussed and the adaptive video bit rate control scheme is described. Experimental results are discussed in Section 3 and related work is explained in Section 4. Section 5 concludes with a summary of results.

2. Congestion Avoidance for Real-time Communication

2.1 Resource Reservation and Congestion Detection

Resource reservation protocols [4, 5] and admission control were designed to avoid the congestion of IP networks. Intermediate nodes manage network bandwidth and respond to the bandwidth demands made by terminals. Intermediate nodes do not accept bandwidth demands that would lead to congestion, so network load is maintained at a proper level. However, existing IP (ver. 4) networks do not have such a resource reservation mechanism.

Without resource reservation, congestion needs to be avoided continuously. This is achieved by detecting congestion in intermediate nodes or terminals and controlling the source bit rate. When the packet queue length in the intermediate node buffers exceeds the maximum level and packets are discarded, the nodes send explicit source quench messages to source terminals over the IP networks. The problem with this scheme is that packet transmission delay becomes too large for real-time communication before congestion is signaled, because intermediate nodes generally have large buffer sizes to cope with temporary traffic overloads. Detecting packet discards at the destination is not suitable for detecting congestion for real-time communication because of the same problem. Terminals can also detect congestion from an increase in the packet transmission delay. This technique is more suitable for measuring the congestion to ensure real-time communication.

2.2 Packet Priority Control and Congestion Control

If IP intermediate nodes could order packet processing according to packet priority, the delay of high priority packets could be decreased. Video and audio packets tagged as high priority are processed earlier than low priority data packets when network congestion occurs. The IP header has a service field that can hold packet priority, but most intermediate nodes cannot process priority. Therefore, packet processing according to packet priority is not possible and the sum of the video/audio traffic and the data traffic on the IP networks needs to be controlled to achieve real-time communication.

TCP offers congestion control scheme to maximize throughput and minimize packet loss. However, real-time video/audio traffic occupies some bandwidth and the delay must be small. Therefore, TCP is not suitable to transport real-time video/audio traffic. I selected UDP for video/audio packet transport but UDP has no congestion control scheme. Therefore, congestion control is by higher layer protocols as described in 2.3.

2.3 Adaptive Video Bit Rate Control by Detecting Congestion

2.3.1 Adaptive Bit Rate Control for a CSMA/CD LAN

If the source terminal can detect congestion by itself, congestion notification is not needed from the destination or nodes. When congestion occurs over a CSMA/CD LAN, the collision packet ratio and packet transmission delay increases. A video communication terminal developed in NTT can detect the number of collisions experienced by the transmitted packet. Because TCP/IP protocols do not have a mechanism to inform the number of collisions to the higher layer, the video communication terminal is provided with such a mechanism. The video communication terminal compares the number of collisions against the predetermined threshold for each connection. When the number of collisions exceeds the threshold, the terminal considers that congestion has been detected. The video communication terminal decreases the video bit rate when congestion is detected, and increases the video bit rate when no congestion is detected over a predetermined period as shown in Fig. 3. The data link layer congestion control scheme (CSMA/CD) is available on a CSMA/CD LAN and the bandwidth of the LAN is relatively large (10 Mbps), so video bit rate control consists of only two stages.

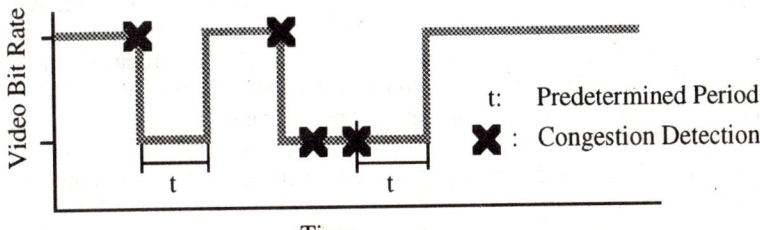

Fig. 3. The video bit rate control over a CSMA/CD LAN

2.3.2 Adaptive Bit Rate Control over connected LANs

The video communication terminal is designed to measure congestion between the source and the destination terminal by themselves. The video destination terminal sends an ICMP echo message and the source terminal immediately replies with an ICMP echo reply message. The video destination terminal measures the increase in the packet round trip delay of echo message to find congestion. After transmitting an echo message, the terminal starts a timer with predetermined threshold (tc) and regards that congestion occurs if the timer expires before the reply message returns. Echo messages are sent at the constant interval of t0. When congestion is detected, the video destination terminal demands that the video bit rate of the source terminal be decreased. As shown in Fig. 4, the video bit rate is held for predetermined period t1 before being reduced again in the face of continued congestion. t1 is set to be several times larger than t0. When congestion is not detected, the video bit rate is increased at the predetermined interval of t2. t2 is larger than t1, for quick decrease and slow increase. The data link layer control scheme is ineffective against congestion over connected LANs and the bandwidth over the connected LANs is sometimes relatively small, so there are more than two video bit rates.

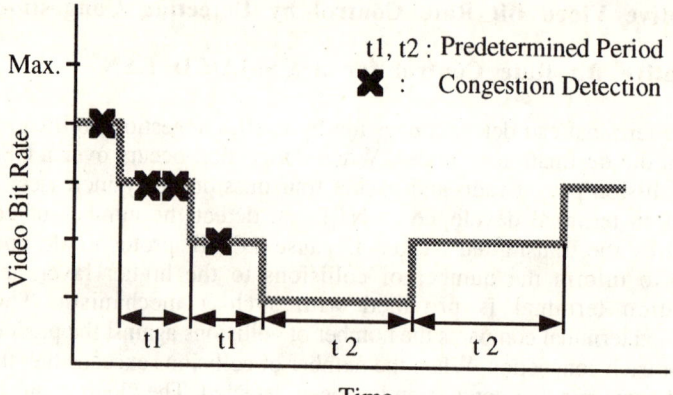

Fig. 4. The video bit rate control over connected LANs

Some rules of changing the video bit rate were added to the rules described below. To suppress oscillation of the bit rate, the bit rate should be decreased immediately by detecting congestion, but not excessively. Therefore, the maximum video bit rate is decreased as soon as congestion is detected as shown in Fig. 5 (point A). When congestion is detected after the video bit rate is increased, the increase is regarded as the one of the causes of the congestion, so the video bit rate is reduced immediately as shown in Fig. 5 (point B). Congestion might continue from (A) or (B) for some time until the intermediate nodes process the packet queues in the receiving buffers, the reduced bit rate is held for t1 regardless of congestion detection as shown in Fig. 5 (A)-(A'), (B)-(B') to avoid excessive bit rate reductions, and congestion detection is restarted.

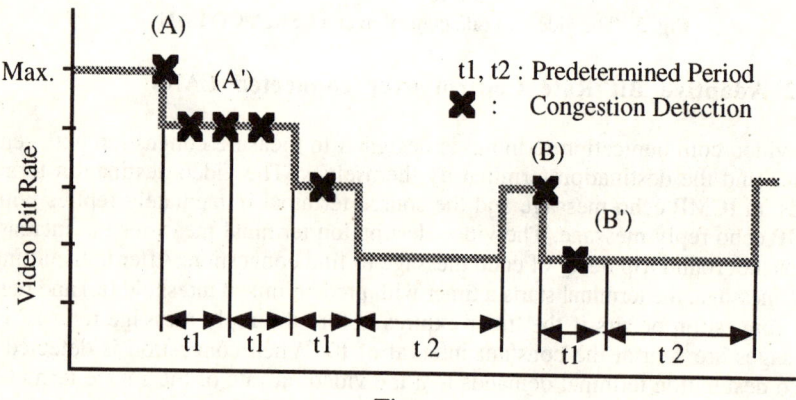

Fig. 5. The video bit rate control over connected LANs

The source terminal could also detect an increase in the round trip delay and control its own video bit rate, but the video communication terminal developed in NTT detects the increase at the destination terminal. This is because the video communication terminal is based on the idea of receiver oriented video bit rate control and the video bit rate control for congestion is merged into the receiving user oriented bit rate control for simplicity.

3. Experiment and Results

The feasibility of adaptive bit rate control was tested in a comparison with fixed video bit rate control using video communication terminals and a traffic generator issuing constant rate packets. Moreover, an experiment was conducted with video/audio traffic and TCP data traffic over connected LANs to show the interaction between adaptive video bit rate control and TCP congestion control.

The transmission delay of actual video/audio packets was measured to estimate the real-time video communication. The data throughput was measured to estimate the efficiency of data transport. The bit rate control scheme is assessed in light of the experimental results.

3.1 Video Communication Terminal used in Experiments

The video communication terminals used in the experiments were a modified form of the commercial product " NTT FM-C100L". Point to point (duplex) video/audio communication is established between the video communication terminals. The video codec and audio codec are based on ITU T recommendations H.261 and G.728 (16 kbps). Video was coded and decoded by hardware boards and the encoder output a CBR video bit stream. The maximum video bit rate was 512 kbps. The video bit rate can be set at one of five stages by changing the coding frame rate and the quantization step size. Video and audio bit streams are assembled into the same packets to decrease the number of the packets and decrease network load. Video-audio packets are assembled and transmitted at predetermined intervals continuously. The video communication terminal always packetizes the audio bit stream even if silent, to avoid cutting off speech beginnings.

TCP is used for communication setup and UDP is used for video audio transport. IP is used for network protocols and the network interface is a 10 Mbps CSMA/CD LAN.

The real-time packet processing capability of the video communication terminal was investigated. The setup of the video terminal was maximized for the capability, i.e. the video bit rate was 512 kbps and the time interval of the packet assembly was 20 ms. Table 1 shows the average packet interval time on the CSMA/CD LAN and the standard deviation relative to the design interval value of 20 ms. The average packet interval time is the same as the design value and the standard deviation is small compared to the interval, so the video communication terminals could realize real-time packet processing on the maximum setup for the capability.

Table 1. Delay and deviation

Average packet interval time	20 ms
Standard deviation	1.1 ms

3.2 Experiment over a 10 Mbps CSMA/CD LAN

A multiport (12 port) hub was used to establish a 10 Mbps CSMA/CD LAN as shown in figure 6. Eight video communication terminals were connected to the hub, and 4 point to point video communication connections were established. A traffic generator added random length data packets to the LAN to simulate data traffic. The data packets varied from 64 to 1500 bytes, the average was 777 bytes long, and were generated at fixed intervals. The generator had one send packet buffer. The LAN analyzer receives the packets on the LAN and stored the packets with received timestamp. The video packet header included a sequence number, so packet discard was found by checking the sequence numbers.

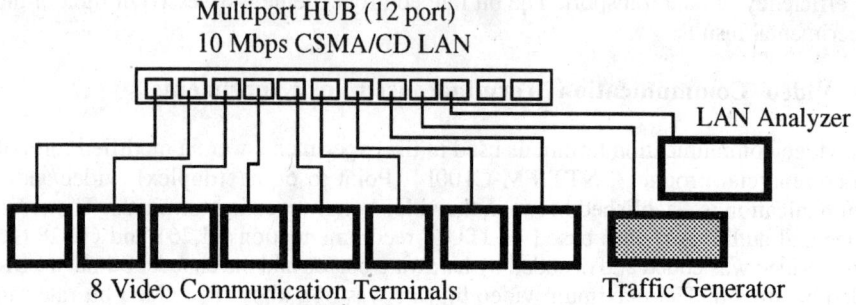

Fig. 6. Experimental configuration.

The video bit rate was controlled to 512 and 256 kbps. The number of collisions indicating congestion was set to 5 and 10 to change the sensitivity of congestion detection. The video packet delay against the 20 ms interval was measured to estimate the real-time communication and the data throughput was measured to estimate the efficiency of the data transport. Data traffic was added to the LAN in four steps:
(A). 0 Mbps
(B). 1.984 Mbps
(C). 5.956 Mbps
(D). 8.926 Mbps

Figure 7 shows video and data throughput as components of the total network utilization rate. With adaptive bit rate control, as total network utilization increased, the video throughput decreased because congestion was detected. When the total network utilization was large, data throughput was larger with adaptive bit rate control than with fixed bit rate control, so the efficiency of data transport was better with adaptive bit rate control.

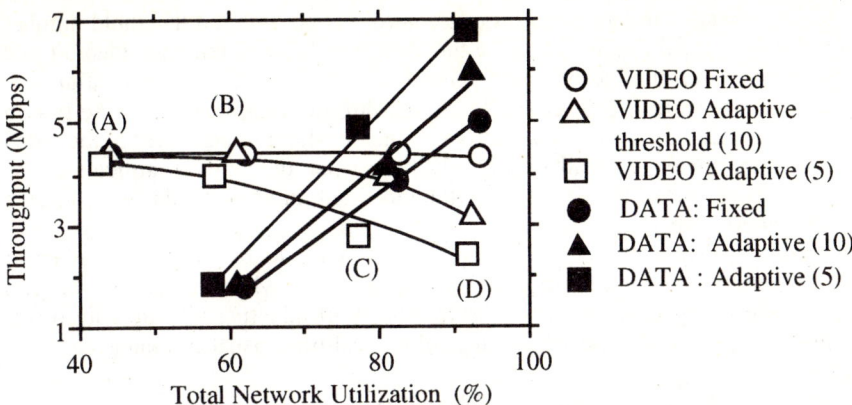

Fig. 7. Video and data throughput

Table 2 shows the number of packets discarded (not transmitted on the LAN) due to excessive numbers of collisions. Total number of packets measured was 16000 and the measurement period was 60 s. Packet discard was lower with adaptive bit rate control, but the discard ratio is very small in all cases.

Table. 2. The number of discarded packets.

Data Traffic	(A)	(B)	(C)	(D)
Fixed	0	0	1	20
Adaptive (threshold 5)	0	0	1	1
Adaptive (10)	0	0	1	1

Excessively delayed packets should be discarded at the receive buffer of the video communication terminal to ensure real-time communication. Figure 8 shows the ratio of packets that were delayed over 100 ms. The ratio decreased with adaptive bit rate control, so adaptive bit rate control effectively supports real-time communication.

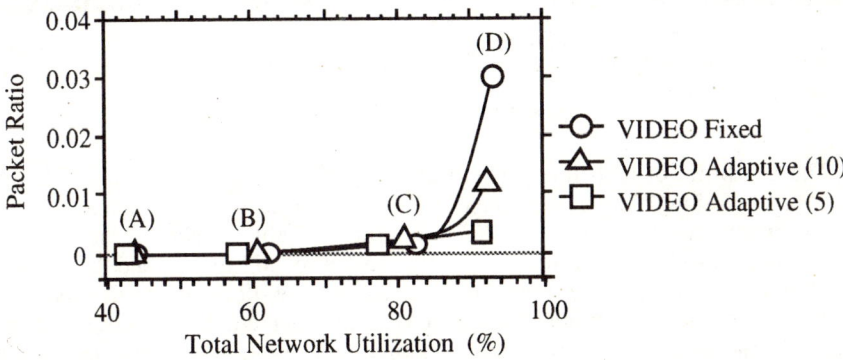

Fig. 8. Packet ration delayed over 100 ms.

The sensitivity of the adaptive bit rate control varied with the threshold number as shown in Fig. 7 and Fig. 8. When the threshold number was ten, the video bit rate was not always controlled to be lower value even when the network utilization was over 90 % as shown in Fig. 7. case (D), so the threshold number should be five to give priority to data transport. However, when the threshold number was 5, the video bit rate was not always controlled to be 512 kbps when the network utilization was under 50 % as shown in Fig. 7 case (A), the threshold number should be 10 to give priority to video quality.

Experiments were conducted using specific work loads, but the number of collisions generally increases over a CSMA/CD LAN when congestion occurs. Therefore, these experiments verify the feasibility of adaptive bit rate control for efficient data-transport and the achievement of the real-time visual communication.

3.3 Experiment over connected 10 Mbps CSMA/CD LANs

Multiport hubs were connected through a 1 Mbps serial line by IP gateways to create connected LANs as shown in figure 9. Six video communication terminals on the LANs executed 3 point to point video communication sessions. Data traffic was simulated by using either the traffic generator to send fixed-rate data or Work Stations to send TCP data. A LAN analyzer captured the video packets and TCP data with received time stamp. The maximum video bit rate was 256 kbps and the rate was controlled in 5 stages. The video packets were assembled and released every 40 ms. The video communication terminal was designed to detect congestion when the packet round trip time exceeded the determined threshold as explained below. The threshold was set to be 100 ms in the experiments and an ICMP echo message was sent every 500 ms.

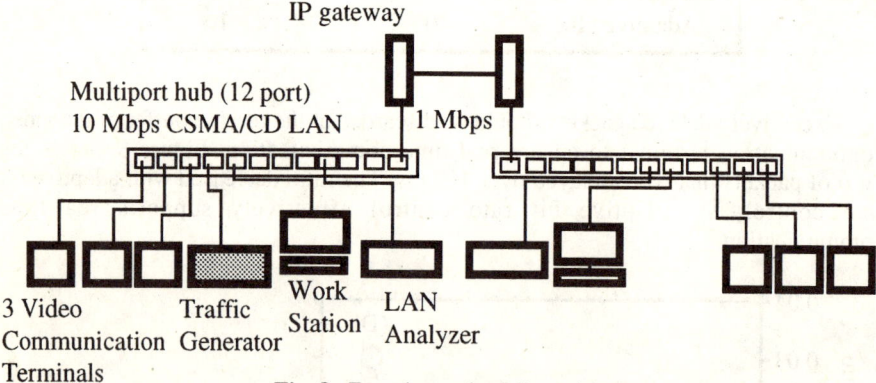

Fig. 9. Experimental configuration.

The capability of real-time packet transmission over the connected LANs was confirmed as follows. The packet round trip time was 4 ms for ICMP echo and reply packets when the packets were the only traffic. Table 3 shows the network throughput over the serial line and the average packet round trip time and the standard deviation of the video packet received times against the 40 ms interval when 1 - 3 point to point video communication sessions were executed. The packet round trip time and standard deviation were small enough that real-time communication was possible when the network has enough large bandwidth compared to the network load.

Table 3. Throughput and delay over the LANs

	1 session	2 sessions	3 sessions
on serial line (kbps)	285.5	570	856
round trip delay (ms)	7	12	21
standard deviation (ms)	1.0	1.2	2.0

3.3.1 Video Traffic and Fixed Rate Data Traffic

A traffic generator was used to add fixed-rate constant-length packets (570 bytes) while the 3 duplex video audio communication sessions were active. The total video and audio traffic was 856 kbps, so the available bandwidth was under 150 kbps over the connected LANs. The traffic generator was controlled to release (I) 298 kbps, and (II) 596 kbps of data traffic to cause congestion. The video bit rates were set at 256, 196, 136, 76 or 16 kbps or fixed at 256 kbps. The rate decrease period (t1) was 1 second and the rate increase period (t2) was 2 seconds, when adaptive video bit rate control was used.

Table 4 shows the packet round trip time and the packet discard ratio in the steady state. The round trip time exceed 1.5 s and packet discard occurred when fixed bit rate control was used. The round trip time under congestion depends on the packet buffer size in the intermediate nodes, but the packet discard ratio can not be ignored for video communication. The standard deviation under fixed rate control was small because the buffers in the nodes were continuously full. When adaptive video bit rate control was used, no packet discard occurred, and both the average round trip time and standard deviation were enough small to realize real-time communication. The results confirmed the effectiveness of the adaptive bit rate control for congestion caused only by the video traffic, because the network bandwidth was constantly insufficient for video traffic in the experiments.

Table.4. Delay and discard

video bit rate control	fixed		adaptive	
data load	(I)	(II)	(I)	(II)
average round trip delay (ms)	1550	1575	51	41
standard deviation (ms)	14	14	34	40
packet discard ratio	0.13	0.43	0	0

Figure 10 shows the change in controlled source video bit rate when the data traffic was 596 kbps and Figure 11 shows the packet round trip delay measured at the destination side. The data traffic was added at about 5 s past from the start of the experiment, and the video bit rate was controlled. The abrupt addition of data traffic caused an excessive increase in the packet round trip delay despite the use of adaptive video bit rate control. However, in the steady state, oscillation of the video bit rate was minimized. The video bit rate control scheme suppressed the oscillation mentioned in 2.3.2 and was found to be effective.

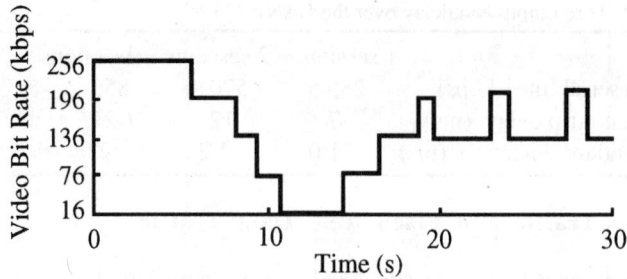

Fig. 10. The change in video bit rate.

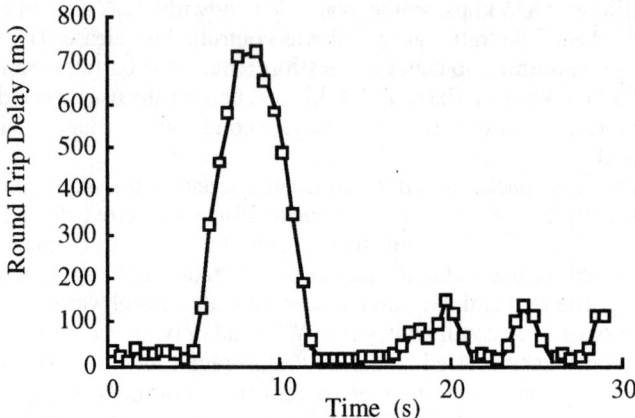

Fig.11. The change in round trip delay.

3. 3. 2 Video Traffic and TCP Data Traffic

The traffic generator released the fixed rate packets without congestion control, but TCP does use congestion control [7], so the interaction between adaptive video bit rate control and TCP congestion control was investigated.

The throughput of TCP data transfer over the connected LANs was measured before the video traffic was added. The right side WS sent 2048 x 512 byte data packets to the left side WS using TCP. The sizes of the send and receive buffers of the TCP sockets were set to 8, 16, and 64 packets to change the TCP window sizes. Accordingly, these values equal TCP window size at the commencement of TCP data transfer. Average data throughputs including data header during data transfer without video traffic are plotted as circles in Fig 12. TCP utilized almost the entire network bandwidth over the LANs regardless of the TCP window size.

TCP data transfers were done while the 3 duplex video audio communication sessions were active. The video bit rate was controlled as described in 3.3.1. Average TCP data throughput for the three buffer sizes of the TCP sockets is shown in Fig. 12. TCP data throughput was increased by using the adaptive bit rate control as compared to using fixed video bit rate control, so the feasibility of adaptive video bit rate control for efficient data transfer was verified.

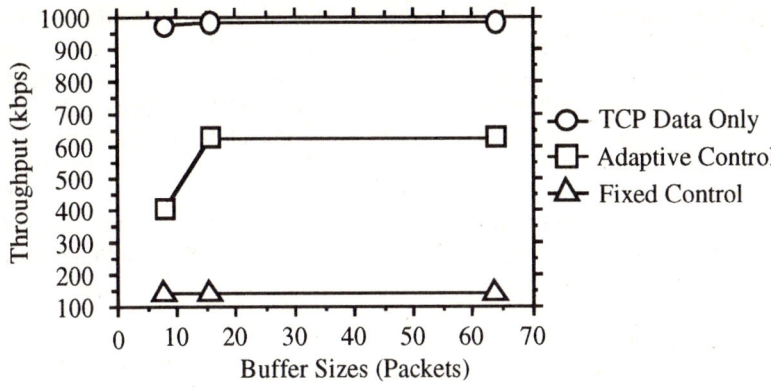

Fig. 12. Average TCP data throughput

Figure 13 shows the average packet round trip delay during TCP data transfer. The average packet round trip delay was small enough to ensure real-time communication even when the video bit rates were fixed and the buffer sizes were large.

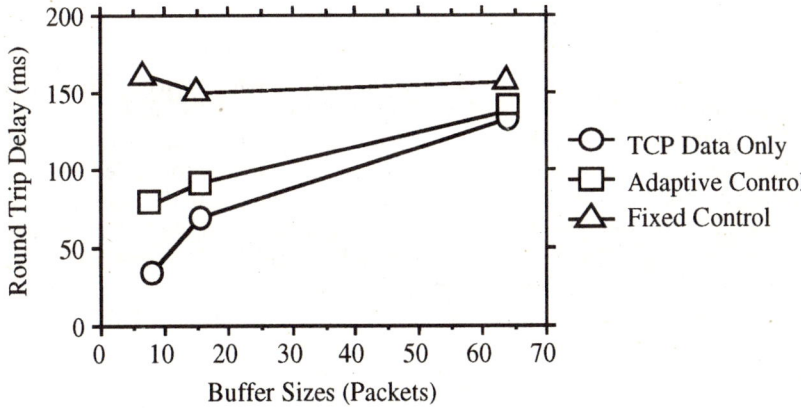

Fig. 13. Average round trip delay

When TCP socket buffer size was 64 packets, the maximum TCP data in transit was 64 packets, i.e. 64 x 512 bytes = 32 kbytes. The available bandwidth for TCP data transfer with fixed video bit rate transfer was under 150 kbps, so the processing time of the 32 kbytes of in-transit data was (32 x 8) / 150 = 1.7 s. The round trip delay shown in Fig. 13 was much smaller than this calculated processing time. Therefore, the changes in the amount of the TCP data in transit were measured and are shown in figure 14, where buffer sizes were 64 packets and the video bit rates were fixed. The amount of the TCP data in transient was calculated as the difference between the last TCP sent sequence number and the last received ack number captured by the LAN analyzer. The amount of TCP data in transient was a maximum of 13 packets and sometimes decreased to 1 packet. When the amount of the TCP data in transient decreased to 1 packet, the 1 packet TCP data was retransmitted data despite the fact that there was no packet discard over the LANs. This retransmission is thought to be caused by the expiry of the TCP retransmit timer. The TCP uses a fast mean update algorithm to compute the mean and the variance in round trip delay and establish the retransmission timeout [7]. In the experiments, the estimation error of the fast mean update algorithm prevented the amount of the TCP data in transient from being increased, so the round trip time did not increase much and adaptive video bit rate control was not necessary for real-time communication. However, if the amount of the data in transit increases by the choice of a different retransmission time out, adaptive video bit rate control is effective in ensuring real-time communication.

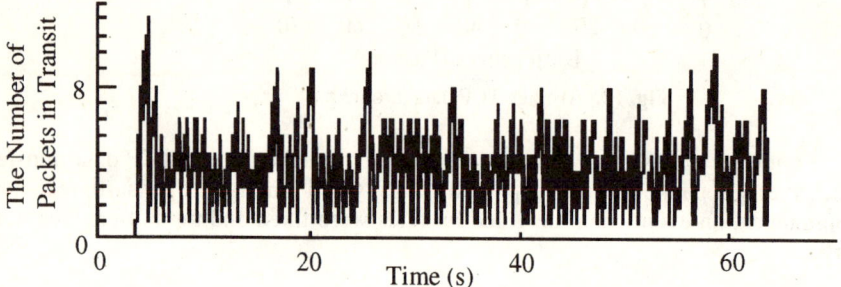

Fig. 14. The changes in the amount of the TCP data in transit.

Fig 15 shows the changes in the video bit rate when adaptive video bit rate control was used and the buffer sizes were 64 packets. TCP data transfer started at about 4 s into the experiment. The packet round trip delay increased with the start of TCP data transfer as shown in Fig. 16 and the video bit rate was controlled. However, the increase in round trip delay was small, which is different from the case of constant rate data packet transfer. Fig 17 shows the changes in the amount of TCP data in transit. The amount did not increase abruptly because the TCP has a slow start algorithm [7] to establish equilibrium. The slow start algorithm prevented an excessive increase in the round trip delay at the start of TCP data transfer.

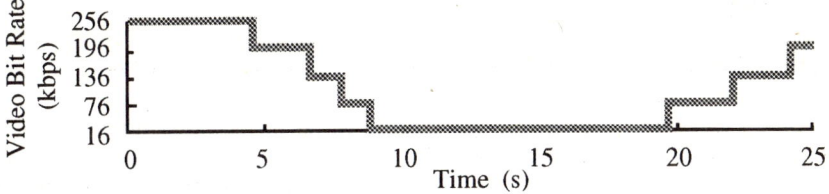

Fig. 15. The change in the video bit rate.

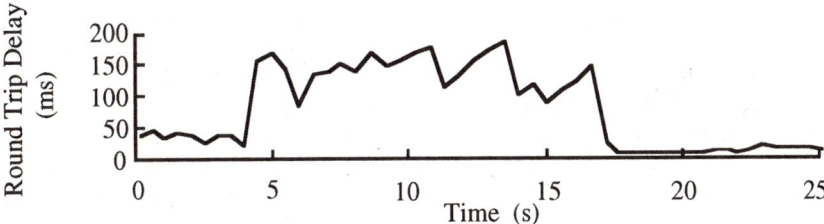

Fig. 16. The change in the round trip delay.

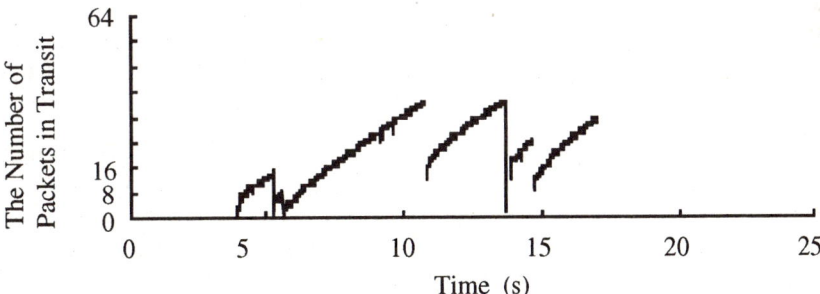

Fig. 17. The change in the number of packets in transit.

From Fig. 15, 16 and 17 the interaction of the adaptive bit rate control and TCP data transfer is as follows. TCP start transmitting data with slow start mechanism, so the TCP data in transit increases. The available bandwidth for TCP data transfer is the rest of the network bandwidth, so the TCP fully uses the rest soon with increasing the data in transit. After using the rest fully, the increasing rate of the data in transit is limited because the slow start mechanism increases the TCP data in transit after receiving ack and the ack rate is limited by self-clocking [7]. The increase rate of the data in transit is limited, so the round trip delay is not increased abruptly. As the TCP data in transit increases, the round trip delay increases and exceeds the threshold. The adaptive bit rate control mechanism detects congestion and the video bit rate is decreased. Thus the available bandwidth for TCP data transfer increases, and the throughput of the TCP increases with increasing the data in transit. The round trip time increases and exceeds threshold consequently, and the same interaction as described above repeats, until the delay does not exceeds the threshold or the video bit rate is decreased to the minimum value.

3.3.3 DISCUSSION

The behavior of bit rate control depends on the amount and the quickness of changes in the bit rate upon detecting congestion.

TCP uses window-based flow control, and halves the window size when congestion is detected. TCP is designed to fully use the network bandwidth to increase the window size to maximize the transport efficiency. Therefore, when congestion is detected, TCP attempts to share the bandwidth with other sessions, by halving the window size [7]. However, real-time visual communication does not use the network bandwidth fully, so the visual communication terminal was designed to gradually control the video bit rate. This change rarely causes excessive oscillation, but it sometimes fails to keep up with abrupt traffic changes as shown in 3.3.1. However, the combination of the gradual bit rate change and the slow start algorithm of TCP prevented an excessive increase in round trip delay as shown in 3.3.2. Only one TCP data transfer was added in 3.3.2, but the slow start mechanism is thought to prevent the excessive increase in round trip delay when the multiple TCP data transfers are done. Therefore, gradual bit rate control is suitable for real-time visual communication over IP networks.

The quickness of bit rate change is determined by the rate increase period and the rate decrease period. The periods should be also decided to prevent excessive oscillation, and to handle abrupt traffic changes. Moreover, the quickness and the amount of the bit rate change should be optimized through the study of human factors. In experiments, the periods were established ad hoc but did prevent excessive oscillation.

The sensitivity of congestion detection depends on the delay threshold. In these experiments the threshold was set to 100 ms and the feasibility of round trip delay based congestion detection was verified. However, the threshold should be finalized by considering some factors. The maximum round trip delay acceptable for real-time interactive human communication is said to be 800 ms [11], so the maximum acceptable transmission delay over the networks is thought to be about 500 ms by excluding the coding and packet processing time in the terminals. Therefore, the threshold might be based on the maximum acceptable transmission delay. However, an excessive large threshold compared to the round trip delay without congestion may delay the detection and cause oscillation, a threshold smaller than the delay without congestion leads detection errors. Furthermore the round trip delay over the networks without congestion usually is not known in advance. Therefore, congestion detection without using the round trip delay is discussed in the following.

The visual communication terminal used in these experiments is based on the idea that the round trip delay should be detected to achieve real-time interactive communication within the acceptable round trip delay. Nevertheless, this congestion detection scheme has a disadvantage, i.e., the round trip time increased even though only one direction was congested and both video communication terminals reduced their video bit rate. Packet discard is a signal of congestion and can be used to detect congestion in each direction. However, the intermediate nodes of IP networks have large buffer sizes to cope with congestion, so the round trip time becomes excessive for real-time communication when packets are discarded. Considering that congestion increases transmission delay, the increase can be used to detect congestion in each direction. Large packet transmission delay values mean late packets, so the increase can be calculated from the increase in the difference between the received interval and the sent interval. The sent interval can be calculated from the difference between the sent timestamp added to sent packets. If packets are sent at the same interval, the timestamp is not needed [9]. The received interval can be calculated from the received time. The sum of the increase in transmission delay can be used to detect congestion without using the round trip delay. The threshold to detect congestion should adapt to a variety of network structures, but the estimation scheme of the threshold is for further research.

Late packets can be detected easily by using the fixed audio decoding rate as follows. The video communication terminal starts decoding the audio signal only after several audio packets are stored in the receive buffer to cope with packet jitters. When congestion occurs, the transmission delay increases and late packets occur, and the number of audio packets in the buffer decreases because the decoding bit rate is fixed. Therefore, congestion can be detected by a strong decrease in the number of audio packets in the receive buffer.

Detecting the increase in the packet delay mentioned above is effective for congestion detection in each direction, however the round trip delay should be measured to confirm the achievement of real-time communication. The measurement of the round trip delay may be low frequency. The measurement just before visual communication session is also useful in confirmation of the achievement of real-time communication.

The IP multicast service is now being used to send video packets to multiple destinations over IP networks. The adaptive video bit rate control scheme proposed in this paper is suitable for real-time interactive communication, so the number of the receivers might be 10 at most. Therefore, large volumes of feedback information [12] do not pass to the source when congestion is detected at the receivers.

The adaptive bit rate control can not be useful in some situations. When the packet delay caused only by the TCP data in transit exceeds the maximum acceptable delay for real-time communication, the adaptive video bit rate control is not effective in real-time communication. If the total traffic of the minimum video throughput exceeds the network bandwidth, congestion must occur in spite of the adaptive bit rate. In these situation, the visual communication sessions should be terminated as showing the state of the network to the user.

4. Related Work

Several source rate control schemes that detect network congestion [6, 8, 9, 10] were proposed for real-time video/audio communication over networks. However, none of the mechanism in these references detects the onset of congestion by measuring the increase in the packet delay to achieve real-time communication. The feasibility of source rate control by finding the packet discard was investigated by using a real system in [8], but few experimental results were reported, and the behavior of the control was not clear. A simulation study was achieved and the feasibility of source control with the addition of data without transport protocol was verified in [6]. However, the scheme in [6] requires that the switches send their buffer occupancy level and service rate to the source which is not possible in present IP networks. The scheme in [9] was very similar to the one introduced in this paper and detecting congestion from the late packets, but its feasibility was not confirmed and congestion was detected at the worse stage to decrease the source rate. The source rate control scheme in [10] was tested in a multicast environment, and its feasibility was verified, but the packet discard rate was used for congestion detection.

The interaction between the source rate control and the congestion control of the data transport protocol is important for adding video traffic to existing packet networks, but the suitable bit rate control scheme was not discussed and the interaction was not investigated.

5. Conclusion

This paper has introduced an effective control scheme that allows real time video sessions to proceed in conjunction with data transfer. Data traffic over IP networks was measured. Average utilization rates were not large, but sometimes data transfer required large capacity. The adaptive video bit rate control scheme proposed in this paper decreases the video bit rate to assign more capacity to the data transfer by detecting the onset of congestion from the increase in packet delay. Experiments proved that the scheme was effective for efficient data transfer and real-time video communication. The TCP congestion control scheme detecting packet discard and the adaptive video bit rate control were found not to conflict in operation and the interaction was discussed. The gradual video bit rate control was found to be suitable.

Even if the network has a resource reservation mechanism, adaptive video rate control is still effective because it simplifies resource reservation and admission control.

References

[1] R. Gusella, "A Measurement Study of Diskless Workstation Traffic on an Ethernet", IEEE Trans. Commun., vol. 38, no. 9, Sept. 1990.
[2] W. E. Leland, M. S. Taqqu, W. Willinger and D. V. Wilson, "On the Self-Similar Nature of Ethernet Traffic," In Proc. ACM SIGCOMM '93 pp. 183 - 193, Sept., 1993.
[3] W. E. Leland and D. V. Wilson, " High Time-Resolution Measurement and Analysis of LAN Traffic: Implications for LAN Interconnection," In Proc. IEEE INFOCOM '91, Bal Harbour, FL, 1360-1366, 1991.
[4] C. Topolcic, S. Casner, C, Lynn, P. Park and K. Schroder. Experimental Internet Stream Protocol, version 2 (ST-II), Internet RFC 1190, Oct., 1990.
[5] L. Zhang, S. E. Deering, D. Estrin, S. Shenker and D. Zappala, "RSVP: A New Resource ReSerVation Protocol," IEEE Network, September, 1993.
[6] H. Kanakia, P. P. Mishra and A. Reibman, "An Adaptive Congestion Control Scheme for Real-Time Packet Video Transport", In Proc. ACM SIGCOMM '93, pp. 20 - 31, Sept., 1993.
[7] Van Jacobson, "Congestion Avoidance and Control", In Proc. ACM SIGCOMM '88, pp. 314 - 329, Stanford, CA, August, 1988.
[8] Michael Gilge and Riccardo Gusella, "Motion Video Coding for Packet-Switched Networks - An Integrated Approach", In Proc S.P.I.E. : Visual Communication, pp. 592 - 602, Nov. 1991.
[9] Luca Delgrossi, Christian Halstrick , Dietmar Hehmann, Ralf Guido Herrtwich, Oliver Krone, Josen Sandvoss and Carsten Vogt, "Media Scaling for Audiovisual Communication with the Heiderberg Transport System", In Proc. ACM Multimedia 93, pp. 99 - 104, Jun. 1993.
[10] Jean-Chrysostome Bolot, Thierry Turletti, "A Rate Control Mechanism for Packet Video in the Internet", In Proc INFOCOM 94.

[11] CCITT SG XV Document, "Considering on acceptable processing delay in the video codec", Document AVC-85, 1 August 1991.
[12] R. Yavatkar, L. Manoj, "Optimistic strategies for large-scale dissemination of multimedia information", In Proc. ACM Multimedia '93, Anaheim, CA, pp. 1-8, Aug. 1993.
[13] T. Sakatani, N. Kanemaki, T. Arikawa and K. Shimamura, "Visual Communication Schemes using CSMA/CD LANs", In Proc IWACA '92, pp. 393 - 399, Munich, March, 1992.
[14] T. Sakatani and T. Tajiri, "A Video/Audio Packet Transmission System over LANs", In Proc the 5 th International Workshop on Packet Video, Berlin, March 1993.

Network Layer Scaling: Congestion Control in Multimedia Communication with Heterogenous Networks and Receivers

Hartmut Wittig
Jörg Winckler
Jochen Sandvoss

IBM European Networking Center, Vangerowstraße 18, D-69115 Heidelberg
Mail: wittig@vnet.ibm.com, {winckler,sandvoss}@dhdibmip.bitnet

Abstract: Because of the admission control schemes in computer networks, in many multimedia communication scenarios it is not possible to get guaranteed quality of service for data transmission along the entire route from sender to receiver. In such cases scaling mechanisms can be used to overcome problems caused by resource congestion. Whenever resource bottlenecks are noticed, the multimedia traffic through these resources will be reduced for a limited time. The scaling scheme described in this paper uses mechanisms in the network layer to enable short delay times between detection of overload and the reduction of multimedia traffic. Network layer scaling allows priority-controlled regulation of multimedia traffic over network resources and adaptive handling of various paths and receivers. The monitoring and scaling strategies in the network layer and a prioritization scheme for multimedia stream packets are described.

Keywords: Quality of Service, Media Scaling, Scheduling, Congestion Control, Heterogenous Receivers, Monitoring, Multicast

1 Introduction

The issue of whether to use guaranteed or best-effort services for real-time multimedia communications remains unresolved within the multimedia community. One group claims that a guaranteed transport service is needed to keep the soft deadlines of real-time multimedia traffic. Prerequisite for this service are resource reservation and scheduling mechanisms within the network system consisting of endsystems, network nodes, and links. The other group says that the real-time multimedia traffic can be transmitted via best-effort connections without guaranteed deadlines. A general approach consists of a mixture of various transmission requirements, as reflected by different service classes and related scheduling algorithms (e.g. [SCZ93]). Multimedia traffic is partitioned into real-time traffic and traffic without real-time requirements. Non-real-time multimedia traffic (e.g., multimedia mail and video downloads) proposes transmission in a best-effort service class without any quality of service (QoS) guarantees; real-time traffic (e.g., live-TV-broadcasting, video conferencing) should be transmitted via guaranteed services.

There are important reasons for a services model which offers both guaranteed and best-effort services for multimedia traffic:

- Different multimedia applications have different requirements on the communication system. As an example, while video-on-demand or audio-on-demand applications (in the next generation of present HiFi equipment) will be very sensitive to quality reductions, it is quite easy to accept small transmission errors in video conferencing scenarios (in the next generation of present phones). According to the QoS requirements, different service classes can offer different service costs.
- A prerequisite for resource reservation is a resource management facility. Different classes of subnets are to be differentiated: those that contain already resource management mechanisms (e.g., ATM, 100BASE-VG-Ethernet), those that can be extended by appropriate mechanisms (e.g., Token Ring, FDDI), and those that do not provide any resource management at all (e.g., Ethernet, Fast Ethernet). According to the local buffering or processing schemes, reservation mechanisms for the local resources can be supplied by routers and communication endpoints. Often the requests for a certain quality of service must be handled by a network consisting of a broad mix of the above resources. If there is one resource in a guaranteed connection which cannot give guaranteed processing bounds, the service cannot said to be guaranteed as a whole.

  Using congestion control, it is not possible to give guarantees, but an integrated transmission scheme resolves access conflicts to overloaded resources by a selective decrease of the overall transferred data. Multimedia streams belonging to a high QoS class can be protected by using the prioritization scheme discussed in this paper.
- Guaranteed QoS for variable bitrate (VBR) data transmission is inefficient. In most reservation protocols a constant bandwidth is required. Therefore VBR transmissions often are considered as constant bandwidth streams with the average overall bandwidth as parameter. However, in the worst case all multimedia streams reach their peek bandwidth at the same moment and the resources may not be sufficient. In this case an uncontrolled loss of packets will occur. There is also a compromise between guaranteed transmission for the worst case and no guarantee at all. A certain base part of the VBR stream is guaranteed using a resource management system. The variable part of the bandwidth requirements is handled with a best-effort service using congestion control mechanisms, e.g., network layer scaling. Resource monitoring detects arising overloads. A controlled reduction of bandwidth requirements down to the guaranteed base parts are used to handle the congestion.

The definition of scaling is derived from [CSZ92], [TTC92], [Upp92]: Scaling is the dynamic adjustment of traffic load to the currently available resource capacity.

Scaling mechanisms can be realized in different layers of the communication stack. The major difference is between network layer and upper layer scaling schemes. All scaling mechanisms above the network layer are on an end-to-end basis (e.g. [DHH93]). They are unable to deal with multicast trees with different receiver requirements. They cannot overcome congestions on single resources along a route. Either the whole multimedia connection with all receivers or nothing will be affected. The responsiveness of end-to-end mechanisms may not be good enough, especially on large networks and internets.

Hoffman et al. have introduced support for dynamically scaled multimedia streams in the network [HSF93]. The advantages of realizing scaling mechanisms on the network layer are:

- In multiparty connections each receiver can be handled individually according to its capacity.
- Each subnet is handled individually according to its bandwidth management functionality.
- If TSDU borders can be identified within the network layer, useless information of corrupted TSDUs can be discarded earlier.
- Smaller feedback loops allow for more precise scaling on a link. Congestions can thereby be resolved quickly.

Several design alternatives for scaling mechanisms in the network layer are presented and a concrete realization is proposed. This paper has 10 sections. Section 2 describes the overall architecture proposed for scaling in the network layer. Section 3 analyzes the monitoring mechanisms to detect congestions. Section 4 contains information about feasible scaling mechanisms that can be used on the network layer. In Section 5 the protocol for dynamic filter adjustment is described. Section 6 surveys the influences of different parameters on the priorities of media streams. In Section 7 an abstract priority transformation scheme between the parameters and the packet priorities is presented. A theme closely related to dynamic network scaling is analyzed in Section 8: the early discard of corrupted data units. Section 9 contains results of a first analysis performed by applying the priority transformation functions to sample video streams. Concluding remarks are given in Section 10.

2 Overall Network Layer Scaling Architecture

A general approach for media scaling communication between a data source and data sink is shown in Figure 1.

The media scaling mechanism can be compared with regulation principles in the control technique. The steering device is the filter. The work of the filter is controlled by feedback messages. In the context of network layer scaling such messages are called scaling messages. In general, a filter is a device which can absorb data packets to minimize resource requirements. In the network layer scaling scheme filters are used to reduce bandwidth by dropping packets of a multimedia stream.

The generation of scaling messages by monitors can be either time- or event-driven. Time-driven monitors send periodic feedbacks, event-driven monitors watch over measurements (e.g., resource utilization) and send the feedback when exceeding a certain threshold. Event-driven scaling messages are recommended because of the fast reaction times.

In general, a multimedia stream can flow from the sender across many network hops to the receiver. Each hop plays the role of potential sink for the previous hop and the source role for the following hop.

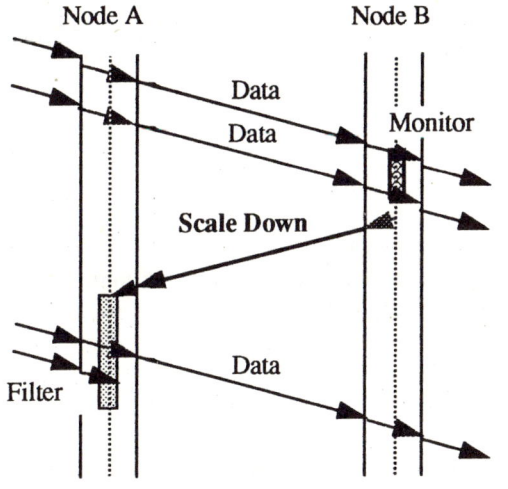

Figure 1: Scaling Facilities in the Network Layer

3 Monitoring of Transmission Bottlenecks

From an abstract point of view, the transmission line of multimedia streams between sender and receiver can be considered as a sequence of resources (see Figure 2). Each active resource (e.g., CPU, network) is characterized by its ability to serve the users requirements (e.g., throughput, delay, reliability). Passive resources (e.g., buffer) decouple the active resources and balance short-term processing differences between neighboring active resources.

3.1 Bottleneck Indicators

The key resources on the transmission line are the network, CPU, and buffers. Each of these resources can become a bottleneck during data transmission. A mid-term capacity problem of one resource results in a capacity problem of another resource — the previous buffer resource. For example, if the network capacity is exhausted the sender cannot place its packets onto the network. This results in an overflow of sender buffers. In this case the detection of overflowing buffers on the sender side is an indication of the bottleneck network. Buffer overflow on the receiver side can be an indication of a CPU bottleneck because the receiver is not able to handle incoming packets.

Table 1 shows an overview over common bottlenecks and their indications at the

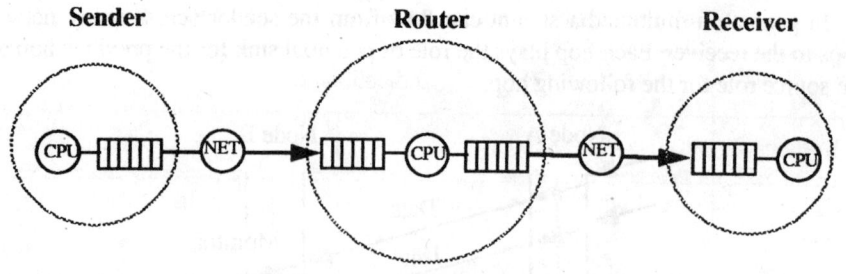

Figure 2: Transmission Line of Multimedia Streams

sending or receiving side. One indication alone is often not enough to determine a resource overload. For example, heavy CPU loads indicates a bottleneck, but the resource cannot definitely said to be overloaded. For example, the heavy CPU load might be caused by non-real-time processes such as computation of off-line video compression task. Therefore it is helpful to monitor CPU utilization consumed by multimedia processes only. The combination of several indications (e.g., high CPU utilization plus lost packets at receiver) can also determine the CPU as bottleneck.

Congestion in the network can first be detected by receivers receiving late packets. If the congestion continues packets are lost. Reduced throughput in combination with the CPU load is also an indication of congestion on the receiver (CPU high) or the network (CPU low). Most common video compression algorithms however, result in variable bit rate streams. So the paper will focus on these. Network congestions can also be detected at the sender side. If the number of packets in the network adapter buffer space rapidly grows, the network is unable to accept them in a timely fashion.

3.2 Monitoring and Measurement

In the following section different monitoring schemes are analyzed: monitoring of delay, throughput, utilization, and loss rate. Main criterion to monitoring mechanisms are their precision and resource consumption. To avoid additional overhead built-in monitoring mechanisms of resources are used (e.g., the CPU usage field of UNIX processes).

Delay Monitoring

Delay monitors observe the processing times of packets. Delay monitoring is suggested for detection of CPU and network bottlenecks. There are different ways of delay monitoring: In the transport layer, each logical data unit is represented by a transport service data unit (TSDU). Each TSDU has an expected arrival time on the next network node. A TSDU arriving later than expected indicates congestion. There are several possibilities to define the expected arrival time of a TSDU. One could, for example, define its value simply as the actual arrival time of the previous TSDU plus the period of the message stream (the reciprocal value of its rate).

Table 1: Indication of bottlenecks at sender and receiver nodes

Bottleneck	Indication at the	
	Sender	Receiver
CPU at sender	• high CPU utilization • lost or late control messages from the receiver • low data throughput	• lost or late packets[a] • low data throughput
CPU at receiver	• lost or late control messages from the receiver	• high CPU utilization • lost or late packets • low data throughput • high receive buffers loads
Network	• high network utilization • high send buffers loads • lost or late indication messages from the receiver • low data throughput	• high network utilization • empty receive buffers • lost or late packets • low data throughput
Buffer	• long-term high buffer utilization without the following resource as bottleneck	• long-term high buffer utilization without the following resource as bottleneck

a. Packet sequence numbers in the header of the network layer and the measurement of inter-frame period allow detection of late and lost packets.

Additionally, the arrival time of the first TSDU of the stream (or any other earlier packet) rather than the arrival time of the previous packet could be used as the basis for the calculation. This helps to avoid false indications of congestions in cases where the previous TSDU happened to arrive early and the current TSDU has a normal delay. The expected arrival time is calculated as the "logical arrival time" of the previous packet plus the stream period. Section 6 describes an early discard scheme which detects TSDU boundaries at the network layer by using a specific start bit in the header of network service data units (NSDUs).

Throughput Monitoring

Throughput monitoring is used to detect CPU and network overloads by watching the packets leaving the resource considered. There are many possibilities to monitor data throughput. Leaky bucket throughput monitoring mechanisms are based on simple packet counters which are incremented by 1 every time a packet arrives and decremented by 1 in fixed time intervals. Another technique is the jumping window mechanism, in which the throughput is measured within a fixed time interval. The moving window throughput monitoring mechanism determines the throughput (*TPT*) of a stream in a given time interval τ_{TPT} by the sum of the length of packets devided by time interval. For example, using interframe-coded video streams this interval can be defined as the duration of one group of pictures. If the monitored throughput falls below a minimum threshold, this indicates problems of previous resources.

Utilization Monitoring

Utilization monitoring can be used for CPU, network, and buffer resources. The utilization of resources can be computed by dividing the currently used parts of a resource by the overall capacity of the resource. For example, the CPU is loaded with time-critical and non-time-critical processes. Time critical processes have a reservation for the processing time and a high scheduling priority. The remaining CPU time is used for non time critical processes. So an overall CPU load near 100% does not necessarily mean that time critical processes cannot get enough time. It is recommended to separately measure the utilization for multimedia real-time communication. Per-process monitoring of CPU consumption is more expensive but allows a more sensitive scaling by noting the different loads of multimedia applications.

Loss Rate Monitoring

Another direct indicator to detect congestion is the rate of packets lost. An assumption needed to enable the detection of lost packets is that all NSDUs of a stream are numbered in sequence. Numbering is done by the sender. The transmission algorithm must keep the sequence of packets through the network. Contrary to the IP transmission of packets, the sequence of packets in most real-time multimedia networks protocols will be retained (e.g., ST-II [DHH94], RTIP [BM91]). If there are gaps in the order of packet numbers, the packets with these numbers are considered as lost packets. The detection of corrupted packets is an indication of network failures, but not a suited for indication of resource bottlenecks. Additionally, the detection of corrupted data packets in each network node requires a time-consuming computation of error-check sequences. Corrupted NSDUs will therefore not be considered.

4 Filters

Our network layer scaling scheme uses filters which reside in the network layer. Filters can be changed in the stream setup phase as well as dynamically at run time. Short term congestion of resources can be managed using filters before the resource. Midterm and long-term changes in filtering scheme should be propagated through the network nodes in direction to the sender; this affects the work of filters on this route.

4.1 Overall Filtering Principle

An example clarifies the principles of filtering for network layer scaling: Figure 3 shows a simple network topology with two streams of a multiparty connection branching in Router B to a target via Router C and to Target D and a point-to-point connection from Sender A to a target via Router C. The monitor functions of Router C detect

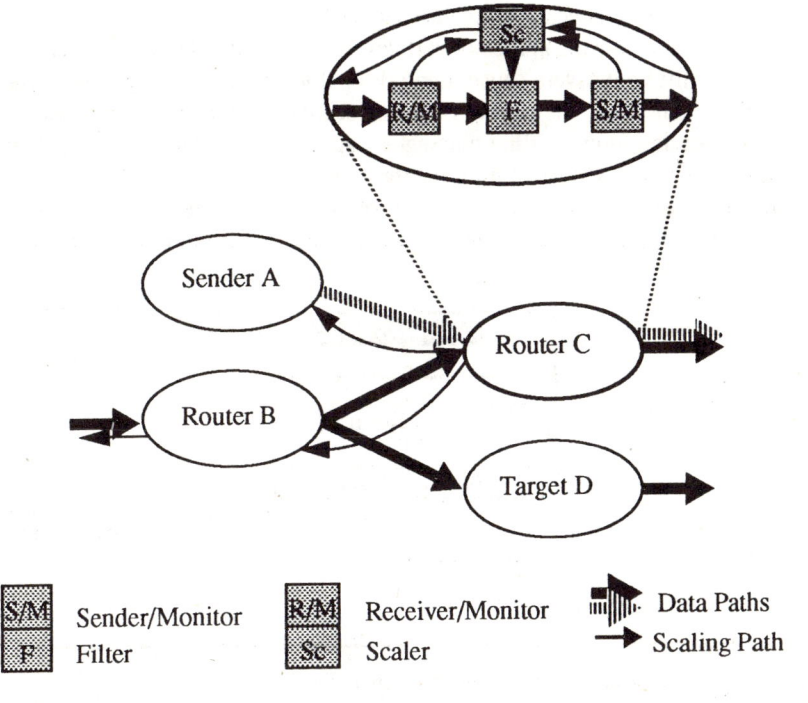

Figure 3: Monitoring and Scaling Devices in the Network Layer Scaling Scenario

situations, in which a resource (e.g., the network between A,B and C) is becoming overloaded and report this to the scaler of Router C. The scaling device[1] (scaler) decides that the congestion can be solved by reducing the traffic in Sender A, Router B and Router C. Router C sends a SCALE_DOWN message to Sender A and Router B. SCALE_DOWN messages are more important than multimedia data, because they protect resources against overloads. Therefore, they should be transmitted with higher priorities than normal data traffic[2]. Both router A and B use a priority scheme to decide which packets should be dropped to meet the new, reduced data rate. The amount to be dropped is defined relative to the maximum transmission rate negotiated

---

1. Scaling devices consists of two basic parts: filters to control data throughput, and decision modules for the filter propagation.
2. Additionally, monitors are adjusted very carefully to report bottlenecks before data is dropped (e.g., the indication is that the CPU load exceeds 90 percent).

in the setup phase of multimedia streams. The calculation of the priority value (up to which PDUs are to be discarded in order to lower bandwidth requirements) is done by a local feedback loop.

The link from Router B to the Target D is not overloaded. So Router B forwards all incoming packets of the stream to Target D (multiparty connection). This results in different QoS in the two branches of the multicast tree.

The reporting of resource bottlenecks to the upper layers of the end systems is the key issue of an integrated scaling scheme. Upper layer scaling mechanisms at sender side and scaling mechanisms in the network layer must cooperate. Independent scaling mechanisms in different layers cause many difficulties, since if many scaling mechanisms overlay each other, the timing of major scaling parameters is impossible. For example, independent downscaling messages can occur at the same time and cause a considerable loss in the quality of multimedia streams.

The reporting to the upper layers (e.g., in Sender A in Figure 3) is done with local scaling messages. Figure 4 is showing the scaling architecture in a sender node. Scal-

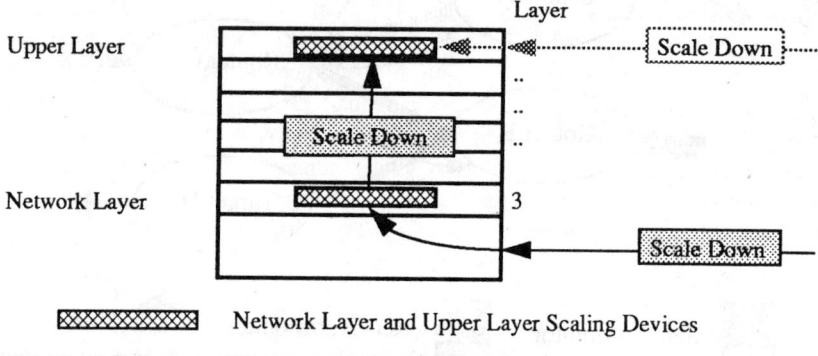

Figure 4: Scaling Architecture in the Sender Node

ing mechanisms in the network layer are suitable to handle bottlenecks on routes to the targets as well as heterogenous receivers. Network layer scaling is based on packet priorities and the discarding of low-priority packets. In the event of long-term resource bottlenecks the scaling message will be propagated in direction to the sender. The scaling message reaches the network layer scaling device in the sender node. If there is a long-term down-scaling, the upper layer scaling devices (e.g. in the application layer) are informed using the "SCALE DOWN" message. By changing the compression schemes of multimedia stream, scaling in the application is more efficient than network layer scaling. For example, the coding scheme can be changed to higher compression rates to relax a network congestion or lower compression rates to relax CPU congestions. Furthermore, the dimension and resolution of the video pictures could be decreased, grey-scale videos presented instead of colored, or the picture rate of a video stream reduced. By changing the packet priorities with regard to the new upper layer scaling parameters in the sender the network layer scaling mechanisms in all hops adapt to changes in the coding scheme.

5 Filter Adjustment Protocol

If a resource bottleneck is detected the monitor sends a scaling message to the previous filter. The scaling message consists of three values (Figure 5).

RELAXATION	DURATION	STREAM-ID

Figure 5: Scaling Message

The relaxation parameter describes the relaxation of the data throughput in relation to a maximum throughput value negotiated between sender and receiver. The relaxation is not described in relation to the actual traffic to prevent unjusted scaling of streams. For example, when the scale down message is related to the actual data rate, a renewed scaling message would affect streams being currently downscaled much more than unscaled streams.

The relaxation parameter is accompanied by a time value which specifies the validity of the scaling message: the duration parameter. When the scaling message times out, the sender can increase the data rate again. If the congestion still continues, the receiver can send further scaling messages to increase the value of reduction or the time value or both. Later scaling messages substitute older ones. If the time value matches the constant PERMANENT, a permanent filter is installed and the scaling message results in a permanent degradation of the data rate. This mechanism is used to handle different receiver requirements.

To gain more flexibility in scaling policies the scaling message holds a list of stream identifiers. If this value is set to ALL, all streams flowing through the overloaded resource (e.g. the succeeding network) are affected by the scaling activities. Otherwise only the streams addressed by the STREAM-ID field will be scaled.

As an alternative to the reduction value, the scaling message can contain a priority mask which specifies the priorities of the packets to be dropped. This allows upper networking layers to describe and propagate static or dynamic filters ([PPA93], [WHD94]). In case of long- and mid-term overload, scaling messages are also forwarded by filters to previous filters.

6 Prioritization of Continuous Multimedia Stream Packets

In case of resource overloads the data throughput must be reduced as fast as possible. The worst case of an overloaded resource is when packets are dropped in an uncontrolled fashion. The amount of damage depends on the type of packet lost. In order to minimize this damage packets should be dropped in a controlled manner using a prioritization scheme which considers the properties of specific streams and packets. The following items have impact to the priority of a stream packet:

- According to the applications and tasks, multimedia streams have different prior-

ities. There are different QoS parameters users of the multimedia transport system can pay for and require to get. As another example, a playback video stream will typically be less important than an unique video stream being recorded.
- In most cases video streams need more bandwidth than audio streams. Comparing the bandwidth requirements of audio and video (see [DHH94]), it can be easily seen that discarding a video picture saves much more bandwidth than discarding an audio sample. Additionally, the human perception of audio errors is much greater than video errors. Therefore it is preferable to discard video packets earlier than audio packets.
- Another influence on the priority of single packets of a stream is caused by the compression technique. Efficient compression algorithms are based on interframe coding techniques (e.g., MPEG-2, MPEG, DVI). As a consequence, the successful decompression of predicted pictures relies on other pictures. Intraframe coded pictures are key pictures and therefore more important. For example, if an I-picture has been dropped, all other pictures belonging to the same picture group cannot be decompressed.
- TSDUs must be split into packets not greater than the maximum transfer unit in the network layer[3]. The priority of a packet relies on its position after splitting. If previous packets have been transmitted successfully, the probability of completing the frame is higher if the last packets of a picture are transmitted.

To derive the priority scheme, the last two points require a more detailed discussion. Figure 5 shows an example MPEG video stream. The stream of the layer being considered has a different layout. A digitized uncompressed video stream consists of a

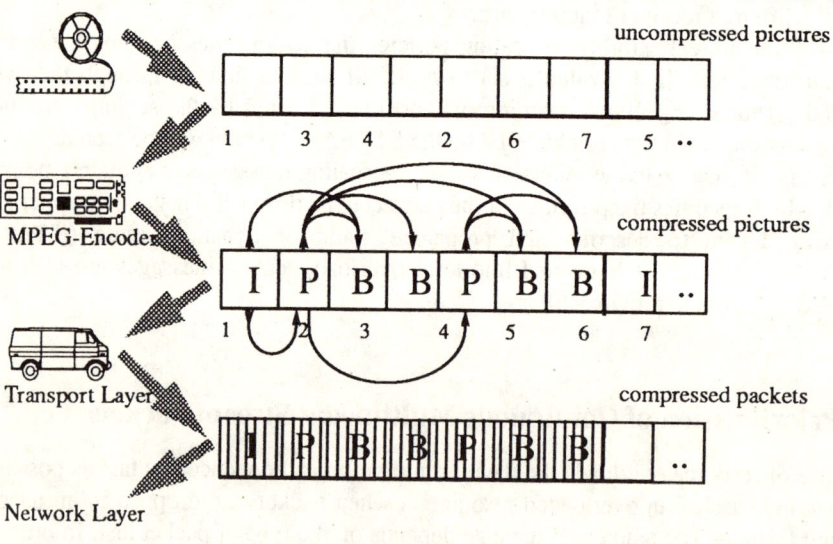

Figure 5: Layout of an MPEG-2 video stream in different layers [MPG93]

---

3. One TSDU contains one picture of an compressed stream.

sequence of pictures (upper part in Figure 5). A compressed video stream containing different kinds of compressed pictures is shown in the middle of Figure 5. The figure distinguishes between three kinds of compressed pictures:

- I: The Intra-coded picture consists of self-contained coded information.
- P: The Predictive-coded picture is the result of a motion compensated prediction and depends upon the past I- or P-picture.
- B: The Bidirectionally predictive-coded picture is a picture which is coded using motioncompensated interpolation from a past and a future I- or P-picture.

A priority scheme of the MPEG pictures can be derived by analyzing the damage to the stream quality when specific picture types are dropped. The greater the gap in the video stream after the drop of a picture, the greater the importance of the picture. Based on MPEG coding the dependence between different kinds of pictures is shown. The compressed video picture can be decompressed only if all other pictures pointing to the picture considered are transmitted correctly. Table 2 is showing which pictures becomes worthless when a specific picture is dropped. The number of the lost pictures is the direct indication of the priority.

As an example, the loss of the P-picture at position 5 causes the loss of two other pictures. The B-picture with number 6 has been compressed using a bidirectional interpolation between the P-picture at position 5 and the next I-picture. This is why both B-pictures with number 6 and 7 cannot be decompressed to the original pictures without this P-picture. Altogether three pictures are lost if P-picture at position 5 is not received. The greatest gap in the video stream is caused by a drop of an I-picture. All other pictures up to the next I-picture cannot be decompressed. Therefore the highest priority (0) is assigned to all I-pictures. B-pictures have the lowest priority (3)[4] because no other pictures are affected by a B-picture loss. The priority of a P-picture depends on its position in the stream. The loss of a P-picture closer to the initial I-picture causes the loss of more pictures than a P-picture at the end of a picture group[5]. Therefore P-pictures are assigned a different priority class between the priority of I- and B-pictures. In the MPEG standard, the length of a picture group is unspecified. To reflect this in this prioritization scheme, enough room has to be left between lowest and highest priority to map the relative position of a P-picture into its priority.

Essentially, this MPEG stream serves to describe the different priorities of the pictures. The same considerations can be made for uncompressed streams or for streams coded with other compression techniques like DVI and MPEG-II. A detailed description of a prioritization scheme of hierarchical compression algorithms like MPEG-II is described in [WHD94]. The main advantage of the prioritization of different kinds of pictures is the minimization of the damage a loss of pictures will cause.

Pictures in the transport layer can consist of many packets in the network layer. Only if all packets of a picture are transferred correctly the picture can be correctly

---

4. In the example the interval of the picture priority is [0;3].
5. A picture group is a series of one or more consecutive pictures intended to assist random access into the sequence. The first coded picture in a group is an I-picture.

reassembled in the transport layer of the receiver. It is better to drop the first packet of a picture (and therefore all other packets of this picture) than the last packet of the other picture (and to loose all already transferred packets of this picture). Increasing priorities solve this problem. The above parameters must be mapped to a certain priority. This priority transformation function will be considered in the next section.

Table 2: Effects of MPEG picture drop

Position	Type of picture	Position of lost pictures	Overall number of lost pictures	Priority (0..highest)
1	I	1,2,3,4,5,6,7	7	0
2	P	2,3,4,5,6,7	6	1
3	B	3	1	3
4	B	4	1	3
5	P	5,6,7	3	2
6	B	6	1	3
7	B	7	1	3

7 Priority Transformation Function

The measurement flowing into the priority scheme is the user-defined stream priority (SP) and the intra-stream priority (ISP), which depends upon the position of the packet. In general, the packet priority (PP) in the network layer is described with

$$PP = f\ (SP,\ ISP)$$

The priority transformation function f maps the SP and ISP parameters into the PP. As a starting point for further discussion, the SP part is expected to be more dominant than ISP. That means, that in comparing two packets with the same position (ISP), the packets belonging to a stream with a higher SP are always transmitted over the network with a higher priority (PP). It is not clear how many different stream priorities are sufficient for the variety of multimedia applications.

The other important part of the packet priority is the kind and position of pictures in the video stream. A proposal for transformation function f is shown in Figure 6.

Each SP gives a certain range to be kept by all packets of this streams. In Figure 6 all packets belonging to a stream can use 5 stages of ISP (4..0). Some examples will clarify the mapping scheme in the priority transformation function.

First, an uncompressed audio stream is considered. Because of characteristics of an

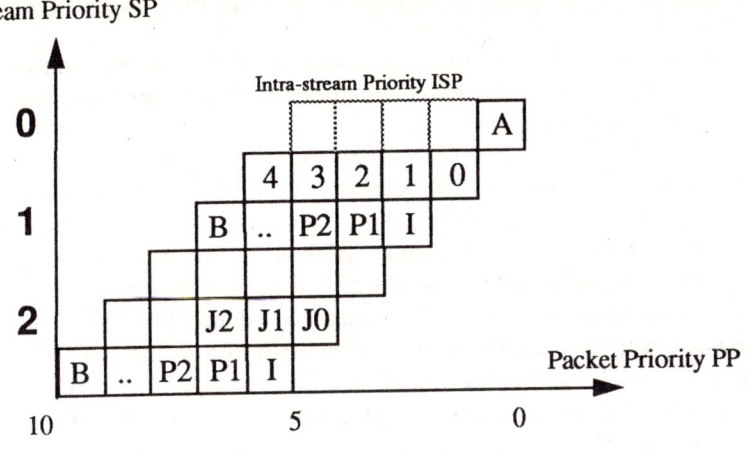

Figure 6: Priority Transformation Function f

audio stream (cf., Section 6) the application specifies the highest SP. The audio stream is not divided into substreams. Therefore the ISP is always zero. All packets of this stream are transmitted with PP=0. In Figure 6 the PP results from an addition of ISP and SP.

A second example is an MPEG stream with SP=2. As shown in Table 2, I-pictures will always get the highest ISP=0 and B-pictures the lowest ISP=4. Because the priority of a P-picture depends upon its position in the picture group, P-pictures at the beginning of a picture group have a higher ISP's than P-pictures at the end of a picture group. To map the different P-picture ISP's to PP's three remaining stages between the highest and lowest ISP (1..3) can be used. If there are no more stages for long picture groups, the P-pictures should be mapped in pairs or triples to the ISP[6]. This transformation function implies that unimportant packets of MPEG streams with higher SP are assigned a lower PP than important packets from a stream with a lower SP. Assuming there is another MPEG stream with SP=5, a packet belonging to an I-picture of this stream has a higher PP than a B-picture of a stream with SP=2.

The last example deals with a Motion-JPEG stream with SP=4. The M-JPEG stream is divided into three substreams using a modulo 3 algorithm. Three JPEG-pictures following each other are assigned to three different substreams (J0..J2). Each substream is assigned a characteristic ISP (0..2). As the result, packets of different substreams are transmitted with different PP's. The picture rate of the M-JPEG stream can thereby scaled.

---

6. The length of picture groups is open in the MPEG standard [MPG93]. In practice most MPEG streams contain one to three P-pictures in a picture group.

8 Detection of Lost Packets and Early Packet Discard

Early packet discard in the network layer is very useful to relax resource overloads. Ramanathan et al. [RRV93] propose mechanisms to discard all remaining packets belonging to a corrupted data unit.

The loss of a single packet of a TPDU implies that this picture can never be decompressed to the original TPDU at the receiver side. Transmission of all packets belonging to corrupted TPDUs are suppressed. Additionally, Table 4 has shown an example MPEG stream in which the loss of pictures inhibits the decompression of other pictures. All pictures depending on the corrupted picture can be dropped.

Losses of packets can be detected by comparing sequential packets numbers. The early discard scheme uses a specific bit field in the packet header called "Start Bit" (SB). This bit is zero for all starting packets of a TPDU. Figure 7 is showing an example MPEG stream containing packets with packet priority and start bit[7]. If a packet loss

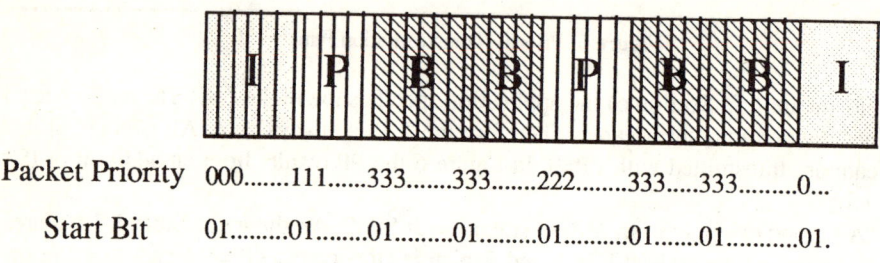

Packet Priority   000......111......333......333......222......333......333..........0...
Start Bit   01.........01.........01..........01..........01.........01..........01.

Figure 7: Example for the early packet discards mechanism

is detected, all packets will be dropped until a packet with the same or a higher priority than the lost packet and with SB=0 will be received. For example, if the second packet of the first P-picture (PP=1; SB=1) has been lost, all packets up to the first packet of the next I-picture (PP=0; SB=0) will be discarded.

9 Results

In this chapter a detailed analysis of streams being transformed using the priority transformation function will be presented. Different MPEG traffic sources are analyzed, and the effects of the priority transformation function and network layer scaling are shown.

The first MPEG stream is an advertisement video clip with many cuts and a high grade of motion. The video stream has an average throughput of 170 kBit/s with a picture resolution of 120*160 pixel. Every picture group of this stream contains two P-

---

7. The packet priority is derived from the prioritization scheme in Section 6.

pictures (picture group format: IBBPBBPBBI). Applying the prioritization scheme to this stream, all I-pictures are assigned ISP=0, all B-pictures ISP=7, the first P-pictures of a picture group ISP=1, and the second P-picture ISP=2. Figure 8 shows splitting the total bandwidth into the different substreams after the transformation. In each diagram the lowest curve describes the bandwidth of I-pictures (ISP=0), the next curve the sum of I- and P1-picture (ISP=0 and ISP=1). The top curve characterizes the total bandwidth of the MPEG stream.

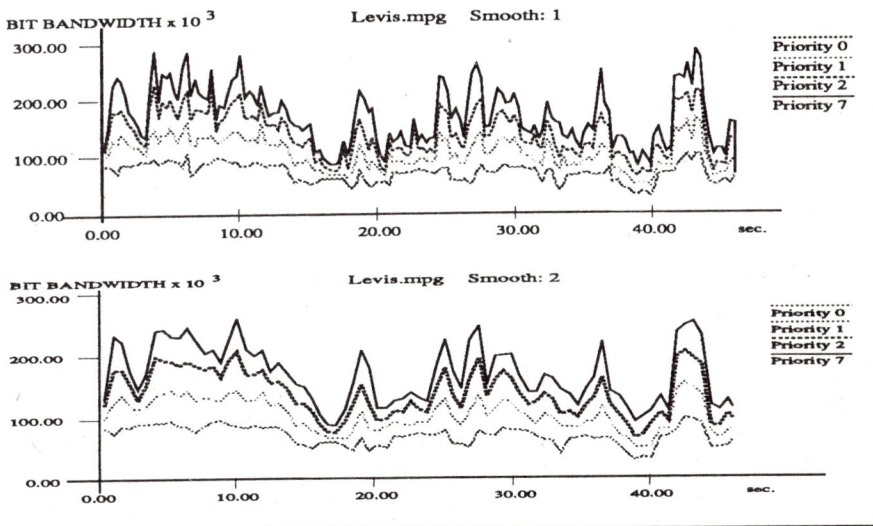

Figure 8: Priority Analysis of Levis.mpg

The upper and the lower diagrams are different because of the smooth factor. In the upper diagram each measurement point is taken at the end of the picture group. This is the smallest unit for throughput measurement of substreams. The second diagram has been smoothed by factor two, this means the throughput has been measured after two groups of pictures. In both cases the high priority substream 0 remains relatively smooth. The reason for this continuity is the independence of I-pictures from the motion parts in the videos. The peak bandwidth of this substream is 110 KBit/s. The peak bandwidth of the whole MPEG stream is 310 KBit/s. Using the presented scaling techniques the bandwidth requests can dynamically be adjusted with a fine granularity down to 30 percent of the peak bandwidth.

Another MPEG stream is shown in Figure 9. The stream is a short scene from "The Simpsons" comic movie. This stream has an average throughput of 710 KBit/s and a resolution of 112*160 pixel. The video stream contains only one P-picture per picture group (picture group format: IBBPBBI). Therefore there is only one substream for the P-pictures (ISP=1). Though in the video there are also cuts and considerable motion, all substreams of this video are very smooth. As already seen in the first example, depending upon congestions the stream bandwidth can be scaled from 840 kBit/s to 250 kBit/s. With scaling mechanisms only low-priority packets are affected at the

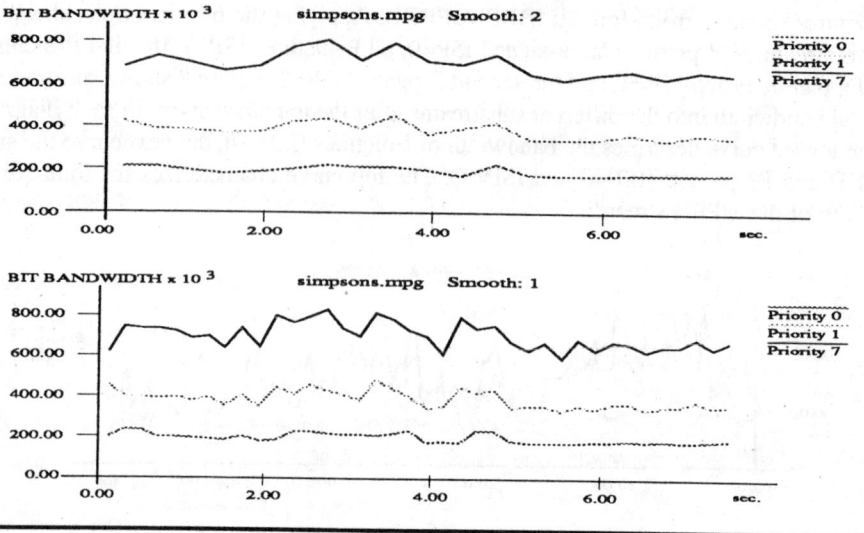

Figure 9: Priority Analysis of simpsons.mpg

beginning of a congestion. Instead of an uncontrolled packet drop which implies the loss of important data the scaling operation affects only low priority parts of the multimedia stream. In many cases this is suitable to smooth temporary peaks of multimedia streams. The main advantage is the well-balanced degree of the QoS of all video streams. It is expected that the loss of single pictures (in case of B-picture loss) in a 30Hz video stream is not noticeable by the human viewer; a 300 ms gap (in case of I-picture loss) cannot, however, be overlooked. A detailed analysis of these topics is underway.

10 Concluding Remarks

A scaling scheme for multimedia streams in the network layer is presented. Network layer scaling enables full adaptive handling of various routes and heterogenous receivers in multi-sender/multi-receiver communication. To resolve resource congestions, packets are not dropped randomly but by specific selection criteria. This reduces damage to a multimedia stream. The mechanisms are specified for video coding techniques like MPEG (I and II) or DVI, but are also useful for audio and intraframe coded video streams like M-JPEG. The key to a feasible scaling mechanism is finding a function that maps different parts of continuous media into logical substreams with different priorities. These prioritization schemes reside in the sender node. Multimedia streams compressed with the MPEG-II codec will offer enhanced capabilities for scaling schemes. In the sender node a flexible prioritization technique takes the temporal, spatial, and frequency scalability of the stream into account and can be used to react to resource congestions in a more intelligent kind. Further work must be done to analyze the effects of transmitting MPEG-II streams over a scaled network. Packet loss is detected in the network layer. To save resources early discard mechanisms drop packets belonging to corrupted and worthless TSDUs as early as possible.

The new scaling mechanisms presented in this paper require a new functionality of the multimedia systems and networks. Monitoring and filtering facilities have to be integrated and protocol mechanisms must be adapted.

Beside the parameters in the scaling message (see Figure 5) there are two other parameters to experiment with:

- Monitoring Interval: Time interval for determination of average loss-rate/delay
- Threshold of loss-rate / delay: Fine tuning of the scaling mechanism

In certain cases it may be useful to allow the service user to change these parameters for an existing connection. CHANGE messages in the network layer protocol can be used to update these scaling parameters. The data packets must contain two header fields to be filled in the sender nodes (see Section 6 and 7):

- Packet Priority
- Start Bit

One important principle of future high-speed networks is minimization of protocol complexity in the network nodes. The presented network scaling scheme is a compromise between the advantages of adaptive network path and receiver handling vs. complexity in network nodes. Priority transformation functions are implemented in the senders of multimedia streams. Monitors for network, CPU, and buffer resources can alternatively be placed in the receiver nodes only, in important gateways, or in all routers. Filters must reside in the network nodes or switches. Further experiments will show suitable network layer scaling topologies.

Further work items are fine tuning of monitoring and scaling mechanisms, e.g., adjustment of monitoring intervals and scaling thresholds. Experiments with different kinds and locations of monitors will follow. An interesting area for network layer scaling is expected in future high-speed networks. In the ATM research community there are many discussions on the prioritization of ATM cells and techniques for admission and congestions control of variable-bit-rate stream traffic over ATM and B-ISDN ([PW93], [YWT93], [SKK93]). The focus of this paper is directed to conventional large networks and internets. However, the network mechanisms appear suitable to be introduced into the ATM layer. In addition to the referenced papers, the network layer scaling technique also proposes prioritization for congestion control with adaptive handling, in the case of multiparty communication.

Acknowledgments

The authors wish to acknowledge the contributions of Luca Delgrossi, Keith Hall, Ralf Guido Herrtwich, Thomas Käppner, Lars Wolf (all from IBM ENC Heidelberg) in discussing and reviewing this paper. Thanks are also due to Peter Renkawitz (University of Mannheim) for supporting the experimentation phase.

References

[BM91] A. Banerjea, B.A. Mah: *The Real-Time Channel Administration Protocol.* Proceedings of 2nd International Workshop on Network and Operating System Support for Digital Audio and Video, Heidelberg, November, 1991.

[CSZ92] D.D. Clark, S. Shenker, L. Zhang: *Supporting Real-Time Applications in an Integrated Services Packet Network: Architecture and Mechanism.* ACM SIGCOMM '92, Baltimore, 1992.

[DHH93] L. Delgrossi, C. Halstrick, D. Hehmann, R.G. Herrtwich, O. Krone, J. Sandvoss, C. Vogt: *Media Scaling in a Multimedia Communication System.* in Proceedings of the First ACM Multimedia Conference, 1993.

[DHH94] L. Delgrossi, R.G. Herrtwich, F.O. Hoffmann: *An Implementation of ST-II for the Heidelberg Transport System.* Internetworking – Research and Experience, Vol. 5, Wiley, 1994

[HSF93] D. Hoffman, M. Speer, G. Fernando: *Network Support for Dynamically Scaled Multimedia Data Streams.* Proceedings 4th International Workshop on Network and Operating System Support for Digital Audio and Video, Lancaster, 1993.

[JST92] K. Jeffay, D.L. Stone, T. Talley, F.D. Smith: *Adaptive Best-Effort Delivery of Digital Audio and Video Across Packet-Switched Networks.* Proceedings 3rd International Workshop on Network and Operating System Support for Digital Audio and Video, San Diego, 1992.

[MPG93] ISO/IEC JTC1/SC29/WG11: *MPEG-2 Systems Working Draft Number 0531.* September 1993.

[PPA93] J.C. Pasquale, G.C. Polyzos, E.W. Anderson, V.P. Kompella: *Filter Propagation in Dissemination Trees: Trading Off Bandwidth and Processing in Continuous Media Networks.* Proceedings of the 4th International Workshop on Network and Operating System Support for Digital Audio and Video, Lancaster, 1993.

[PW93] D.W. Petr, J.S. Wineinger: *End-to-End Priority Cell Discarding Analysis for ATM Networks.* IEEE Infocom, 1993.

[RRV93] S. Ramanathan, P.V. Rangan, H.M. Vin: *Frame-Induced Packet Discarding: An Efficient Strategy for Video Networking.* Proceedings of the 4th International Workshop on Network and Operating System Support for Digital Audio and Video, Lancaster, 1993.

[SCZ93] S. Shenker, D.D. Clark, L. Zhang: *A Service Model for an Integrated Services Internet.* Internet-Draft, October 1993.

[SKK93] H. Saran, S. Keshav, C.R. Kalmanek: *A Scheduling Discipline and Admission Control Policy for Xunet 2.* Proceedings of the 4th International Workshop on Network and Operating System Support for Digital Audio and Video, Lancaster, 1993.

[TTC92] H. Tokuda, Y. Tobe, S.T.-C. Chou, J.M.F. Moura: *Continuous Media Communication with Dynamic QOS Control Using ARTS with an FDDI Network.* ACM SIGCOMM 92, Baltimore, 1992.

[Upp92]) P. Uppaluru: *Networking Digital Video.* 37th IEEE COMPCON, 1992.

[VHN93]   C. Vogt, R.G. Herrtwich, R. Nagarajan: *HeiRAT: The Heidelberg Resource Administration Technique, Design Philosophy and Goals.* Kommunikation in verteilten Systemen, Munich, 1993 (also published as IBM Tech. Rep. No. 43.9213, 1992).

[WHD94]   L. C. Wolf, R.G. Herrtwich, L. Delgrossi: *Filter for Reservation-Based Internetworks.* submitted to ACM Multimedia, 1994.

[YWT93]   K. Yamazaki, M. Wada, Y. Takishima, Y. Wakahara: *ATM Networking and Video-Coding Techniques for QOS Control in B-ISDN.* IEEE Transactions on Circuits and Systems for Video Technology, Vol.3 No. 3, June 1993.

Resource Requirements for VBR Mpeg Traffic in Interactive Applications

Maher Hamdi
Pierre Rolin
Yvon Duboc
Marc Ferry

ENST de Bretagne, Departement Réseaux et Services Multimédias
BP 78 35512 Cesson Sévigné, France
Email: {Maher.Hamdi, Pierre.Rolin}@enst-bretagne.fr
Tel: (33) 99 12 70 23

Abstract

Resource Allocation for variable bit rate video traffic is still a crucial problem due to the complex traffic characteristics of compressed video. This paper presents the results of a traffic study showing the effect of video contents on the bandwidth needed by interactive multimedia applications to transfer video in a loss-free context. Mpeg compression has been applied to four movies with different characteristics. Useful statistics on the VBR traffic are presented. Probability density functions are drawn and discussed. The network bandwidth needed for each sequence is drawn as a function of the application interactivity. Our main results show that for moderately interactive applications, the real time property costs very low while it is dramatically expensive for highly interactive applications. We also prove that the bandwidth required by a video sequence depends strongly on the scene contents and vary with a ratio up to 3.5.

Keywords: *VBR video traffic, Mpeg, Interactive applications, Network bandwidth and memory, Jitter.*

1 Introduction

The emergence of video compression techniques has stimulated the introduction of video facilities in computer networks and the rise of new multimedia applications. The quality of service (QoS) required by multimedia applications may need some performance guarantees at the transport layer of the underlying network. To guarantee the required quality, effective resource allocation schemes are necessary and have to be performed in the network as well as in the application context. Such mechanisms are particularly critical when variable bit rate (VBR) traffic such as digital video coding has to be considered. One major problem is the need of some prior knowledge of the VBR traffic. It allows the network to predict the connection behaviour in order to prevent congestion. The mechanisms proposed in [5] [8] [9] are based on VBR traffic characterization using analytical traffic models. For large scale broadcasting video applications, specific mechanisms such as traffic superposition can be used taking benefits of the statistical properties of the large number of video streams being transported together by the network. The aim of our study is limited to video streams that are likely to be handled by multimedia software in conventional workstations and a point to point architecture so that no superposition technique can be performed. In this case, users need to estimate bandwidth and memory requirements for a single video stream according to the desired QoS.

The resources needed by a coded video stream are estimated according to the VBR traffic characteristics which depend themselves on two main factors: the coding algorithm and the scene contents. In this study, we use Mpeg [3] as the compression algorithm. In [10] the authors studied the effect of the coding parameters on traffic characteristics using one video sequence and a layered Mpeg coding scheme. However we believe that scene contents have a great influence on the traffic characteristics and resource reservation. Our work aims to the study of the effect of the movie type (scene contents) on the generated traffic. This gives insights to the user on how far the network conditions are affected by the kind of the video sequence. In order to cover different types of video, four sequences (10 min. long each) have been selected for this study.

Bandwidth and memory required by a VBR video stream depends on the QoS of the related multimedia application. Multimedia conferencing systems need some real time properties to be guaranteed by the network because of their interactivity. In this paper, we study the effect of delay restrictions on the bandwidth needed in a loss-free context. We use a single stage limited capacity multiplexer to model network resources. The delay/bandwidth trade-off in the case of VBR video traffic is studied with real measures; the greediness, in terms of bandwidth, of real time video applications is highlighted.

The paper is organized as follows: section 2 presents the the four sequences' characteristics as well as the coding parameters used for Mpeg compression. Results presented in section 3 include the frame by frame bit rate, probability density functions and some elementary statistics on the resulting VBR traffic. In section 4 we investigate the bandwidth required by each sequence according to the delay limitation. Our conclusions are presented in section 5.

2 Video Sources and Encoders

2.1 Video Sources

In this section, we focus on the description of the video sequences and the coding algorithm. Since multimedia applications are not restricted to any specific type of video, four different sequences are considered in our study corresponding to four different types of video. Many studies have been done in this field but most of them have concentrated on one type of video sequences [5] [8] [9] [10] [12]. Conference type sequences were used in [5] [8] [9] [12]. In [11], four general video sequences have been compressed and concatenated totalling 6000 frames used to perform simulations. In this study we chose four typed video sequences with different subjective characteristics. We fixed the duration of each sequence to 11 minutes in order to have realistic results for bandwidth dimensionning and significant statistics and density functions. The first sequence, *Video Clip* is a commercial music video clip that shows a strong image complexity, a wide range of colours, rapid scene changes, scene cuts, camera zooms, pans and almost no fixed plans. The second sequence *The Spitting Images* is a TV program (puppet-show) where a group of marionettes are actively moving within a moderately changing scene. The third sequence, *Mayotte* is a kind of TV reporting showing natural sceneries of Mayotte's island, images have natural colour structure with slow motion and no frequent scene change. The last sequence *Pierre's Conference* is a teaching session showing the whole body of a teacher speaking with few gestures of hands and shoulders in a fixed plan. This sequence represents conference type video. Table 1 summarizes the subjective characteristics of the four sequences. Images are captured using a Parallax Board Card on a SUN Station 10, this equipment allows a capture rate of about 15 images per second with a resolution of 384x288 (near the CIF format) and 24 bits per pixel coding. The image resolution is suitable for workstation based multimedia applications. Note that these characteristics of the video sequences do not apply to broadcasted TV quality or other specialized video equipments (VCR..) because of the limited temporal and spacial resolutions. The 10,000 images generated by each sequence are then converted to the YUV 4:1:1 format before being compressed.

Criteria	Video Clip	Spitting Images	Mayotte	Pierre's Conference
Scene Change Frequency	High	Medium	Low	Low
Similarity between successive plans	Low	Medium	Medium	Very High
Image Quality	Good	Very Good	Good	Medium
Color Complexity	High	Medium	Medium	Low

Table 1. Subjective characteristics of the four video sequences

2.2 VBR Mpeg Coding

The compression algorithm used in this study is the one specified by the joint international commitee ISO-IEC/JTC1/SC2/WG8, known as the Mpeg-1 algorithm [3] [7]. It was defined for video streams having an average rate of about 0.9 to 1.5 Mb/s.We believe that this algorithm will be frequently used by LAN based multimedia services since it was designed to support a large number of applications [13]. Some software realizations are already in use within the Internet public domain and allow to playback stored compressed video. Hardware implementations are needed however to build interactive video applications.

Mpeg video compression allows temporal redundancy to be exploited via motion compensation which is one of several techniques of inter-frame coding. Spatial redundancy is reduced by the Discrete Cosine Transform technique combined with a visually weighted adaptive quantization (see [3] for details). When an image is coded without reference to other images, it is called intra-frame coding. With Mpeg compression, all images are not coded identically. The input video is divided into groups of pictures (GoP) starting with an intra-coded image (called image I for Intra) followed by and arrangement of images of type P (Predictive) or B (Bidirectional) which are compressed using inter-frame coding. B images are coded with reference to a past image and a future image of type I or P. In principle, the best compression ratio is achieved for B frames. The quantization step is controlled by a parameter called the factor of quantization which acts on the compression ratio at the cost of some image quality degradation. It can be dynamically changed to perform rate control on the output bit stream. These encoding parameters (GoP length, factor of quantization,..) have to be fixed for the Mpeg encoder depending on image quality requirements as well as average bit rate restrictions caused by the capacity limitation of the transport system. They are also responsible for an extra-delay due to inter-frame coding (image buffering into the encoder and the decoder's memory), this delay has to be considered according to the application delay restrictions.

Parameters	Value
Image Resolution	384x288
Quantization Factor	8
Frames between I pictures	5
Frames between P pictures	2
Frame sequence as to be displayed	...IBBPBBI...
Rate Control	None

Table 2. Mpeg encoding parameters

Since we are not interested in studying the effect of encoder parameters on the generated traffic (this has been studied in [10] [11]) the four sequences have been compressed using only one parameters configuration. (see table 2.) This configuration could be considered as non-optimal for the coding of some sequences but it allows a real comparison of the generated traffic according to a single parameter: the movies contents.

The software implementation of Mpeg-I encoder used is the one developed by the Portable Video Research Group at Stanford University [6].

3 Density Functions and Statistics

3.1 Video Traffic Modelling

Video traffic are characterized by a periodic packet generation and variable packet lengths. The packet size is equal to the number of bits in the corresponding coded image. Let's denote by P_i the length of the ith packet and by p the generation period. In our case p is equal to 66 ms (15 images/s). We consider a slotted time axis where p is the duration of the time slot. We assume packet P_i is entirely generated during slot $i-1$ and transmitted during slot i. This assumes at least one frame buffering in the coder which is reasonable when using the Mpeg coder. We define the instantaneous rate of the video stream as the following time dependant function:

$$r(t) = P_i/p \text{ with } ip < t < (i+1)p \qquad \text{(Eq. 1)}$$

This definition allows us to consider the pure VBR traffic without any disturbance resulting from multiplexing or queuing. Note that from a practical point of view, equation 1 assumes that packet transmission is smoothed over one frame period to avoid very high instantaneous rates. Heeke [4] found that such smoothing mechanisms increase statistical multiplexing gain.

One major problem in video traffic studies is to approximate the $r(t)$ function by an analytical model that allows queuing systems resolution or numerical simulations. Several models have been proposed in the recent literature especially in the context of congestion control for ATM based networks. In [1] the authors show that most of these models are unsatisfactory and give references to some of them. Our approach is to provide results and measurements based on real video traffic so that accurate conclusions can be drawn.

3.2 Results

For each sequence, the $\{P_i\}$ series are measured at the coder's output. The $r(t)$ functions are easily calculated using equation 1 and are plotted in Fig. 1. Table 3 presents the statistics performed on the $r(t)$ functions.

Sequence	Average Rate (Mb/s)	Min Rate (Mb/s)	Peak Rate (Mb/s)	Variance (Mb/s)^2	Standard Deviation (Mb/s)	Squared Coefficient of Variation	1st order Auto-Correlation	Burstiness
Video Clip	1.106	0.143	5.033	0.422	0.649	0.3444	0.3360	4.55
Spitting Images	1.026	0.253	3.072	0.368	0.606	0.3496	-0.2482	2.99
Mayotte	0.505	0.198	1.597	0.087	0.295	0.3421	-0.3150	3.16
Pierre's Conference	0.581	0.245	1.468	0.168	0.410	0.4991	-0.4194	2.52

Table 3. Elementary statistics on Mpeg frame length

Average rate is defined as the average of $r(t)$ over the 10,000 time slots.
Peak rate is defined as the highest value reached by $r(t)$. This value is generated by the largest frame after compression.
Min rate is the lowest value of $r(t)$.
Burstiness is defined as the peak rate to mean rate ratio.

From table 3 we see that the average rate caused by *Video Clip* is twice higher than the one caused by *Mayotte*. This shows that the effect of the movie contents on the efficiency of true VBR coding algorithms (without rate control) is by no means small. *Mayotte* has the lowest average rate, followed by *Pierre's Conference*. This argues the fact that low rates can be generated by different types of movies and not exclusively by video conference sequences. Results found in [5] [8] and many other papers which were relative to conference type video may probably be used for slow moving high redundant sequences such as *Mayotte*. The *Video Clip* sequence gives the highest average rate caused by the complexity of its images and the very low gain performed by the inter-frame coding.

Peak rates vary from 1.5 to 5 Mb/s among the four sequences. Obviously these values could be reduced if the transmission were smoothed over more than one frame time but this introduces additional delays in buffering. Hard real time video applications cannot suffer such delays and are forced to transmit data at their peak rate. In this case the application should not use the same bandwidth when it switches between sequences of different types. The trade-off delay/bandwidth is studied in the next section.

For the squared coefficient of variation, the value 0.34 is common to almost all sequences so that one can consider that it is not dependant on the video contents. The sequence *Pierre's Conference* has the highest value in spite of the regular form of its instantaneous bit rate as shown in Fig. 1.

The instantaneous bit rate $r(t)$ is plotted in Fig. 1 as a function of time. *Pierre's Conference* has the most regular shape. One can easily notice three horizontal strips corresponding to the I, P and B images generated by Mpeg coding. This shows the efficiency of the inter-frame coding since the bit rate generated by B frames is lower than the one generated by I and P frames. This separation can be clearly seen in *Mayotte* and *Spitting Images*, but is not perceptible in the case of *Video Clip* whose traffic seems to be strongly bursty.

The distribution of the compressed images size for the four sequences is presented in Fig. 2. *Pierre's Conference* curve has a special interest as one can appreciate the properties of the Mpeg coding algorithm. The three separated peaks relative to the Mpeg frame types (I,P and B) show the efficiency of the motion compensation technique when performed on a sequence with very high temporal redundancy. One can modelize such a traffic by superposing three constant bit rates with different arrival periods. In [5] the authors characterized a video tele-conference sequence by a gamma distribution and raised the question of the coding algorithm effect on the frame length distribu-

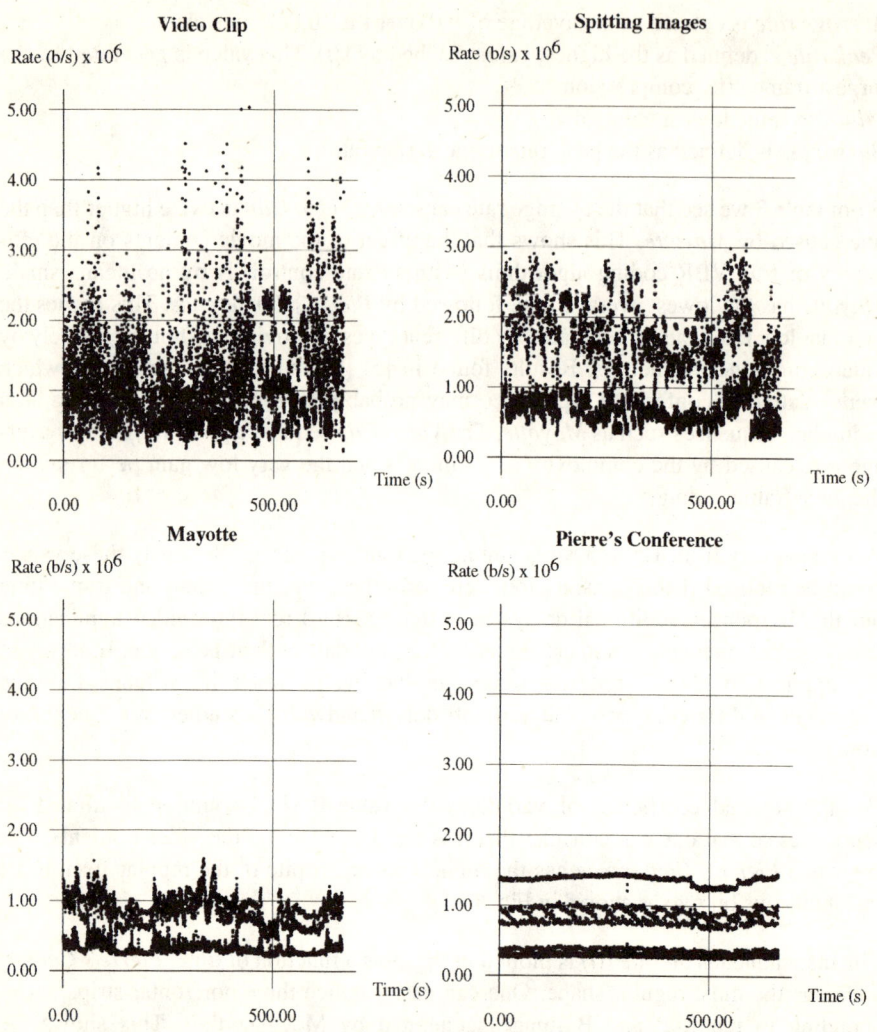

Fig. 1. Instantaneous bit rate

tion. From Fig. 2 we see however that when both intra and inter-frame coding are used, gamma distribution (or any other bell-shaped distribution) do not fit conference type compressed video.

In *Mayotte*'s density function we can easily recognize the peak relative to B frames whereas P and I frames are represented by a bell-shaped peak. The lack of the P frames peak results from the simple image structure which provides a high spacial redundancy. *Mayotte* and *Pierre's Conference* have the same range of frame lengths (up to 100 kbits). In *Spitting Images* the I frame length reaches 200 kbits and can go up to 300 kbits/s in the *Video Clip* sequence. It appears clearly that the latter has the lowest

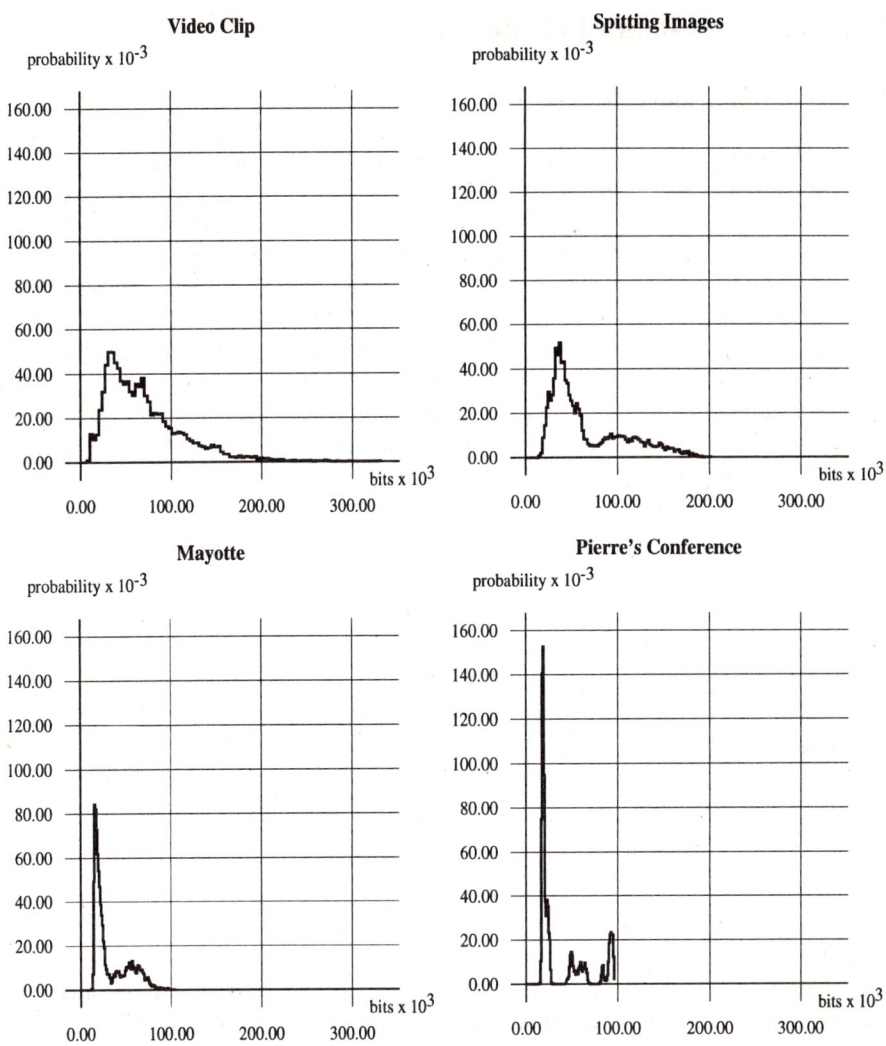

Fig. 2. Distribution of frame length

temporal redundancy. The density functions of these four sequences plotted in Fig. 2 show that VBR traffic characterization is very dependant on the video sequence studied as well as on the coding algorithm. A combined classification according to the scene contents and to compression techniques seems to be necessary to provide reliable video traffic models. The curves shown in Fig. 2 and the statistics in table 3 are useful in video traffic characterization studies.

4 Memory and Bandwidth Requirements

4.1 Video Applications Interactivity

This section considers distributed video applications built according to the client/server model. The server provides video images to the client which displays them on the screen. The server can be a video camera combined with a video coder, or simply a storage device where movies are already compressed and stored. A network is used to transfer video data between the server and the client. Video-conference and video-on-demand are examples of distributed multimedia applications handling video. The QoS of such applications is expressed by several performance parameters like image loss probability, image quality, network and decoding delays... In our study, we are not interested in coding and decoding parameters, we focus our attention on network related QoS parameters. The latter depends on the network resources and the properties of the carried traffic. In sub-section 4.2 we study the relation between network resources, expressed in terms of buffer and bandwidth, and QoS parameters, expressed as the application interactivity defined hereafter.

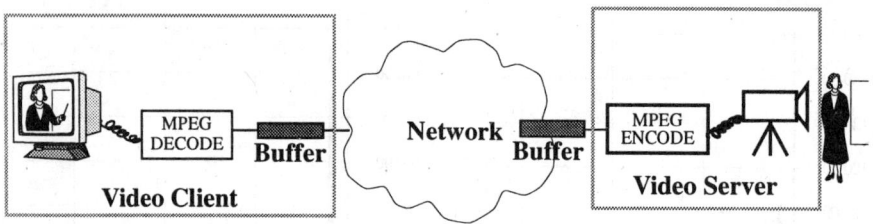

Fig. 3. A distributed video application

Video traffic can be transferred in different ways depending on the application QoS requirements. Short non-interactive video sequences (where the network delay is not important) can be transported by classical file transfer services. In this case no delay bounds are guaranteed by the network, re-transmission mechanisms can be used so that the data is delivered error-free. In the case of delay-sensitive applications, the lag between the video client and the video server has to be fixed at the connection establishment and guaranteed by the network using adequate resource reservation algorithms.

For instance, in video-on-demand applications, the client sends a request to the server asking for the transmission of a certain movie and has to wait for the network response. In this case the application is said to be interactive. Video-conference applications are also interactive since they allow real time conversations between users. We define the interactivity of multimedia applications as the requirement of a bounded response time between the client and the server such as defined above.

In general, buffers are needed at the two ends of the video connection as shown in Fig. 3. At the server side (the sender), a multiplexer is needed for buffering frames at the network interface. Since images experience variable delays in the network because of their variable length and/or the sharing of the network with other traffics, a buffer is needed at the client side (the receiver) and has to be filled before the decoder starts consuming the compressed images. This buffer has to be properly dimensioned to ensure that the decoding process will not be broken. In other terms, this buffer is necessary to absorb the jitter introduced by network delay variation. Delays experienced by video traffic are those introduced by the two buffers in addition to the propagation delay. The latter does not depend on the network bandwidth nor on the traffic properties, thus it is not taken into consideration for the remainder of this study. We define the response time of an interactive video application as the delay introduced by the two buffers of Fig. 3. The response time is directly related to the network bandwidth and the traffic variable rate and can be considered as a measure of the application interactivity. In the next sub-section we analyse the relation between response time and network bandwidth using the traces of the four compressed sequences presented in section 3.

4.2 Buffer Analysis and Results

We consider that video data are not multiplexed with other kinds of traffic. The study is limited to a one stage multiplexing modelling the network access. A network channel with constant capacity is allocated to the video stream, this can be considered as a virtual circuit in an ATM network interface, a dedicated router or N-ISDN connections...

At the server side

The queuing system under consideration is almost[1] D/G/1/K. The model is illustrated in Fig. 4.

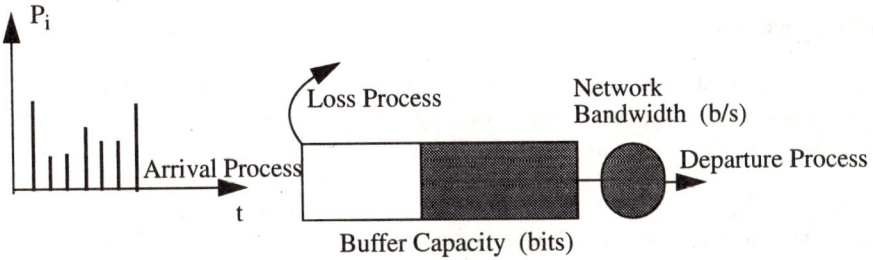

Fig. 4. Network resources: memory and bandwidth

---

1. The only difference is that in the true D/G/1/K, K means the system capacity in clients whereas in our case it is seen as a memory capacity in bits.

Let c denote the server capacity in bps and M the buffer capacity in bits. Let $\{X_i\}$ be the time series, defined at the begining of each time slot, where X_i denotes the number of bits in the buffer just after the input of the entering packet of size P_i. The $\{X_i\}$ process is built as follows:

$$X_{i+1} = \min(M, P_{i+1} + \max(0, X_i - cp)) \quad \text{with } X_0 = P_0 \quad \text{(Eq. 2)}$$

The statistical analysis of $\{X_i\}$ allows the computation of two important QoS components: the frame delay and the frame loss processes. Despite the fact that most multimedia applications dealing with video are tolerant to some loss of images, results presented herein are computed on a basis of no frame loss. These results have to be considered as an upper bound function of the resources required with respect to the model of Fig. 4. Under this assumption, equation 2 can be written as:

$$X_{i+1} = P_{i+1} + \max(0, X_i - cp) \quad \text{with } X_0 = P_0 \quad \text{(Eq. 3)}$$

The buffer memory needed (say M) to avoid frame loss is then:

$$M = \text{maximum}(X_i)_{\text{over i}} \quad \text{(Eq. 4)}$$

M is dependant on c and the $\{P_i\}$ series and represents the size in bits of the network buffer that has to be reserved to avoid overflow.

We denote by $\{l_i\}$ the process of departure instants from the multiplexer and by $\{t_i\}$ the process of arrival instants. We have:

$$t_{i+1} = t_i + p \quad \text{and} \quad l_i = t_i + X_i/c \quad \text{with } t_0 = 0 \quad \text{(Eq. 5)}$$

The delay experienced by video packets in the network interface buffer is bounded by:

$$D = M / c \quad \text{(Eq. 6)}$$

At the client side

The decoding system consumes one compressed frame per time unit p. Since the $\{l_i\}$ process is not a periodic one, some buffering is necessary to ensure that the decoding process will not be broken as it was highlighted in sub-section 4.1.

Let's denote by k_i the instant at which the decoder begins consuming (decoding) the ith frame. We have: $k_{i+1} = k_i + p$. As the propagation delay is neglected, the condition of the presence of each frame in the decoder buffer before its decoding starts is expressed as:

$$\text{maximum}(l_i - k_i)_{\text{over i}} = 0$$

using equations 4, 5 and 6 we obtain:

$$k_0 = D \quad \text{(Eq. 7)}$$

The decoding process should begin D seconds after the first frame has been encoded and put into the network buffer so that we ensure that all frames will be present in the decoder's buffer before their decoding time. In other terms, D is the buffering time needed (at the client side) to absorb the jitter introduced by the variable service time of frames due to their variable length. From equation 7, it results that D is the upper bound of the delay introduced by the client buffer. However, the total delay for a video packet in the two buffers is exactly equal to D since we can write $k_i\text{-}t_i = D$ (for each i) as a direct result of equation 7. The receiver play-point is D seconds later than the sender play point. According to the definitions given in sub-section 4.1, D is a measure of the connection interactivity.

From equation 6 it seems clear that the interactivity property is in a trade-off with the

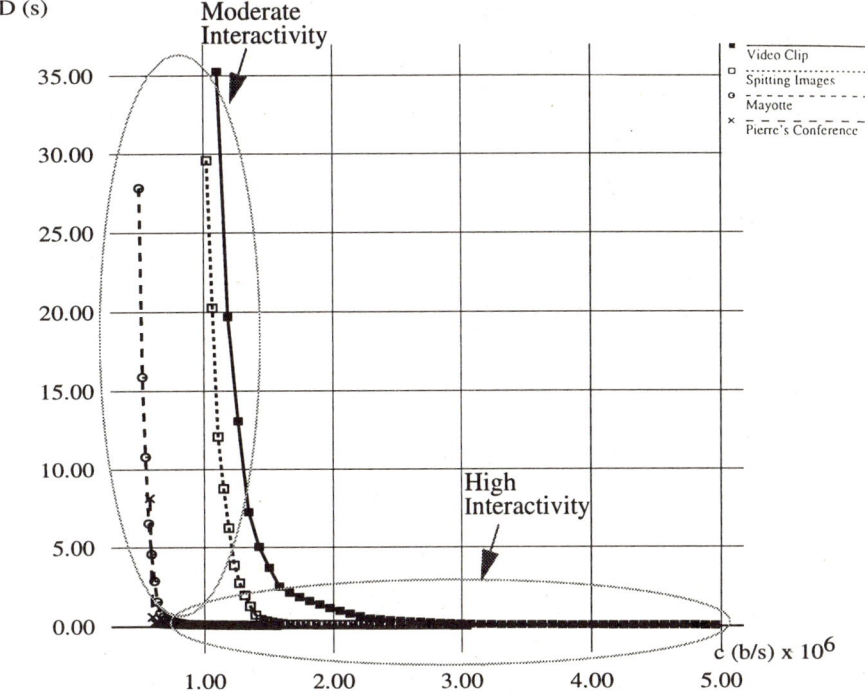

Fig. 5. Delay vs bandwidth

bandwidth reserved by the network. In order to study this compromise, we have drawn D as a function of the bandwidth c using numerical computations of M given by Eqs. 3 and 4. Curves are shown in Fig. 4. The first observation is that if the application tolerates a latencys of few seconds, the bandwidth required is about 0.5 Mb/s for *Mayotte* and *Pierre's Conference*, and about 1.5 Mb/s for *Spitting Images* and *Video Clip*. These values are close to the average bit rate of each sequence. On the other hand, if the delay must be around the frame period (high interactivity), the bandwidth required is close to the peak rate. The vertical part of the curves (corresponding to moderately interactive applications) shows that we can drastically improve the response time of the application by slightly increasing the network bandwidth. For instance, *Mayotte* and *Spitting Images* require 25 kb/s (70 kb/s for *Video Clip*) more bandwidth to improve their response time by 5s. Moderately interactive applications are then encouraged to reduce their response time since it costs very low in terms of network bandwidth. On the other hand, for highly interactive applications (response time less than one second), the bandwidth grows dramatically as the response time draws near the frame period. In Fig. 6 (which is the same curve of Fig. 5 plotted with logarithmic scale for Y axis) we can see that to improve their response time by 30 ms, *Mayotte* (respectively *Spitting Images* and *Video Clip*) needs 0.5 Mb/s more bandwidth (respectively 1 Mb/s and 1.5 Mb/s). The real-time property cost is very expensive for highly interactive video applications. Remind that these conclusions are relative to both the VBR traffic properties which characterize the open loop Mpeg coding and the QoS assumptions we made in sub-section 4.1 (loss-free and jitter cancellation).

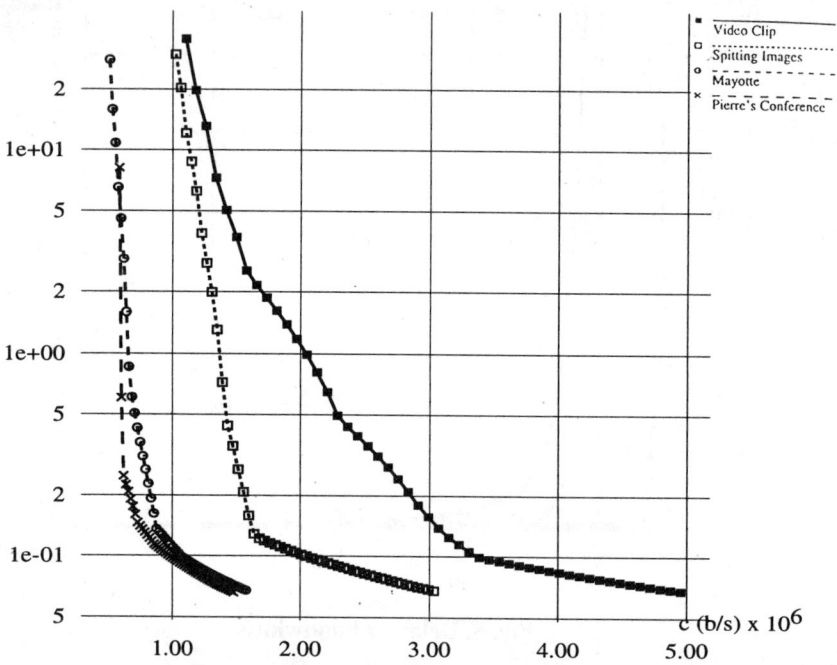

Fig. 6. Delay vs bandwidth -y axis in log scale-

The next observation is about video contents effects on network bandwidth. Let $C_{Video\ Clip}(D)$ (resp. $C_{Mayotte}(D)$) denotes the inverse of the function plotted in Fig. 6 which is relative to the *Video Clip* sequence (resp. *Mayotte* sequence). As a measure of the effect of video contents on the network resources we define the following ratio:

$$E(D) = \frac{C_{VideoClip}(D)}{C_{Mayotte}(D)}$$

$E(D)$ is calculated using the curves of Fig. 6 and is plotted in Fig. 7 which shows that $E(D)$ varies from 2.3 to 3.4 depending on the response time of the video application. These values show that video contents, if ignored, can have great effects on the efficiency of resource reservation mechanisms. If we consider applications such as video-on-demand, switching between sequences of different types may require an extra-bandwidth reservation or release unless a worst case policy is applied which leads to an inefficient resource utilisation. In the ATM context, resource management protocols such as FRP/DT developed at Cnet Lannion [2] seem to be attractive for interactive video applications which handle different kinds of movies.

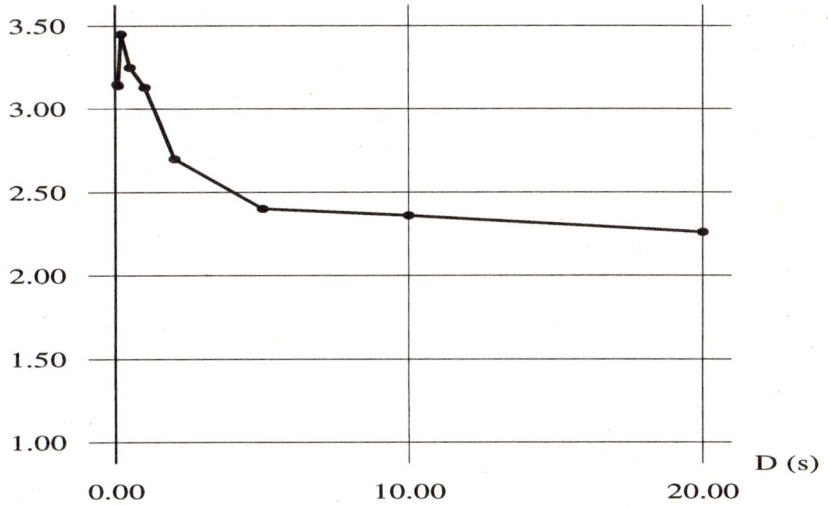

Fig. 7. Movie contents effect on bandwidth

5 Conclusion

Variable bit rate traffic generated by multimedia applications is still a crucial problem for network resource management. Adequate reservation mechanisms have to be performed by the network to provide guarantees for the quality of service required by the application. In this paper, we studied memory/bandwidth requirements for Mpeg traffic in interactive multimedia applications. Four video sequences of different kinds have been digitized and compressed using the Mpeg coder. Instantaneous bit rate and densities have been plotted and other statistics have been presented and discussed. One of our observations was that the efficiency of the inter-frame coding is very relative to the scene contents and has great effects on the resulting traffic shape. Another observation was that models presented in the literature for video-conference traffic and based on a bell shaped density function seem not to be adequate when both intra and interframe coding are performed. A combined classification according to the scene contents and to compression techniques seems to be necessary to provide reliable video traffic models.

The second part of this paper dealt with the network resources needed by Mpeg VBR traffic when the application requires bounded delays. A single stage multiplexing was studied using real traces of the four compressed movies. A particular attention was paid to the jitter caused by the variable length of video packets and its effects on the network bandwidth and memory. Curves showing the bandwidth/delay trade-off have been plotted. Our conclusion is that moderately interactive applications can easily improve their response time by slightly increasing their bandwidth. For some sequences, 5% (500 to 525 kb/s) more bandwidth saves up to 30% of the end to end response time (from 15 to 10 seconds). On the other hand, for highly interactive applications (delay less than one second), the bandwidth grows dramatically as the response time draws near the frame period. In this case, up to 50% (1 to 1.5 Mb/s) more bandwidth is required to improve by 30% the response time (from 100 to 70 ms). Another observation concerning the scene contents is that movies with complex structure and fast moving pictures can require up to 3.5 times the bandwidth required by another conference typed video sequence.

Acknowledgements

The authors would like to thank Mr. H. Afifi and Miss D. Bourges for their helpful comments.

References

[1]   J. Beran, R. Sherman, M. S. Taqqu, and W. Willinger. Variable-Bit-Rate Video traffic and Long-Range Dependence. *Accepted for publication in IEEE Transactions on Communications*, 1992.

[2]   P. E. Boyer and D. P. Tranchier. A reservation principle with applications to the ATM traffic control. *Computer Networks and ISDN Systems*, 24:321–334, 1992.

[3]   D. Le Gall. Mpeg: A Video Compression Standard for Multimedia Applications. *Communications of the ACM*, 4(34):305–313, April 1991.

[4]   H. Heeke. Statistical Multiplexing Gain for Variable Bit Rate Video Codecs in ATM Networks. *International Journal of Digital and Analog Communication Systems*, 4:261–268, 1991.

[5]   D. P. Heyman, A. Tabatabai, and T. V. Lakshman. Statistical Analysis and Simulation Study of Video Teleconference Traffic in ATM Networks. *IEEE Transactions on Circuits and Systems for Video Technology*, 2(1):49–59, March 1992.

[6] A. C. Hung. The PVRG-Mpeg Software Codec v1.0. *Stanford University, Available by anonymous ftp from havefun.stanford.edu:pub/mpeg*, 1991.

[7] International Organization for Standardization ISO Mpeg. Coding of Moving Pictures and Associated Audio for Digital Storage Media at up to 1.5 Mbits/s. November 1991.

[8] B. Maglaris, D. Anastassiou, P. Sen, G. Karlsson, and J. D. Robbins. Performance Models of Statistical Multiplexing in Packet Video Communications. *IEEE Transactions on Communications*, 36(7):834–844, July 1988.

[9] M. Nomura, T. Fujii, and N. Ohta. Basic Characteristics of Variable Rate Video Coding in ATM Environment. *IEEE Journal on Selected Areas in Communications*, 7(5):752–760, June 1989.

[10] P. Pancha and M. El Zarki. Bandwidth Requirements of Variable Bit Rate Mpeg Sources in ATM Networks. In *INFOCOM'93*, pages 902–909. IEEE Computer Society Press Los Alamitos, California, March 1993.

[11] D. Reininger, D. Raychaudhuri, B. Melamed, B. Sengupta, and J. Hill. Statistical Multiplexing of VBR Mpeg Compressed Video on ATM Networks. In *INFOCOM'93*, pages 919–926. IEEE Computer Society Press Los Alamitos, California, March 1993.

[12] P. Sen, B. Maglaris, N.-E. Rikli, and D. Anastassiou. Models for Packet Switching of Variable-Bit-Rate Video Sources. *IEEE Journal on Selected Areas in Communications*, 7(5):865–869, June 1989.

[13] W. Tawbi, F. Horn, E. Horlait, and J.B. Stéfani. Video Compression Standards and Quality of Service. *The Computer Journal*, 36(1):43–54, 1993.

Transmission of MPEG2 Applications Over ATM Networks

Teresa Andrade[1] and A. Pimenta Alves[2]

[1] JNICT/INESC
[2] FEUP/INESC, Largo Mompilher 22, Apartado 4433,
4007 Porto Codex, Portugal

Abstract. ATM, the Asynchronous Transfer Mode cell-based switching technique, was finally recognized in 1992 by the relevant International standardization entities, as the switching and multiplexing solution for implementing the future Broadband Integrated Services Digital Network (B-ISDN). Because of its service-independent characteristics, ATM will provide B-ISDN with capabilities for integrated support of a wide range of services in a very flexible and efficient way. Transmission of multimedia services in the B-ISDN is currently being addressed by several Study Groups of ITU-T and ETSI. In addition to identifying the required network capabilities and to defining the functional model of multimedia services in the B-ISDN, the possible alternatives for performing the multiplexing operation of the different components of a multimedia application are also being evaluated. The MPEG2 standard (ISO/IEC 13818), which is now in its final phase of specification, is being accepted worldwide as the appropriate coding technique to implement almost all types of video applications and multimedia services involving good-quality full-motion video. It seems therefore important to evaluate the alternatives for performing the multiplexing operation of the elementary components of an MPEG2-based multimedia application over ATM networks. In fact, different Study Groups of both ITU-T and ETSI, as well as the ATM Forum, are now evaluating alternatives for mapping MPEG2 streams into ATM cells.

1 Overview of the MPEG2 Standard

ISO and IEC in collaboration with ITU-T and ITU-R, have created joint working groups with the purpose of specifying universal standards in several communications domains, including audio-visual digital communications. One of such groups is the ISO/IEC JTC1/SC29/WG11 known as the Moving Pictures Experts Group (MPEG), which was established in 1988 with the purpose of defining a coding algorithm suitable for storage and retrieval of moving pictures and associated audio in Digital Storage Media (DSM) with bit rates not exceeding 1.5 Mbits/s.

The first draft proposal for this ISO standard (MPEG-1) was produced in September 1990. The work of MPEG1 reached the status of International Standard in 1992 under the designation of "IS-11172, Coding of Moving Pictures and Associated Audio for Digital Storage Media at about 1.5 Mbits/s", universally known as the MPEG1 Standard. In 1991 MPEG started a second phase of work with the scope of developing a standard with a wider range of applicability rather than just storage and retrieval in DSM, offering much higher picture resolutions. This second phase of work is known as MPEG-2 (ISO/IEC 13818) and has already produced a generic video coding algorithm recently standardised by ITU-T as the H.262 recommendation. The MPEG2 standard defines a hierarchical digital television system with standard, extended and high definition television picture resolutions (SDTV, EDTV, HDTV) at bit rates around 4, 10 and 15-20 Mbit/s.

The MPEG2 standard (ISO/IEC 13818) is divided into four parts: MPEG-Systems (ISO/IEC 13818-1), MPEG-Video (ISO/IEC 13818-2, ITU-T H.262), MPEG-Audio (ISO/IEC 13818-3) and MPEG-Compliance Testing (ISO/IEC 13818-4).

The *Video part* specifies the coded bitstream format for high quality digital video. In order to cope with the requirements of different classes of applications, a set of profiles has been defined offering different picture quality resolutions. In addition to being compatible with MPEG1, it supports interlaced video formats, higher picture quality, several picture resolutions including 4:3 and 16:9 as well as other facilities for HDTV. MPEG2 is a generic standard, which means that it is application and transmission medium independent, offering a wide set of tools from which it is possible to extract different subsets in order to fulfill the requirements of specific applications. For instance, to implement hierarchical coding, MPEG2 identifies different forms of video coding scalability using the following basic scalability tools: data partitioning, SNR scalability, spatial scalability and temporal scalability. Hierarchical coding is useful in application areas such as video telecommunications, video on ATM, interworking of video standards, HDTV with embedded TV, etc. Scalability enables different decoders to reconstruct different versions of the same video source by using sub-sets of the total encoded bitstream. The total bitstream is structured in two or more layers with a base layer and a number of enhancement layers. Hierarchical coding could also be achieved with the simulcast technique where a number of independently coded versions of the same video source are transmitted to several receivers. However, this technique althougth very simple, doesn't provide an efficient management of bandwidth and transmission resources as the use of scalable tools does. When these tools are used separately - basic scalability - the video stream is divided in two different layers referred to as the lower layer and the enhancement layer. If combinations of the basic tools are used - hybrid scalability - the number of layers may raise up to three.

In basic *spatial scalability* the generated layers have different spatial resolutions. The lower layer provides the basic spatial resolution of the video source and can be decoded by itself to regenerate a basic version of the encoded video source, while the enhancement layer carries additional information to upgrade the basic version to full-spatial resolution images. *Spatial scalability* is the appropriate tool to be used in applications where inter-working of video standards is necessary as well as in video broadcasting applications - it allows a smooth transition to the HDTV system maintaining compatibility with existent standard TV systems, it enables the implementation of a graceful degradation mechanism by which, under poor reception conditions, the picture resolution of a higher layer may temporarely drop to that of the lower layer, it may provide resilience to transmission errors by sending the information of the base layer through a channel of better performace or by improving the error correction mechanisms only for that layer, etc.

In *SNR scalability*, the two layers present the same spatial resolution but different video qualities of a single video source. The lower layer provides the basic video quality while the enhancement layer carries the information which when added to the lower layer allows reconstruction of a high-quality version of the video input. *SNR scalability* also provides high degree of resilience to transmission errors for the same reason stated above for spatial scalability and is specially suited to be used in applications requiring a minimum of two quality levels of the same video source. This technique is therefore intended to be used in HDTV applications with embedded TV, in video services with multiple qualities, in video services over ATM transmitting the lower layer in a high-priority channel, etc.

When two transmission channels are available, the suitable scalable tool to use is *data partitioning*. With this technique the encoded information of the total bitstream is divided in two streams according to their relative importance in the reconstruction process - the more critical data such as headers, motion vectors and DC coefficients are grouped together and sent through the channel with better error performance, while the less critical data (higher DCT coefficients for example) is sent through the other channel with worse error performance. In *data partitioning*, unlike the other two basic scalable tools previously described, neither layer may be decoded by itself to regenerate a version of the original video signal.

In basic *temporal scalability* the video stream is divided in two layers which present different temporal resolutions. The lower layer may be decoded by itself to regenerate a video signal presenting a basic temporal rate, corresponding to the temporal resolution of the actual TV systems. The enhancement layer, when decoded and temporally multiplexed with the decoded information of the lower layer, allows reconstruction of a full temporal resolution version of the video source. Only sophisticated systems of the future, will be able to regenerate and display such a full temporal resolution video signal.

The *Audio part* specifies multi-channel audio coding, supporting up to five full-bandwidth channels (right, left and central channels and two background channels) and two low-frequency channels and up to seven multilingual commentary channels. It also supports coding of high-quality conventional stereo and mono channels at bit rates not exceeding 64 Kbit/s.

The *Systems part* specifies the method of combining into a single stream, one or more video and audio elementary streams previously coded according to parts 2 and 3 of the standard and other type of non-MPEG2 data. The video and audio coding algorithms when applied to video and audio information produce compressed elementary streams, which are packed with system information thus generating Packetized Elementary Streams (PES). The MPEG2 Systems Stream is the result of multiplexing PES belonging to different sources into one stream having one of the two possible formats specified in the Systems part of the MPEG2 standard - The *Program Stream* and the *Transport Stream*.

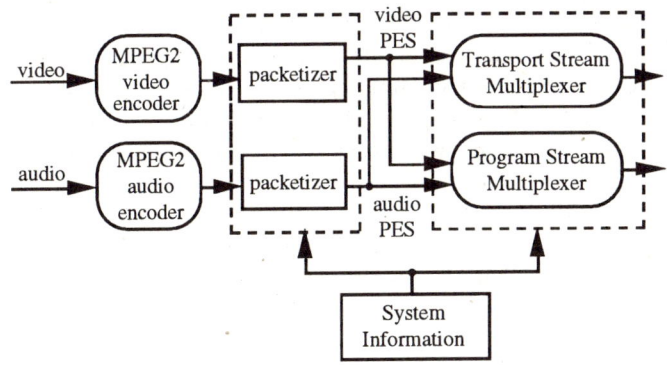

Fig. 1. Simplified overview of ISO/IEC-13818 scope

A *Program Stream* is a combination of Packetized Elementary Streams which share a common time base, as for example a coded video sequence and its associated sound. Its syntax is quite similar to the one specified in the Systems part of the MPEG1 standard (ISO/IEC 11172-1). The *Program Stream* is structured in packs which convey a variable number of variable length PES packets. This stream format was designed to be used specially in applications implemented over error-free environments.

The *Transport Stream* syntax is used for multiplexing one or more digital programs (video + associated audio) and individual elementary streams, which may have their own individual time-bases. The *Transport Stream* can be seen as a sequence of fixed length packets, each of which carries information belonging to one single coded source. The fact that transport packets have a relatively small fixed length, improves interoperability among applications and makes the *Transport Stream* Syntax the most appropriate to be used over error- prone environments, thus presenting the broader range of application.

PES level can be seen as an intermediate format between Program Stream and Transport Stream and it consists of variable length packets with a payload field conveying coded information belonging to one single source, preceeded by an header holding control information associated to that stream such as timing, priorities, scrambling, etc. A PES packet header starts with a 32-bit start code field, which also identifies the stream to which the payload belongs to. It may also include the time-stamp fields referring to the first access unit which starts in that PES packet.

According to the Transport Syntax, each PES packet (header + payload) is divided into 184-byte long segments to fit the payload field of the fixed- length packets. These packets are 188-byte long, which means that a 4-byte header with system information preceedes the payload field. The transport packet header includes a 13-bit field for elementary stream identification (PID). Each PID identifies at most one elementary stream, which means that each transport packet carries information belonging to only one single source.

The contents and characteristics of the Transport Stream are described in the Program Specific Information (PSI) tables, which are sent in transport packets of known PIDs (reserved and selected). These tables carry the information which allows the receiver to correctly demultiplex and present programs [3]. They identify for instance, the PIDs of the transport packets which carry the PCR for each program, the associations of elementary streams to form programs (through the PIDs of the transport packets carrying those elementary streams), etc. When a change occurs in the contents of the Transport Stream being described, new PSI data must be sent to allow the receiver to update its own locally reconstructed PSI tables and thus to continue presenting programs correctly. Timing information is carried at the system level by means of a common 27 Mhz system clock - the System Time Clock, STC. Samples of this time-base are encoded in the Program Clock Reference (PCR) field in the headers of transport packets, and are used at the receiver to lock an initially free-running 27 Mhz local clock, thus regenerating the time-base for each program. Synchronization between elementary streams is achieved through the inclusion of time-stamps fields (Presentation Time Stamps, PTS and Decoding Time Stamps, DTS) in the headers of PES packets.

The MPEG2 standard is expected to cover a large field of applications such as Digital Terrestrial Television Broadcasting (DTTB), Direct Broadcasting Satellite (DBS), Satellite and Electronic News Gathering (SNG and ENG), distribution by cable or fiber, distribution and interactive applications on ATM networks, retrieval systems on DSM, disc or tape, etc. This means that MPEG2 will be used in almost every application involving the transmission of video and sound, independently of the delivery mechanism being used. MPEG2 will also allow a smooth transition to the HDTV broadcasting system, providing compatibility with existing standard TV services.

2 Channel Adapters for MPEG2 applications over ATM networks

Support of multimedia applications based on the MPEG2 standard over ATM-based networks, involves consideration of five major aspects. These are:
  * multimedia multiplexing methods;
  * AAL protocols to use;

---

[3] a program is formed by elementary streams which share a common time-base, i. e., which refer to the same PCR clock.

* provision for error detection and/or correction mechanisms;
* clock recovery methods;
* CDV strategies.

This document deals only with the first point, as it identifies possible solutions for the implementation of ATM channel adapters for MPEG2 applications based on different multiplexing methods and discussing their relative performance. The analysis performed, demonstrated that the multiplexing solution adopted has a considerable impact on the options to be made regarding the other four issues.

According to the level at which the multiplexing operation of the different components of the multimedia application is performed, the ATM Channel Adapters (ATM CA) may be basically divided into two major groups: *user-multiplex* and *channel-multiplex* ATM CAs.

2.1 Class 1: *user-multiplex* ATM channel adapters

In this class of adapters, the multimedia stream is seen by the network as being one single monomedia stream, which means that the different types of information are conveyed through the same virtual channel and are equally assigned the same transmission parameters, as for instance the Quality of Service parameter. The channel adapter performs a transparent mapping of the MPEG2 Transport bitstream into ATM cells, not taking into account its multimedia nature nor taking advantage of MPEG2 Transport header functionalities.

Nevertheless, it is possible to envisage two scenarios within this group of ATM CAs, as depicted in figure 2. In the first one, some external control information is either provided to the ATM CA by the MPEG2 multiplexer, or the channel adapter performs some sort of bitstream parsing to obtain that control information, identifying the beginning of the MPEG2 Transport Stream (TS) packets. In this scenario it is possible to align cell boundaries with TS packets, assuring in this way that each cell will convey only one single media (figure 2. a). This may bring advantages regarding the performance of the error detection/correction mechanisms that may eventually be used. Also in this case, the ATM CA could make use of the transport priority flag present in the TS packet header, to assert the CLP bit in the ATM cell header. This case will be referred to as case A of the set of ATM channel adapters that will be described in this document.

On the other hand, the channel adapter may simply receive the MPEG2 multiplexed bitstream and slice it into ATM cells without taking into account nor cell or TS packet boundaries (figure 2. b). This scenario will constitute case B.

To achieve alignment between cell boundaries and TS packets and take advantage of the functionality of the TS packet header, the input interface format of those channel adapters connected to the output of the MPEG multiplexer, could accept MPEG2 coded data in one connector and status information in a different one. The status connector might include control signals provided by the MPEG2 TS multiplexer, such as a packet synchronous signal and a transport priority flag.

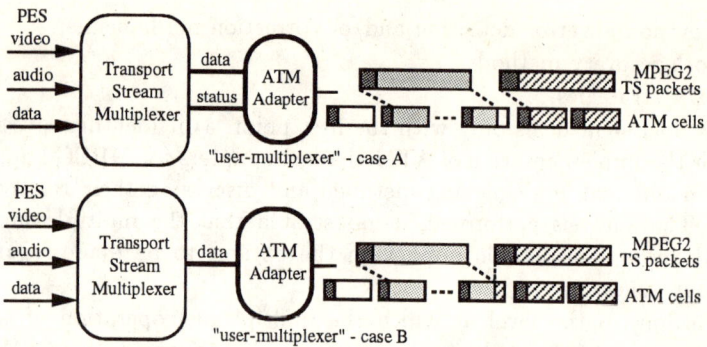

Fig. 2. *User-multiplex* ATM CA

2.2 Class 2: *channel-multiplex* adapters

In the second group of adapters, the *channel-multiplexer*, the different monomedia information streams are treated individually. Within this class of CAs, it is possible to foresee several distinct situations, according to the channel adapter input interface format and to the sub-level at which the multiplexing operation is internally performed. This will lead us to the eigth cases listed in table 1.

Table 1. Possible cases belonging to the *channel-multiplexer* class

	Channel multiplexer			
interface format		internal sub-level for performing the multiplexing operations	network resources	case
multiplexed stream	PID selection	AAL	1 VC	C'
		ATM	n VCs	C
	PES selection	AAL	1 VC	D'
		ATM	n VCs	D
separated streams	output from elementary encoders	AAL	1 VC	E
		ATM	n VCs	F
	output from PES encoders	AAL	1 VC	G
		ATM	n VCs	H

According to the input interface format, two basic alternatives can be identified. These depend on the existence or not of the MPEG2 Transport Stream multiplexer. The ATM CA may be connected to the output of the MPEG2 TS multiplexer, then performing some demultiplexing operations to recover individual media, which leads to cases C and D, or it may receive the elementary streams in separated channels (cases E, F, G and H). This means that according to the input interface format, there will be two distinct types of ATM CAs: one which input is connected with the output of the MPEG2 TS multiplexer and another which precludes the use of the MPEG2 TS multiplexer, receiving the elementary streams separately.

Each of these two scenarios can be further divided into two distinct cases. In the first scenario, where the input of the ATM CA is connected to the output of the MPEG2 TS multiplexer, two distinct modes of operation were then identified. The difference between them relying on the amount of demultiplexing processing functions performed over the MPEG2 Transport Stream.

The channel adapter of case C, the *PID switch* assigns transport packets to VCI values according to the value of the PID field present in every transport packet header. Because each PID value identifies at most one elementary stream, ([1] ... *transport packets of one PID value carry data of one and only one elementary stream.*), mapping the PID field onto VCI values will assure that different elementary streams will be transmitted in distinct virtual channels. Thus, if the channel adapter performs a simple separation of transport packets in distinct virtual channels according to the value of the PID field, then the *PID switch* channel adapter is obtained (figure 3. a).

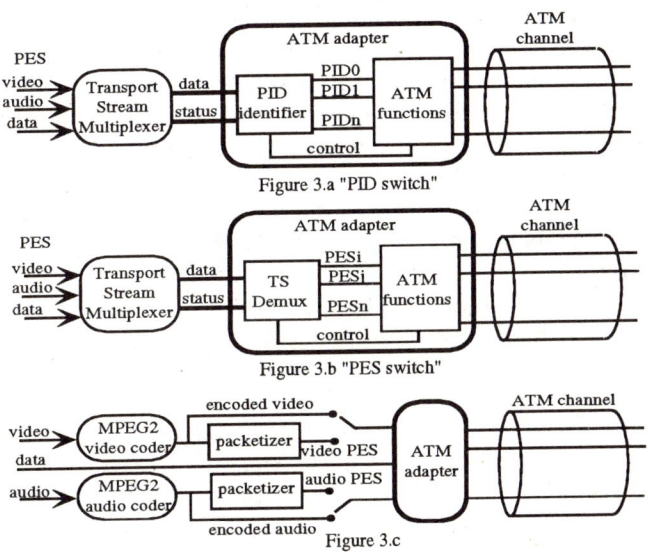

Fig. 3. *Channel-multiplex* ATM CAs

The channel adapter of case D, the *PES switch*, has a very similar mode of operation except that it parses the MPEG2 multiplexed bitstream down to a level where individual packetized elementary streams may be regenerated. Because PES packets carrying data from different elementary streams are identified through the *stream id* field present in every PES packet header, this type of adapter assigns different VCI values for each PES according to the value of the *stream id* field (see figure 3. b).

In the second scenario, where the ATM CA receives the elementary streams in separated channels, the interface can be done directly at the output of individual encoders or it may be performed at the PES level. These two situations are depicted in figure 3. c.

Regardless of the input interface format, this kind of adapters may either use a single Virtual Channel (VC) for all different types of monomedia streams, or instead may use separated VCs to transmit each elementary stream. These two situations correspond to the cases of performing the multiplexing operation of the individual media at the AAL layer or at the ATM layer.

3 Comparative analysis

A comparative analysis based on several transmission performance parameters and user/application requirements, has been performed over six[4] of the eigth options above identified for the implementation of an ATM channel adapter for MPEG2 applications. Table 1 summarizes the result of this comparative evaluation.

Hardware complexity

Channel adapters belonging to the *user-multiplex* class are the easier to implement, specially if no attempts are made to align cell and TS packets and to use control signals conveyed in the TS packet headers. In the *channel- multiplex* group, additional complexity is required for multi-channel management when the elementary streams multiplexing operation is performed on a VC (Virtual Channel) basis. However, if media/VCIs mapping tables are used, implementation of this kind of channels adapters can be made quite simple. Regarding the two channel adapters that implement some MPEG2 demultiplexing functions, it seems that the *PID switch* alternative is simpler because it only requires identification of a single transport header field, performing the ATM functions over the original transport packets (transport packets remain unchanged as they are segmented into ATM cells).

Bandwidth efficiency

Bandwidth resources are more efficiently managed by the *channel-multiplex* adapter which receives the different media separately, because the MPEG2 Systems multiplexing layer is not included. At the first sight, the *PES switch* approach seems to be equally efficient because it presents the same small overhead weight - transport packet headers are chopped off before performing ATM functions. However, if buffering of a number of payload transport packets carrying data from the same PES is not made, this advantage will easily be canceled

---

[4] Although the use of a single VC has been referred above for the scenario where the MPEG2 TS multiplexer was present, those cases (C' and D') won't be evaluated because they have little meaning: it wouldn't be logical to waist processing power in demultiplexing operations to recover individual media from the MPEG2 multiplexed stream to multiplex back into one single channel those recovered streams and transmit them in one single VC.

because the payload field of transport packets is not by itself an integer multiple of the cell payload field length (and for performing VCI multiplexing, each cell must carry only one type of elementary information). This means that in order to improve bandwidth efficiency in this type of adapters, additional buffering requirements would have to be provided, which would also increase the end-to-end delay.

Standardization/compatibility

Those solutions which are directly connected to the output of the MPEG2 TS multiplexer and those that foresee the use of multiple VCs for multimedia components (which is aligned with proposals presented within the standardization bodies), are the methods which provide the highest degree of compatibility. However, as far as it concerns this parameter, there is a clear distinction among the channel adapters which receive the MPEG2 multiplexed bitstream: channel adapters from the *user-multiplex* class as well as the *PID switch* support full functionality of the MPEG2 Systems Layer, while the *PES switch* solution does not. This means that the *PES switch* method only provides compatibility at the level of the input interface format but, just like the channel adapters which receive the elementary streams in separated channels, does not provide compatibility with MPEG2 receivers at the receiving-end because it does not transmit all the information of the MPEG2 System layer.

Media synchronization

The channel adapters which receive the different media separately, prior to the MPEG2 Systems multiplexing layer, will have to provide additional mechanisms for media synchronization as well as for clock recovery functionality. They will possibly have to make use of the mechanisms that will be provided by the ATM technology. Instead, those solutions which include the MPEG2 Transport Multiplexer, are already provided with the necessary tools for assuring media synchronization and clock recovery at the receiver. Those tools are included in the MPEG2 Systems layer syntax by means of time-stamps and a system clock encoded in specific fields of the Transport Stream syntax and of the PES syntax.

Flexibility

The alternatives that deal separately with the different elementary streams provide more flexibility to the system, as transmission parameters may be individually assigned according to the characteristics of each media. On the contrary, and as an example, in the *user-multiplex* class of adapters, all media must be assign the Quality of Service parameter of the most demanding medium.

Robustness to errors/Influence of cell loss

The *user-multiplexer* channel adapter is the least resilient to errors because it can not identify the elementary stream where the error has occurred. Also, the occurrence of a cell loss may affect several media. In the *channel-multiplex* class of adapters, one cell loss will influence only one elementary stream and it will be possible to provide separate error correction mechanisms for each type of information.

4 Conclusions

Table 1 presents a summary of the discussion performed above. It clearly indicates that the *channel-multiplex* class where the MPEG2 multiplexer is present, particularly the *PID switch* method, constitutes the best solution.

Table 2. Comparative analysis

	Method					
	user-multiplexer with MPEG2 multiplexer		*channel-multiplexer* with MPEG2 multiplexer		*channel-multiplexer* without MPEG2 multiplexer	
parameters	case A (with alignment)	case B (without alignment)	PID switch	PES switch	VC mux	SAR mux
bandwidth efficiency	-	+	-	+	++	++
standardization	+	+	++	++	++	
MPEG2 compatibility	++	++	++		-	-
media synchronization	++	++	++	+	-	-
error resilience	-	-	++	++	++	-
hardware complexity	+	++	+		-	
flexibility	-	-	+	+	+	-

References

1. ISO/IEC: Information Technology - Generic Coding of Moving Pictures and associated Audio information - Part 1: Systems ISO/IEC DIS 13818-1 (1994)
2. ISO/IEC: Information Technology - Generic Coding of Moving Pictures and associated Audio - Part 2: Video, Recommendation H.262 ISO/IEC DIS 13818-2 (1994)
3. ITU-T SG XVIII: IVS Baseline Document (1992)
4. ITU-T SG XV: Status Report on ATM Video Coding Standardization, Issue3 (1992)
5. ITU-T SGXV: Multimedia multiplexing method for audiovisual communication Doc. AVC-370 (1992)
6. ISO/IEC JTC1/SC29/WG11: ATM mappings of Transport Layer Packets (1993)
7. ISO/IEC JTC1/SC29/WG11/MPEG93/645 Preliminary studies into ATM mapping of MPEG2 Transport Layer packets (1993)
8. ATM Forum/93-1016: Recovery of the MPEG2 system clock over ATM (1993)
9. ATM Forum/93-976: ATM multimedia System Structure (1993)
10. ISO/IEC JTC1/SC29/WG11/MPEG93/750 MPEG2 Transport Packet transmission over ATM (1993)

11. ATM Forum/93: MPEG in ATM networks and a proposal for a VBR Video Adaptation Layer: AAL6 (1994)

Vocabulary and abbreviations: Access Unit - coded representation of an elementary stream presentation unit. In the case of video, an access unit is the coded representation of a picture. In the case of audio, is the coded representation of an audio frame. *ALL* - ATM Adaptation Layer. *ATM* - Asynchronous Transfer Mode. *CA* - Channel Adapter. *CLP* - Cell Loss Priority. *MPEG* - Motion Picture Experts Group. *Packetizer* - a functional block which builds packets from a raw elementary stream, including control information in the packet header. *PES* - Packetized Elementary Streams. *PID* - Packet IDentification. *PS* - Program Stream. *QOS* - Quality of Service. *TS* - Transport Stream. *VC* - Virtual Channel. *VCI* - Virtual Channel Identifier.

Protocols for Multimedia Conferencing - An Introduction to the ITU-T T.120 series

W J Clark,

Centre for Human Communications, BT Labs, Martlesham Heath, Ipswich, IP5 7RE

Abstract. For the past 5 years, ITU-T (formerly CCITT) has been standardising a suite of protocols for Audiographic and Audiovisual conferencing for use over a range of networks. The standards form a basis for a range of real-time interactive services such as Reservation & Conference set-up, Conference information, Chair controls, File transfer, Group editing, Whiteboards, Multiple pointers on screen, Multiple windows, Access to Databases.

A key feature of these standards (known collectively as the T.120 suite) is that they have been designed for multipoint use from the outset, thus making possible a wide range of facilities which have previously been restricted to point-to-point operation. The paper will describe some of the background to the standards, and give an overview of the major features and attributes.

1. Introduction

The MultiMedia Communications Forum (MMCF) has identified the lack of generic system support as the primary technological factor holding back the deployment of realistic, large scale, distributed multimedia applications. They suggest that there are two basic technologies required to make feasible such support: an appropriate transport service for communications needs and a set of generic multimedia services to provide a framework for application development. The ITU T.120 suite of protocols, which has been given the colloquial name of MultiLayer Protocol (MLP), addresses both these technology areas, by providing both the necessary transport service for a range of networks, and a number of generic services. It is intended that the generic services can be utilised by means of appropriate Application Programmer's Interfaces (API) which will give the necessary structures to allow Vendors to provide enhanced applications.

One of the key features of the T.120 suite is the provision of multipoint working. With more organisations being located on multiple sites, and the increasing formation of specialist working groups there is a clear need to provide communications systems and applications which offer more than just the traditional point-to-point communication. The term usually applied to multisite communication is "Multipoint" and this can be provided either by means of network embedded equipment, or as a feature of a particular network topology [1,2]. Within this paper, emphasis will be given to Multipoint Conferencing, but the data protocols described can also be applied to a number of other services, indeed the T.120 protocol can communicate and manage all forms of telematic/data media between two or more multimedia terminals.

2. Multipoint Communication

Multipoint conferencing has the objective of enabling a number of users at remote locations to interact in as natural a way as possible. Since existing Wide Area Networks (WANs) and equipment provide primarily point-to-point connections, any multipoint system must work in such an environment.

Within a WAN, two possible network topologies can be considered. The first, a mesh network (Figure 1) allows a connection from each site to every other site. It has the disadvantage that as the number of locations increases, the number of point to point connections increases even more rapidly.

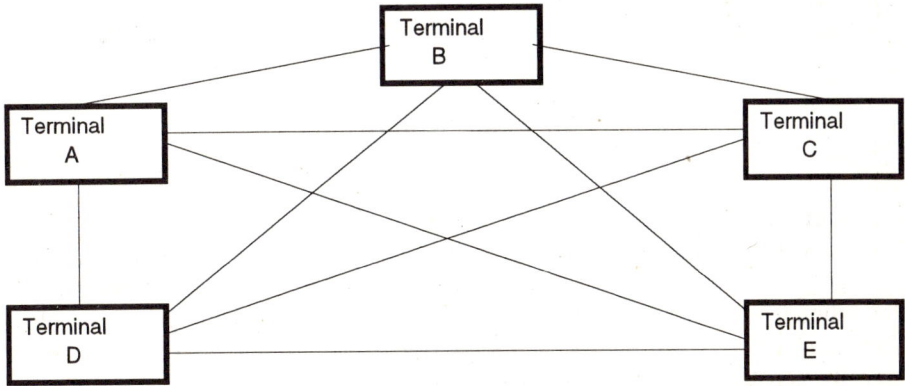

Figure 1 - Mesh Network

The Star network (Figure 2), as an alternative, provides a single connection from each site to a central node. From this node, information can be selected, processed and distributed to all other points of the star. This gives a reduction in the required number of links compared to the mesh configuration. A key element of the star is the extra equipment which is needed at the hub, and this is given the generic name of Multipoint Control Unit (MCU).

From the T.120 standards viewpoint, although no constraint is placed on the configuration of the physical connections between terminals, practical systems might be connected to one star point as in Figure 2 or connected to others in a chain.

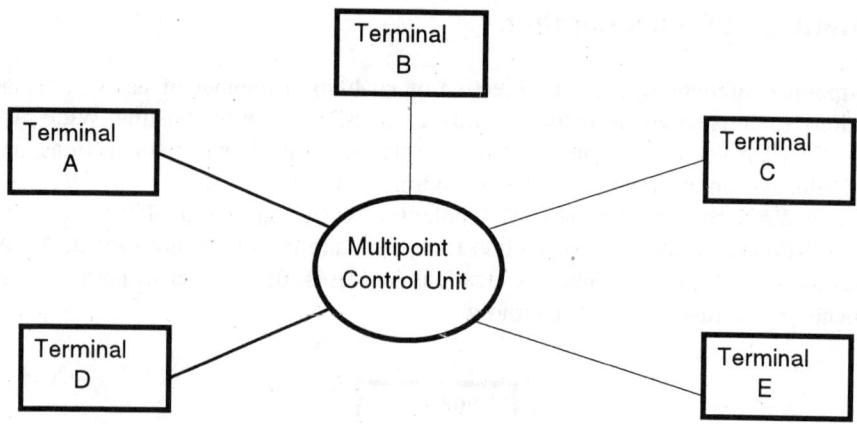

Fig 2 Star Network

In a conferencing environment, three types of information can be distinguished. These are real time audio and video as used in a videoconference, together with data. Although they are used together, they can be conceptually regarded as three separate information types which each has its own special requirements. In Figure 3, the three information streams are shown multiplexed together within a common stream. Typically for ISDN connections, the ITU-T Rec.H.221 would be used for the multiplex structure.

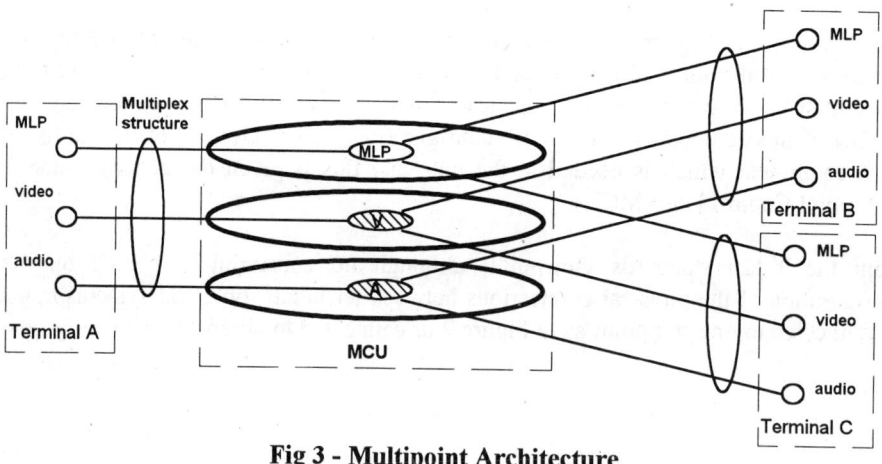

Fig 3 - Multipoint Architecture

It may be asked why a new protocol is needed for multipoint operation. Figure 4 shows two terminals connected point-to-point; in this situation, existing data protocols make use of the bi-directional nature of the link to provide error correction.

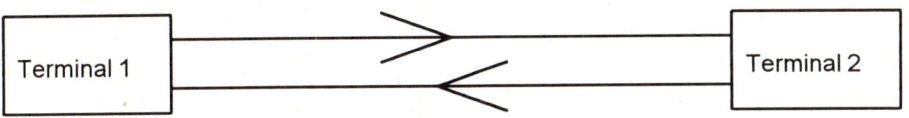

Figure 4 - Point-to-point Duplex Protocol

In a multipoint environment, as shown in Figure 5, either each transmitter has to offer error correction facilities to every receiver, or alternatively the MCU must provide the link error control on a link by link basis. The latter strategy is used by the MLP and this leads to the concept of a point-to-point transport protocol, above which are placed the multipoint protocols; this is explained in further detail in Section 3.

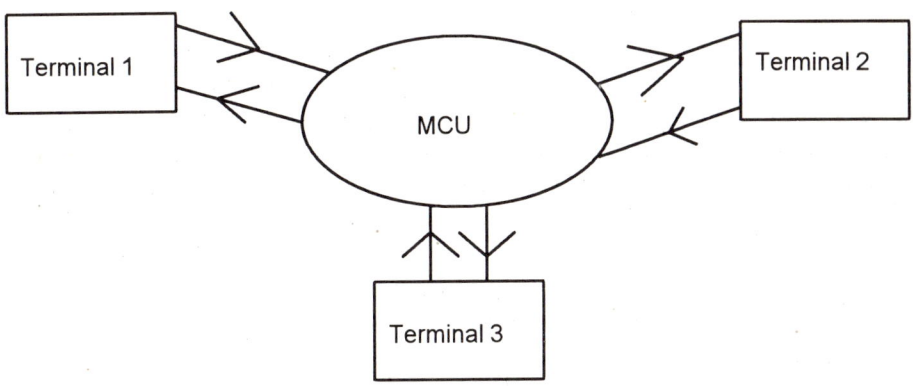

Figure 5 - Multipoint Duplex Protocol

3. The T.120 Protocol Suite

The T.120 protocol suite [3,4,5,6,7] defines the protocol stacks for a number of different networks including PSTN, ISDN, public or private switched data networks and other transport protocol stacks for LAN and ATM are currently being defined. The protocol stack is designed to provide efficient real-time interactive services where speed and timeliness of delivery are critical.

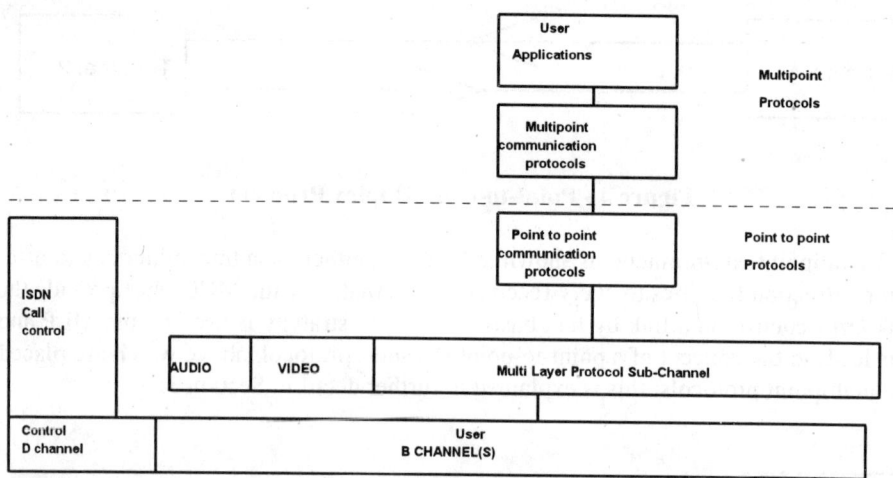

Fig 6 - Basic MLP Stack for ISDN

Fig. 6 shows how the overall protocol stack is mapped within an ISDN B-channel and how it may be considered in terms of specific point-to-point and multipoint layers. By these means, existing point-to-point protocols can be used as part of a multipoint protocol.

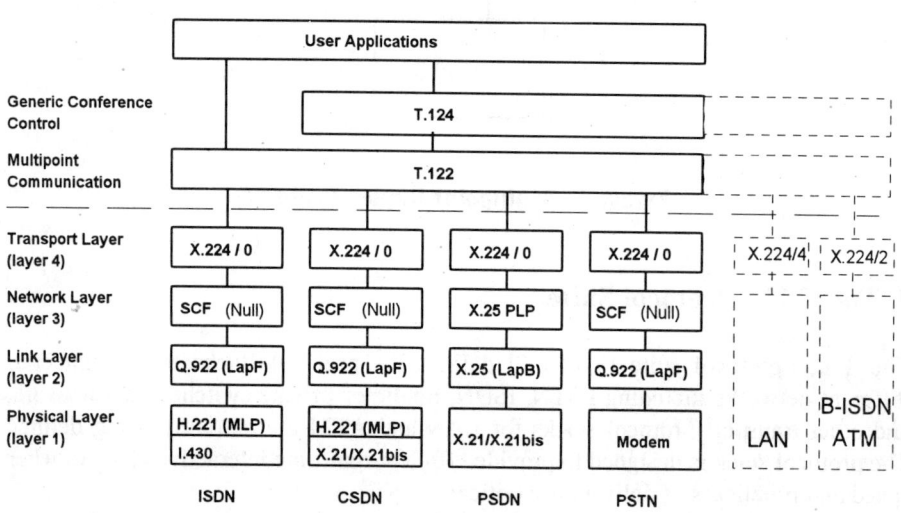

Note: CSDN = Circuit Switched Data Network
PSDN = Packet Switched Data Network

Fig. 7 MLP Communication Profile

Fig. 7 shows the communication profile structure for a range of different networks, and the following sections will describe the individual layers in some detail.

The purpose of the standardised layers of the protocol is to support, within a common environment, user applications which can be provided by a number of vendors. Layers 1 to 4 accord with similar layers in the standard OSI model, whilst the higher layers provide specific multipoint facilities.

3.1. Layer 1 - The Physical Layer

This layer varies according to the exact physical medium used for transmission. Typically for ISDN use, ITU-T Rec. H.221 is used which defines particular sub-channels for audio, video and data within a number of ISDN B-channels. Note that call set-up and control for ISDN is handled externally to the T.120 stack.

3.2. Layer 2 - The Link Layer

The link layer makes use of LAPF (Link Access Procedure Framed - mode bearer services) and provides the reliable point-to-point links on which the multipoint service is based. LAPF provides an interleaving of multiple data streams to form a multiplex and thus provides for multiple logical links and a non-blocking priority scheme. Error and flow control are provided on each of the links. LAPF was chosen for the MLP protocol as its extensions over the earlier LAPD provide symmetry in the user to network interface and allow for direct user to user inter working and support for frame relay and switching services.

The LAPF protocol is based around HDLC (High-Level Data Link Control) frames which use synchronous framing and the layer below to provide a method of clocking and synchronising the transmission and reception of data. The protocol relies on the technique of bit stuffing to ensure frame delimiters are unique and cannot occur within the data stream.

3.3. Layer 3 - The Network Layer

This layer has functionality only in the control plane and is a null layer with regard to data. The function of the SCF (Synchronisation and Control Function) is to manage the establishment and release of network connections within the T.120 channel. As described above, LAPF supports multiple logical links that correspond to Data Link Connections (DLCI). Acting through layer 2 management the SCF manages all DLCI assignments. It does this in communication with its peer SCF over the reserved management channel DLCI zero.

3.4. Layer 4 - The Transport Layer

Currently the Transport Layer consists of Rec. X.224 class 0. Rec. X.224 allows a common interface to be defined between the various point-to-point network profiles and the lowest common Multipoint layer, MCS.

An additional reason for providing this functionality is the possible future need to provide segmentation to reduce the layer 2 frame latency to an acceptable level. Since both high priority/interactive data, and low priority bulk data must share the services of the link layer, frame latency becomes important in large conferences which include a number of MCUs, where there may be 8 or more tandem protocol stacks between a sending and a receiving terminal.

3.5. The Multipoint Communication Service

The Multipoint Communication Service (MCS) is a protocol which takes groups of point-to-point transport connections from the layers below and maps them seamlessly together to present a multipoint environment to the layers above, thereby defining a multipoint data delivery service. The MCS services are defined in Rec T.122 and the MCS protocol is defined in Rec T.125. Much of the power and flexibility of MCS is derived from the fact that it provides its services in a manner that is independent of the underlying network connections, thus allowing portability and interconnection between different network types.

Segmentation may also be accomplished by the MCS layer, above layer 4, and this is the preferred location for segmentation. However, segmenting at MCS incurs a much larger packet overhead, particularly for smaller Layer 2 frames.

MCS provides mechanisms to send data by the shortest route to all or to a subset of the multipoint group; in addition a Uniform Sequenced Data Transfer service is provided that ensures that data transmitted simultaneously from several sites arrives at all receivers in the same order. Tokens are provided to both allocate resources to particular applications and to enable signalling and synchronisation between applications.

Referring to Figure 8, the MCS functionality is provided by the "MCS Provider". Each MCS Provider connects to another via an "MCS Connection". The set of MCS Connections between MCS Providers is known as an "MCS Domain" which is analogous to a conference in the higher layers. There will an MCS Provider for each node connected to the Domain and MCS provides a hierarchy that will ensure that only one of the Providers will be assigned the 'Top Provider' role and act as the resource provider for the Domain.

An application user attaches to a domain and receives a "User Identifier". Once a domain has been set up users may join one or more "Channels". A channel connects all the terminals which have expressly joined it. Any terminal within the Domain may send to that channel and all terminals joined to the channel will receive from it. A number of types of channel can be used:-

- Multicast Channels, open to all and used to send data to all or a subset of users
- Private Channels, where the channel manager can decide who is allowed to join the channel.

These are further divided into:

- Static channels which are pre-defined and allocated specific functionality.
- Dynamic channels which may be used as unique address identifiers allowing specific locations or functions to be explicitly identified and addressed. They may also be used to form sub groups allowing data sent to that channel to be received by all group members.

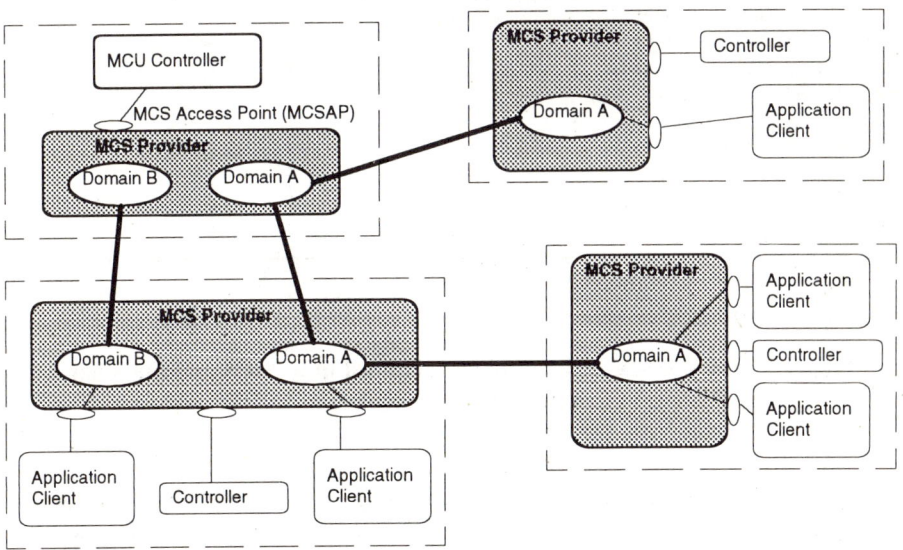

Figure 8 - A four site network with multiple Domains showing the relationship of MCS Providers, MCS Connections and MCS Domains.

In the example of Figure 8, some of the MCS Providers have multiple connections and some have multiple attached users. Some MCS providers reside on MCUs, whilst others reside on terminals. A "Controller" application, within the MCU or terminal builds each MCS connection by means of the "Connect Provider" service.

3.6. Generic Conference Control

The Generic Conference Control Protocol (GCC) is defined in Recommendation T.124. GCC resides in the MLP protocol stack above the MCS layer and it is MCS that defines the Multipoint delivery mechanism used by GCC.

GCC provides the high-level framework for conference management and control, addressing the requirements of audio graphic, audio-visual terminals and Multipoint Control Units (MCUs). It encompasses generic functions such as conference establishment and termination, managing the conference and application databases, remote actuation, conference conductorship and bandwidth control. GCC also provides co-ordination between the real-time, and non real-time aspects of a Multipoint conference.

Figure 9 below presents an overview of the scope of the GCC Recommendation showing its relationship with MCS and other applications.

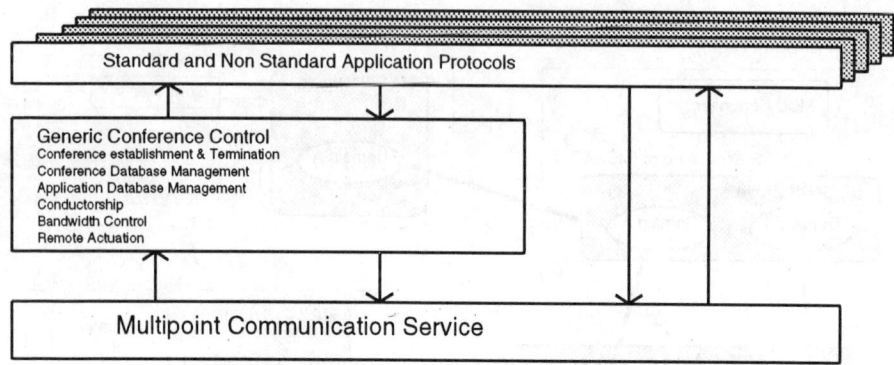

Figure 9 - GCC functionality and its relationship to MCS and the other application protocols

The application layer comprising GCC and MCS supports additional interactive applications and conference capabilities, which may include both standardised and non-standardised applications.

4. The Services and Application Protocols Provided by the T.120 Protocol Suite

As well as the services provided by MCS and GCC, the ITU-T SG8 is standardising a number of application interworking protocols. The aim is not to stifle innovation of new applications but rather to provide a framework which can be used by software developers to ensure that a degree of interworking can be achieved. Five areas are currently being defined and these are described in the following sections.

4.1. Audiovisual Control

This standard (T.AVC) sets out the procedures to be used for management of the real-time speech and video channels within the conference. Such topics as selection of particular video views and control of microphones and audio mixers are included.

4.2. Multipoint Binary File Transfer

This standard (T.MBFT) addresses the need for an interactive file transfer protocol for use within conferences and will in the future provide mechanisms which allow for automated file distribution and retrieval. The standard imposes no requirements on the type of data which can be transmitted, files being segmented into fixed size protocol data units to which sequence numbers are added. This enables the application to recover from link disconnections or a packet loss. As well as sending files other features include provision for terminals to request a file from another node, and to minimise file transfer traffic during an interactive conference, files may be preshipped.

4.3. Still Image Transfer

This standard (T.SI) provides for the transmission of still images of various types as well as annotation or other operations. A number of simultaneous workspaces can be defined, and each operation and image referred to a particular workspace. The shared workspace or "Whiteboard" is one of the most common conferencing tools and a number of Vendors have produced packages for this application.

4.4. Facsimile

Under study at the time of writing, this standard will provide multipoint facsimile.

4.5. Further Standardised Media

Consideration is being given to standardising a number of additional media. These include:

- T.Sound for the encoding of audio objects of various types.
- T.MPTV for the encoding and transmission of moving-video objects conforming to MPEG.
- T.MH for the transmission of MHEG management information and AVI scripts.

5. Using the T.120 Protocol Suite

In parallel with the standardisation work, vendors are developing various toolkits which can be used to build applications using the T.120 protocol suite. An appropriate way in which this can be done is by defining specific Application Programmer's Interfaces (API) which decouple the hardware and MLP implementation from the new software applications. One such proposal is shown in Figure 10.

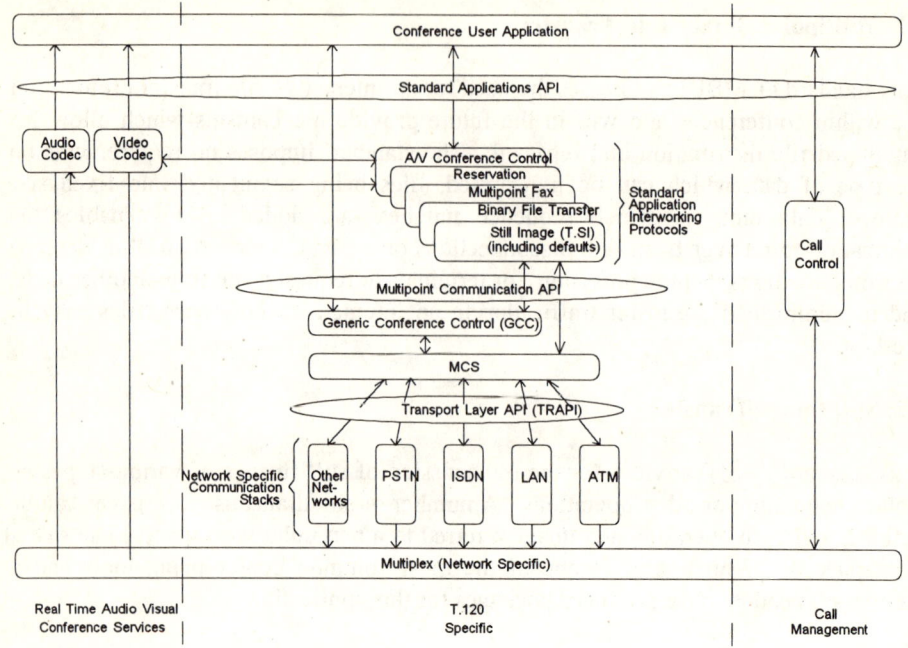

Figure 10 - Proposed APIs using the T.120 protocol suite

6. International Groups using T.120

As well as ITU-T SG8, three other international groups are immediately concerned with the T.120 protocol suite. The MCCOI (Multimedia Communications Community Of Interest) was established in 1993 to bring together the key players from across the whole of the multimedia communications industry, namely Telcos, manufacturers and software developers; current membership stands at 45. Its primary objective is in developing and verifying through practical trials the interworking capability of videotelephony and multimedia communications systems based on ITU standards including H.320 and T.120 series. Its vision is to establish the ability for any customer to equip themselves in the near future with real time Desktop Multimedia Communications Systems that will allow them to interact with anyone around the world, with full capability at the touch of a button.

CATS (Consortium for Audiographics Teleconferencing, Inc.) was also established in 1993 to promote the creation of international standards for audiographics conferencing and to form the critical mass of vendors and users necessary to promote market demand for standards based products.

ITCA (International TeleConferencing Association) also has a working group which is specifically looking at the T.120 series from the users' viewpoint.

7. CONCLUSIONS

The paper has given a brief description of the T.120 protocol suite. The standardisation process is almost complete, and with the backing of industry, the T.120 standards are expected to find widespread acceptance. The next part of the process is to further define the relevant APIs, to specify additional features within the T.120 series and to carry out international interworking trials.

8. REFERENCES

[1] Mason T I & Lewis D E: "Multipoint Videoconferencing", IEE Colloquium, Video Conferencing, Has The Time Come, Dec 1987

[2] Clark W J: Multipoint Multimedia Conferencing as an ISDN Application, IEEE Communications Magazine, May 1992

[3] ITU-T Draft Recommendation T.120, Introduction to the Audio graphics and Audio-visual Conferencing Recommendations - in development

[4] ITU-T Recommendation T.122 (1993), Multipoint Communication Service for Audio graphic and Audio-visual Conferencing

[5] ITU-T Recommendation T.123 (1993), Protocol Stacks for Audio graphics and Audio-visual Teleconference Applications

[6] ITU-T Draft Recommendation T.124 (1993), Generic Conference Control Service Definition

[7] ITU-T Recommendation T.125 (1994), Multipoint Communication Service Protocol Specification

A Platform for Multimedia Telecooperation Bridging Endsystem Heterogeneity

Gabriel Dermler[1], Thomas Gutekunst[2], Edgar Ostrowski[3], Nelson Pires[4],
Thomas Schmidt[5], Michael Weber[5], Heiner Wolf[6]

University of Stuttgart[1]
IPVR
<dermler@informatik.uni-stuttgart.de>

Swiss Federal Institute of Technology[2]
TIK
<gutekunst@tik.ethz.ch>

Technical University of Berlin[3]
PRZ
<ostrowski@prz.tu-berlin.d400.de>

INTERSIS Automação[4]
<M4240@eurokom.ie>

Siemens AG[5]
ZFE ST SN 21
<schmidt@dfki.uni-sb.de>

University of Ulm[6]
Distributed Systems
<wolf@informatik.uni-ulm.de>

Abstract

Joint Viewing and Tele-Operation Service (JVTOS) is an advanced teleservice allowing distributed users to work in a collaborative fashion with multimedia. JVTOS offers services for multimedia collaboration across high-speed networks and is aimed at running in heterogeneous workstation environments comprising different hardware platforms and different operating and window systems. JVTOS comprises facilities for session management, floor control, multimedia application sharing, telepointing, and audio/video communication. This paper outlines the design of JVTOS on different platforms.

1 Introduction

Synchronous computer-based collaboration is one of the new multimedia services made possible by the advances in network and endsystem technology. Its promise is to replace physical conventions of people by an environment allowing humans to travel and cooperate electronically. This paper presents a system competing to establish a flexible cooperation environment and to overcome the frontiers of proprietary hard- and software: JVTOS.

This Joint Viewing and Tele-Operation Service enables synchronous cooperative work by enabling a JVTOS user to share privately owned applications with other users. This implies that each user is presented with the same view of the application output and that the application is controllable by any of the users. JVTOS aims at providing a system level platform which allows the sharing of arbitrary existing single-user applications. In addition, JVTOS supports cooperation by a picturephone and a telepointing facility.

JVTOS has two outstanding features. Firstly, JVTOS realizes the sharing concept not only for text/graphics application output as it is done in e.g. Timbuktu or the known X multiplexers (e.g. SharedX, Xmux, XTV) [1] but also for multimedia applications such as multimedia authoring systems or film editors. Secondly, it realizes the described platform in a heterogeneous workstation environment comprising different hardware and different operating and window systems. Specifically, JVTOS supports SUN Sparcstations with SunOS, Siemens-Nixdorf workstations RW420 with IRIX 4.x, the Apple Macintosh with MacOS, and the IBM PC with MS-Windows.

JVTOS is issue of work package 4.2 within RACE 2060 project CIO [2].

2 Service Description

JVTOS is a new telecooperation service for high-speed networks [3], [4], [5]. It is structured into a set of four user-level services: *Session Management*, *Application Sharing*, *Picturephone* and *Telepointing*.

The *Session Management Service* is the major control of the entire JVTOS teleservice. It administrates and runs sessions. Sessions are the frame in which a collaboration takes place. The Session Management Service offers a variety of admission and floor control policies to accommodate different ways of cooperative work. JVTOS sessions are dynamic in that they allow users to enter or leave ongoing sessions.

The *Application Sharing Service* allows cooperation-unaware single-user multimedia applications to be shared among several heterogeneous workstations. The terms "cooperation-unaware" and "single-user" denote that the applications were actually constructed for a single user only and hence are not aware of being run in a group context. Multimedia applications may handle text and graphical information, still pictures, moving pictures (video and animation), and sound. To maintain the single-user behavior of shared applications, floor control mechanisms are used to determine which user is allowed to direct input to a shared application. JVTOS allows simultaneous sharing of more than one application.

The *Telepointer Service* allows a session participant to move a telepointer in shared windows being visible to all other session participants. The Telepointer Service also distributes the floor holder's mouse pointer.

The *Picturephone* offers desktop video conferencing and thus allows the session participants to communicate audiovisually with each other.

2.1 Session Management

In the context of a session, users may be invited by the session chairman or they may request to join or leave an existing session. The chairman can assign and revoke the floor for shared applications. The *Session Management Service* (SMS) coordinates all these operations necessary from the start of a session until its end. It acts as the mediator between the users and the involved services by providing a set of operations which are grouped into core session management (e.g. open/close the session), participant management (e.g. invite a participant), floor control (e.g. assign/revoke the floor) and service management (start/terminate a service).

2.2 Multimedia Application Sharing

A telecooperation environment requires *joint viewing*. This allows multiple users, each on his own computer workstation, to view and interact with an application. A possible solution is to build a new set of *cooperation-aware applications* which explicitly support this requirement. The most critical problem with this approach is that users would be limited to the use of only special cooperation-aware applications.

Application sharing is another solution. It exploits properties of the operating/window system to allow joint viewing with unmodified applications. Applications and terminals are the basic entities in the application sharing service. *Application* means a cooperation-unaware single-user multimedia application running on a computer, *terminal* denotes the set of a user's input/output facilities such as display, keyboard, mouse and audio/video input/output devices. Users access shared applications through their terminals. The *Multimedia Application Sharing Service* (MASS) is the JVTOS service interconnecting shared applications and user terminals. It allows all session participants the joint viewing of an application and directing input to the application. The participants may use different hardware platforms and window systems, namely: *X11R5* on *SUN/SNI* workstations, *QuickDraw* on *Mac*, *MS-Windows* on *PC*.

2.3 Telepointing

Communication between JVTOS users is enhanced by globally visible pointing tools. The telepointer service is to monitor movements of pointers owned by a user and to display these movements locally and at remote user sites. Each user may own a set of telepointers. In addition, the service tracks the mouse pointer of the floor holder and mirrors it on remote displays in the corresponding shared windows.

2.4 Picturephone

JVTOS includes a Picturephone which provides interpersonal audiovisual communication. It supports two-way audio and two-way video streams between all participants of a JVTOS session. For each session participant status information is shown in a dedicated window indicating connectivity, visibility, and audibility. Each user can individually select the streams to be presented.

3 Implementation concepts

3.1 Basic JVTOS Structuring

The structuring of the JVTOS design covers two dimensions. In the *functional dimension* it was decided to realize different services of JVTOS as independent modules (Fig. 1). This approach ensures future expandability and reusability of JVTOS parts in the context of other CSCW-oriented work.

Session Management is the entity which provides information concerning user participation (names, addresses, roles) to all other JVTOS services. Each of these services is otherwise responsible for managing all service-related objects contributing to cooperation in a session. For instance, the MASS creates, announces and shares applications for JVTOS users. Similarly, the Telepointer and the Picturephone modules are responsible for managing their respective means of cooperation. The Session Management also coordinates the set-up and termination phases of JVTOS services.

A second decision led to a symmetrically *distributed architecture* (Fig. 1). One benefit of this is that each JVTOS system is autonomous, i.e. it can connect to another JVTOS system without using any centralized "conference" servers. A second benefit is that processing load is distributed among the participating machines, thus avoiding potential performance bottlenecks when increasing the number of participants. Also, using direct communication links between modules of the same service can be expected to be significantly faster than communication via central hops.

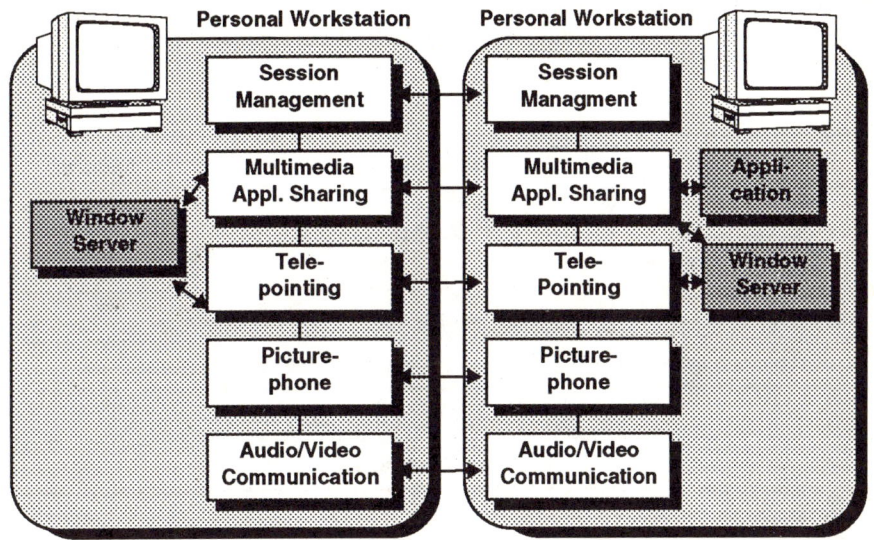

Fig. 1. Basic JVTOS Structure

3.2 Heterogeneity

One key goal of JVTOS was to bridge heterogeneity across different computer platforms. *Procedural and format heterogeneity* apply to the peculiarities of each platform's procedures and formats to generate, transfer and display data. A common exchange protocol for data transfer had to be established. For text/graphics the X protocol was selected. Non-X endsystems (Mac, PC) employ local translators (located within the MASS) to translate their proprietary local I/O calls into or from X protocol elements.

For audio and video control a control protocol was developed which is used by the MASS when detecting local audio or video I/O calls. For exchanging video data the JFIF format is currently being used, while for audio 16-bit linearly coded samples at 22 kHz are assumed. Though otherwise possible, JVTOS is currently supposed to support these mandatory "JVTOS" formats.

A second aspect of heterogeneity relates to *performance*. JVTOS allows each user to select the data it wants to present on its endsystem. This includes the freedom to select the shared applications a user wants to view and the audio and video sources of the picturephone. By making selections each user can tune the selection of viewed data to its endsystem performance.

3.3 Audio/Video Handling

JVTOS incorporates an Audio Video Communication Service (AVCS) module. Its main purpose is to transfer audio/video data from a source port to a set of corresponding remote sink ports ([3]). Thereby, the AVCS shields all aspects of end-to-end transfer from AVCS clients (the MASS and the PP). These aspects relate to possible transport level multicast support, synchronization and in the future possible data format conversions and quality scaling. The AVCS concept also makes JVTOS easier to adapt to new audio/video hardware and control software since it concentrates most of the related functions.

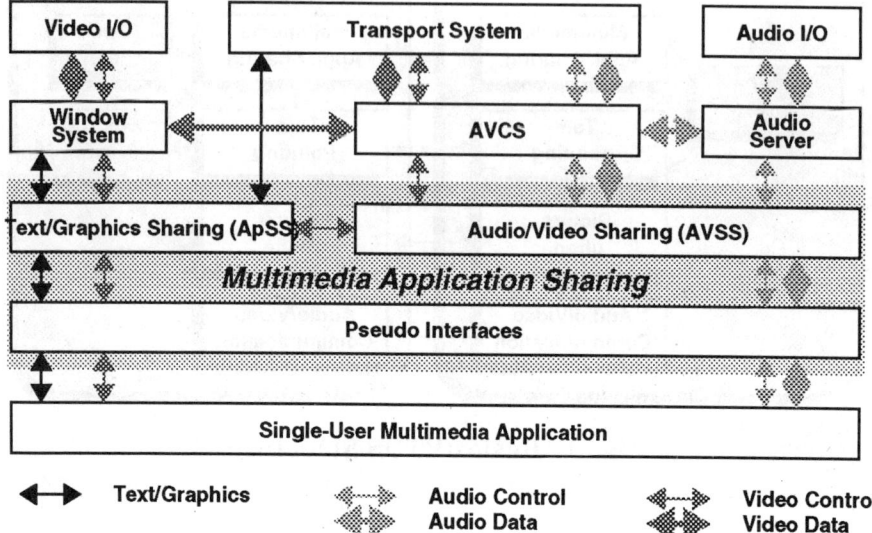

Fig. 2. Audio/Video Sharing Data Flow

The AVCS provides an API allowing the MASS and PP to export a source port and to import a previously exported port (connect it to local output port). This paradigm allows AVCS clients to select those streams a user is interested in and which can be displayed by a user's machine. Figure 2 shows the functional structure of JVTOS with regard to sharing audio and video generated by an application.

4 Platform Specific Aspects

UNIX. The UNIX operating system provides a pre-emptive multitasking environment for the active applications. Thus the JVTOS services can be implemented as a set of concurrently executing UNIX processes. So each of the blocks in Figure 1 is implemented as at least one process. The processes communicate using available UNIX interprocess communication facilities.

MAC. The Macintosh operating system provides a cooperative multitasking environment for active applications. Applications determine on their own how much time they use and how much processing time remains for other applications and background processes. Application do not get a granted part of the processing time. JVTOS is used like a system level service. Thus, JVTOS will run in the background, while a shared application is in the foreground. If JVTOS were an application, the multitasking system would not guarantee processing time for JVTOS and its services. Hence, the JVTOS services are implemented as interrupt driven modules.

PC. On the PC platform, JVTOS is realized as a set of four different processes. The Picturephone, the session management module, and the Windows to X translator (WinX) which has two processes, the interceptor and the translator. These processes communicate using the standard MS Windows interprocess communication protocol Dynamic Data Exchange DDE which allows applications to communicate and "share" memory objects.

5 Summary and Conclusion

This paper describes the implementation of a system supporting cooperative work. Unlike similar available systems, JVTOS does this support in a heterogenous and multimedia environment. Heterogeneity is present in the flavor of the used platform and applications. JVTOS runs on four different hardware platforms using different operating and window systems: SUN/SNI (Unix/X11), PC (MS-Windows) and MAC (QuickDraw).

Shared applications may be initiated on a workstation of any type. Such applications use the window system of the workstation they run on, but it is still possible to interact with such applications using a workstation of any of the other types. Multimedia support includes sharing of multimedia applications, with the implication that another element of heterogeneity has to be taken care of, namely the handling of different kinds of multimedia toolboxes associated with the different platforms.

Since JVTOS integrates different endsystems over different high-performance network technologies the domain of telecooperation is opened to a wider group of potential users and thus potential collaborators.

6 References

[1] J. E. Baldeschwieler, T. Gutekunst, B. Plattner: "A Survey of X Protocol Multiplexors". *ACM Comp. Comm. Review, Vol. 23, No. 2,* New York, pp. 13-22, 1993.

[2] W. Bauerfeld: RACE-Project CIO (R2060): Coordination, Implementation and Operation of Multimedia Tele-Services on Top of a Common Communication Platform; *Intl. Workshop on Advanced Communications and Applications for High Speed Networks '92,* pp. 401 - 405, 1992.

[3] G. Dermler, T. Gutekunst, E. Ostrowski, F. Ruge: "Sharing Audio/Video Applications among Heterogeneous Platforms" accepted at the 5th IEEE COMSOC Workshop Multimedia '94, Kyoto, Japan, May 1994.

[4] G. Dermler, T. Gutekunst, B. Plattner, E. Ostrowski, F. Ruge, M. Weber: "Constructing a Distributed Joint Viewing and Teleoperation Service in a Heterogeneous Workstation Environment". *Proceedings, 4th IEEE Workshop on Future Trend of Distributed Systems in the 1990's, Lisboa, Sept. 93*

[5] Thomas Gutekunst, Bernhard Plattner: "Sharing Multimedia Applications among Heterogeneous Workstation". *Proceedings, Second International Conference on Broadband Islands.* Athens, 1993.

A Scheme for Multimedia and Hypermedia Synchronization[*]

Nikos B. Pronios, Theodoros Bozios
Development Programmes Department, INTRACOM S.A.
P.O. BOX 68, Peania 19002, GREECE, e-mail: {npro, tmpo}@intranet.gr

Abstract. In this paper we present a synchronization scheme that addresses in an integrated manner the complex issues of synchronization required by Structured Multimedia and Hypermedia Presentations. Taking into account the capabilities of both the overlaying multimedia application layer and the network-supported Quality-of-Service (QoS), we describe a new multimedia synchronization service residing at the communication system, between the network transport services and the multimedia application layer. The use of this service performing coarse inter-stream synchronization is optional.

1 Introduction

The introduction of multimedia computing and communications has added another dimension to the already complicated problem of synchronization in communications, that of *Multimedia Synchronization*. The multimedia synchronization can be classified as *Single Continuous Medium (or Intra-Medium) Synchronization*, *Multiple Media (or Inter-Media) Synchronization* like lip synchronization, *Conditional Synchronization* when the presentation of a medium is linked to the satisfaction of a condition and *Multipoint Synchronization* when the presentation events occur almost at the same time at different locations [1], [2]. The need for multimedia synchronization arises by the introduction of *Distributed Structured Multimedia and Hypermedia Presentations*[1] containing *Continuous Media (CM)* (e.g., video, audio) and having *multipoint* requirements.

Three basic strategies can be used for multimedia synchronization [3] [4] [5]: the *Synchronization Markers (SMs)* transferring timing information, the addition of *Synchronization Channels (SCs)* which periodically transmit timing and synchronization information, and the *Multiplexing* of the streams transferring media with temporal relations, onto a single logical channel. Much work has been recently done in the area of multimedia synchronization, addressing specific synchronization aspects, at the user-presentation (e.g., MHEG [1] and MPEG system level [6]),

---

[*] The work described here is being carried out as part of the RACE II R2008 EuroBridge project.

[1]These Presentations imply the existence of a spatiotemporal script for the user-presentation.

network level (e.g., Internet Real-Time-Protocol (RTP) [7]) and dealing with the modeling of the multimedia synchronization problem, without sufficient emphasis on the identification of Multimedia Synchronization Performance measures.

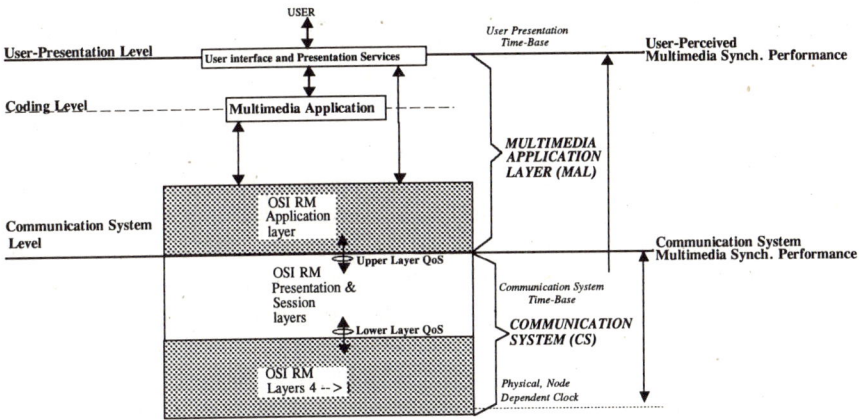

Fig. 1: The Unified Layered Framework for Multimedia/Hypermedia Synchronization [2] used in the proposed synchronization scheme

The proposed scheme differs from others in that it addresses the complex forms of synchronization required by Multimedia/Hypermedia Presentations, within the three-layer framework proposed in [2] and depicted in Fig.1, in an integrated manner. It uses SMs and it approaches the synchronization needs from the user-presentation point of view, taking also into account existing or forthcoming standards in the coding and user-presentation level. To address synchronization problems encountered during the retrieval and presentation of structured presentation documents, it uses the functionality of the MHEG standard as a baseline. The above mentioned framework, distinguishes the responsibilities and synchronization performance measures in three layers, the bottom two layers comprise the *Communication System (CS)* and the upper layer is the *Multimedia Application Layer (MAL)*. The focus of this paper is the multimedia synchronization service performing *many-to-one* (i.e., media sources at different nodes) inter-media synchronization at the upper layer of the CS.

The rest of this paper is organized as follows: In section 2, we provide a short description of the proposed synchronization scheme. We present the design objectives and the basic operation, as well as the proposed new synchronization entities. In section 3, we provide certain conclusions and directions for future work.

2 Proposed Multimedia and Hypermedia Synchronization Scheme

The CS should be capable of supporting the transfer of monomedia objects constituting a multimedia/hypermedia object from different logical channels. The transfer through different channels creates problems in maintaining the temporal relations between these objects. Due to the stochastic nature of network environment, even when we have network resource reservation and admission control, we can assume that the total network delay is non-deterministic. Data elements traveling through different logical channels may arrive in the destination with different offsets creating synchronization problems at the presentation level.

The user-perceived degree of synchronization depends on the combination of CS and MAL. The proposed multimedia/hypermedia synchronization scheme is based on the idea that the CS's role in multimedia synchronization is to provide a *coarse* (lower level) synchronization before the MAL can perform the *fine* (higher level) synchronization. Thus, the CS should forward the media streams, belonging to the same structured presentation, to the MAL with asynchronisity levels that the MAL can handle. This basic concept exists in *all* synchronization schemes, since fine synchronization (or *tracking*) can be achieved at the cost of reduced dynamic range, that is the range of operation of the specific synchronization mechanism. Since wide synchronization range results in reduced resolution-matching capabilities, a coarse synchronization (or *acquisition*) stage is used prior to the tracking.

2.1 Design Objectives

The scope of the proposed scheme, and the associated service, is to bridge the synchronization capabilities of the underlying networks and the overlaying MAL. This is achieved by performing a coarse many-to-one inter-media synchronization at the upper layer of the CS. The basic design objectives of this scheme, are:
1. Compliance/use of existing multimedia/hypermedia document architectures.
2. Utilization of QoS formalism that is becoming prevalent in the high performance communication networks and services.
3. Distinction of the synchronization problems into those caused by the CS and those caused by other systems involved (e.g., DBMS, OS, playout devices).
4. The CS synchronization mechanism will deal with the CS's problems, and will perform a coarse inter-media synchronization, leaving the fine-synchronization issues to be dealt by the MAL synchronization mechanism.
5. Correlation of CS and user-presentation events. The differences that can exist between the relationship of bit rate and presentation rate are correlated with the incorporation of SMs at two different levels, using the concept of *Synchronization Group (SG)*, i.e., the connections to be synchronized.
6. The optional use of the CS's synchronization mechanism. The need of using this service is determined by the MAL taking into consideration the performance of the underlying networks as well as the MAL capabilities.

Presentation Time Stamps (PTSs)
The PTSs correspond to the user-presentation level, and are inserted by the MAL in the media streams during the "original" presentation of the media (which then can be stored or transferred for presentation to another node). Their values correspond

to the duration of medium presentation measured from the start. The PTSs are incorporated in the media streams, are stored along with it, and are not visible by the network. The positions and the frequency in which the PTSs are inserted in the media streams, are specified by the application taking into account possible requirement during the presentation of this object. They correspond to the MHEG *timestones* and can be used for fine inter-media and conditional synchronization.

Network Time Stamps (NTSs)

The NTSs are the SMs that are inserted in the media streams by the CS at the sources and are checked by the CS at the destination. They do not correspond to a logical unit from the user presentation point of view (e.g., video frame). They are *sequential numbers* inserted in order to detect/correct inter-media synchronization problems caused by the CS. The NTS value change rate is specified by the MAL and imposes a lower limit on how often it is possible to detect and correct synchronization problems in the receiver, and, consequently, how accurate the synchronization can be. The MAL decides for the NTS change rate in a medium stream according to the significance of the information it contains for the user, taking also into account the underlying networks and the synchronization problems they may cause, the available resources (e.g., buffers), etc.

2.2 Basic Operation

In the proposed scheme the role of the CS is to maintain the synchronisity levels between the media streams belonging in a synchronization group between the levels specified by the MAL. During a multimedia presentation, all the *Elementary Presentation Units (EPUs)* [8] with a specified relation between their PTSs numbers must be displayed appropriately. The basic operation of the scheme, is:

- The MAL at the destination determine the media to be synchronized and their QoS characteristics. In case of stored media, they estimate also the retrieval times and rates for each medium.
- The CS *assumes* that the retrieval times and rates are correct and attempts to maintain the temporal relations of the media streams when these streams are delivered to the MAL at the destinations as they were requested.
- At the sources, the CS synchronization mechanisms change NTSs values in the rate specified by the MAL of the destination.
- At the destination, the CS detects synchronization problems caused by network, by examining the NTSs.
- According to the MAL policy, the CS synchronization mechanisms can buffer the data from the streams whose NTSs have arrived until all the NTSs are collected. It can then pass the data up to the MAL with aligned NTSs.
- The MAL examines the PTSs to detect other synchronization problems and takes corrective actions before presenting the objects.

The scheme introduces two new entities, one at the *MAL*, the *Multimedia Control Entity (MCE)*, which is related to the PTS dependent actions, and one at the *CS*, the *Synchronization Manager & Monitor (SMM)* that handles the NTSs (see Fig 2.).

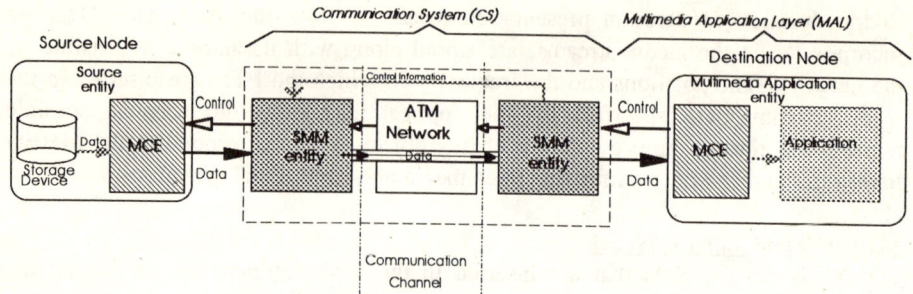

Fig. 2. The Proposed Synchronization Scheme: New Synchronization Entities

Multimedia Control Entity (MCE)

The MCE is an MAL entity. Its role is to support the different types of distributed multimedia applications. It has interfaces and interactions both with the multimedia application and the CS for the provision of multimedia synchronization. This entity incorporates certain functionalities from the MHEG engine, as well as an extended set of functionalities for the interchange of the multimedia/hypermedia (MH) objects through heterogeneous networks. How "strict" the synchronization can be depends on the number of PTSs inserted in the media stream by the MCE during the storage phase, how often the MCE evaluates these PTSs in the presentation phase, and the asynchronicity levels that the presentation processes can resolve. The basic functionality of the MCE is:

- Parsing, encoding, interchanging, interpreting of MH objects.
- Evaluation of the presentation status of MH objects.
- Fine inter-media and conditional synchronization using PTSs.
- Determination of the required channels' QoS characteristics based on Media type, Encoding Method, and Presentation attributes.
- Scheduling of MH objects retrieval and *serialization* [1].
- Support of "*graceful degradation*" enabling an application to conform to the current environment resources (including network resources).
- Decomposition of the composite MH object in its monomedia components for possible parallel processing and transmission through different communication channels at the source for display in the destination.
- Support of pre-defined synchronization policies which take into account the *human media perception properties* [9].
- In case of continuous synchronization problems, QoS re-negotiation or message to the MCE at the destination to change the media retrieval rates.

Synchronization Manager & Monitor (SMM)

The *SMM* is a new CS layer entity which operates on *Elementary Presentation Units (EDUs)* [6] performing NTS insertion at the sources and NTS value checking and removal at the destination. It is dedicated to handling the problems caused only by the CS, assuming initially that the information is inserted synchronized in the network, and attempting to maintain this synchronization when it delivers the information to the MCE at the destination.

The SMM receives from the MCE the SG, the multimedia object with the most significant and meaningful synchronization information, from the application's point of view within the SG *(the Synchronization Master Object (SMO))*, the degree of the significance of synchronization information of each medium *(Synchronization Priority (SP))*, the synchronization tolerance level *(Sync$_{level}$)*, and the accepted number of consecutive times Sync$_{times}$ that the Sync$_{level}$ can be exceeded [10]. The SMMs at the sources change NTSs values in rates specified by the SPs. The NTS values will be monitored by the SMM at the destination. The SMM will inform the MCE about synchronization problems and, according to application policy, it may try to make corrective actions. Depending on this policy, when all the EDUs with the same value in their NTSs are collected by the SMM, will be forwarded to the MAL. The SMM can be built using those of the functionalities of the RTP protocol that provide support of the various forms of multimedia synchronization. The SMM could also use directly the services of the ATM AALs.

3 Conclusions

In this paper we presented an architecture and associated mechanisms supporting the multimedia synchronization needs, addressing both the CS and the MAL. Currently we are in the process of simulation of the SMM. The workplan leads to performance analysis for the CS's synchronization support, using CS-oriented measures.

References

1. "Coded Representation of Multimedia and Hypermedia Information Objects," ISO/IEC JTC1/SC29/WG12, MHEG Committee Draft, June 1993.
2. "Multimedia and Hypermedia Synchronization: A Unified Framework," N.B. Pronios, Th. Bozios, 2nd IWACA, Heidelberg, Germany, 26-28 September, 1994 .
3. "Extending OSI to Support Synchronization Required by Multimedia Applications", Michel Salmony and Doug Shepherd, IBM European Networking Centre.
4. "A Continuous Media Transport and Orchestration Service," Campbell et al, OSI 95 project, August 92.
5. "Synchronizing the Presentation of Multimedia Objects - ODA Extensions -," Petra Hoepner, Multimedia Workshop, Stockholm, April 1991.
6. "Coding of Moving Pictures and Associated Audio, " ISO13818-1: CD, March 1994.
7. "RTP: A Transport Protocol for Real-Time Applications," S. Casner and H. Schulzinne, October 1993.
8. "Multimedia Synchronization: The role of the Communication System," Th. Bozios, N. Pronios, "BROADBAND ISLANDS: Towards Integration," ELSEVIER 1993, pp 151-172.
9. "Multimedia Synchronization Techniques: Experiences Based on Different System Structures," R. Steinmetz, MULTIMEDIA' 92, California April 1992, pp 306-314.
10. Multimedia Synchronization Functionality in the EuroBridge Platform," RACE 2008 EuroBridge project, Th. Bozios, N. Pronios, March 1994.

Integration of Existing Applications into a Conference System

Dieter Riexinger and Kathrin Werner

IBM European Networking Center, Vangerowstr. 18, D-69115 Heidelberg
Mail: {dieterr, kwerner}@vnet.ibm.com

Abstract: An important aspect of computer supported cooperative work is the exchange of data during a conference. Usually, data is stored in data bases or managed by applications which are no conference applications and cannot be shared easily among conference participants. How can this data be used during a conference? In this paper we point out that an ideal solution for this problem is the integration of conference aware systems and conference unaware applications. We describe our experiences with the extension of a public domain sharing system for X applications into a universal sharing component which can be easily integrated into any conference system.

1 Introduction

Workstation conference systems facilitate meetings with several people located across the building or even across different countries without leaving the office. The participants of a computer supported conference save travel time and travel expenses. Another important advantage of such a system is the access conference participants have to their documents throughout the conference. This data may be held in large data bases on a mainframe or locally on workstations. In general, data is created by tools that are not developed for conference situations. To exchange this data among conference participants there must be a possibility to import data into conference applications or to integrate the applications themselves, that are used to process the data.

Our goal is the extension of a conference system which provides a set of special conference applications to facilitate the integration of already existing applications. As a flexible conference system must run over several platforms, sharing of applications of different window systems such as X11, Presentation Manager, MS-Windows or MacOS must be supported. As a first prototype we extended IBM's Person to Person conference system to allow sharing of X applications. To realize this prototype we used parts of the XTV [1] system.

In this paper we describe our experiences with the integration of Person to Person and XTV. First we give an overview of the possibilities to exchange data between participants of a computer supported conference. In chapter 3 we introduce the Person to Person conference system and describe XTV. We motivate why we have chosen XTV to be integrated into the conference system in chapter 4. In chapter 5 the changes made in XTV and its conversion into a sharing component for integration into any conference system are presented. Finally we sum up our results and give an outlook on future work.

2 Information exchange in a computer supported conference

An important aspect of computer supported conference systems is the ability to synchronously exchange information. Information can be presented by a conference participant to other conference participants for viewing. Sometimes participants are allowed to annotate or even to modify presented information. Usually, the information to be exchanged is stored in files or database systems and are processed by applications that are not written for conference situations. To change this data within a conference and to have the changes available afterwards data must be exchanged in its native format, e.g., a WordPerfect document is made available within a conference without changing the format of the document.

In this chapter different approaches for exchanging information within a conference are examined with regard to two aspects:

- Data exchange format
  Data can be exchanged in its native format or must be converted into a format known by the conference system.

- Data access provided to the conference participants
  Conference participants are allowed to change data or there is only a view of the information presented.

In general, changing data is only feasible if the applications being used to process this data can be integrated into the conference and shared among the participants. That way, data can be exchanged in its native format. Since these applications are not developed for conferencing and do not notice if they are running in a conference context they are called conference unaware. Distribution of data to conference participants must be controlled by an external component and is transparent to the application. The external component decides whether the conference participants should have direct data access or only a data view. Based on a floor policy the right to modify the information under discussion can be passed. A disadvantage of this approach is its lack of flexibility. Since the semantics of operations and data is unknown outside the application, adaptation to possible conference requirements is not feasible. Data exchange is based on the strict "What You See Is What I See" principle. The advantage of sharing conference unaware applications is that users are already familiar with the tools used. Moreover, they can easily present their data without any conversion.

To avoid the "What You See Is What I See" restriction the application has to be conference aware, meaning that it recognizes the existence of a conference situation. Then it is able to decide which data will be distributed at what time to which participant according to the conference requirements. In general, conference aware applications cannot exchange data in its native format if they have not been used to produce this data. Data has to be converted into a format readable to the conference application. A typical example for a conference aware application providing unrestricted data access is a shared editor. Participants can not only manipulate the text processed but can also save a copy on their local system. In table 1 these applications are classified to require data conversion. If an application only allows viewing of data, like chalkboard applications, conferees can possibly annotate but not change the objects under discussion. Desired changes must be done by a single participant who has access to the application which is used to process the data. Modified data has to be imported again into the conference application. A disadvantage of conference aware applications is that information that must be exchanged, usually has to be converted into a format readable for the conference application. The advantage is that such applications can be developed to meet special conferencing demands like the "customer-expert" problem in kiosk applications.

Both, conference aware applications and conference unaware applications, have pros and cons. A comprehensive conference system should support both approaches. It should facilitate the development of new groupware applications and also the integration of already existing applications. That way the user is not only provided a flexible, powerful tool but also synergy effects will be exploited concerning conference management, audio/video support, remote pointing, etc.

3 Person to Person

Person to Person is an IBM software product, referred to here as P2P. P2P is a CSCW system that supports users to share and exchange information by way of simple text messages, files, application screens, graphics, images and motion video. The system consists of several applications. They are: Call Manager, Address Book, Chalkboard, Talk, File Transfer, Clip, Video and Stills Capture. Running collaborative applications in a network environment requires sophisticated communication facilities. This includes different types of networks, namely LANs, WANs and even asynchronous modem links. The network infrastructure must be driven by appropriate communication protocols. The P2P communication subsystem handles different network and communication protocols transparently for the user. It supports multiple communication adapters and can be configured for multiple communication protocols like NETBIOS and TCP/IP. Connections over ISDN and modems are supported.

Usually, system infrastructure is heterogeneous. For an efficient usability in such an environment the interoperability of a conference system is mandatory. P2P has been designed to support different operating systems, namely OS/2 and Windows. A technology prototype for AIX has been developed at European Networking Center. Conferences can be set up transparently between conferees across the borders of hardware platforms, operating systems and window systems. The interoperability is achieved by using a system independent protocol for the exchange of data.

If a P2P conference is established using various networks and communication protocols, it is not necessary that all conferees are attached to the same kind of network and run the same protocol. Each user can use the communication subsystem of his choice (if supported by Person to Person). In the situation where system boundaries concerning the networks and/or protocols will be crossed there must be one node that is able to support both networks and protocols. This node will act as a communication router. An interesting aspect here is that the router node can be a workstation that is not actively participating in the conference. Then it is called a passive node.

The chalkboard application is the prime means by which participants of a P2P conference can share data. Chalkboard offers also a method to integrate data of conference unaware applications. This functionality is called mirroring. The contents of any application window can be loaded as a background image into Chalkboard. After a window is mirrored and is part of the conference an additional button is added to its window panel. If this button is pressed another copy of the window contents is mirrored into Chalkboard and distributed. Optionally the user can select periodic mirroring and specify the time between mirroring of the selected window into chalkboard.

The mirroring function makes P2P a flexible tool to exchange information, but this kind of application integration provides only a view of the data. For example, if a text is to be discussed in a conference the original text cannot be changed, the remote participants cannot make input to the original application, and no scrolling of the text is possible.

4 A Sharing Component

The ideal sharing component should enable the sharing of single user applications of all kind of window systems like X11, Presentation Manager, MS-Windows, MacOS or others. In the first step we restricted the demand on the sharing component to support only the X Windows system. In the following a list of requirements is developed a sharing component has to fulfill to be a candidate for integration into the Person to Person conference system.

4.1 Requirements on a sharing component

The X window system can be characterized as a network oriented window system. The application program and the input and output devices are separated by a communication system. X applications are well suited to be shared by inserting a process into the connection between X client and X server. This process intercepts all messages and forwards them properly. In Figure 1 two possible architectures for X Protocol Multiplexers are shown, the centralized and the replicated approach. In the first, only one sharing component is running connected to all participating X servers. In contrast, in a replicated architecture there is a sharing component for each X server. A message from the X client is sent to the local X server and distributed to all other sharing components.

X clients interact with the X server by creating, changing and destroying so-called resources, which are referred to by identifiers. Since all resources will be created on each participating X server they will have different identifiers. Therefore a request message cannot simply be distributed to all X servers without any modification. Resource identifiers contained in requests have to be translated to identifiers referring to the appropriate resource on the addressed server. In the centralized approach this translation has to be performed for each participating X server by the sharing component which that way can become a bottleneck for larger conferences. In the case of replicated sharing components, requests can be distributed without any changes. Each component converts the resource identifiers only once. The replicated architecture both has the potential to distribute processing load and enables the use of multicast communications.

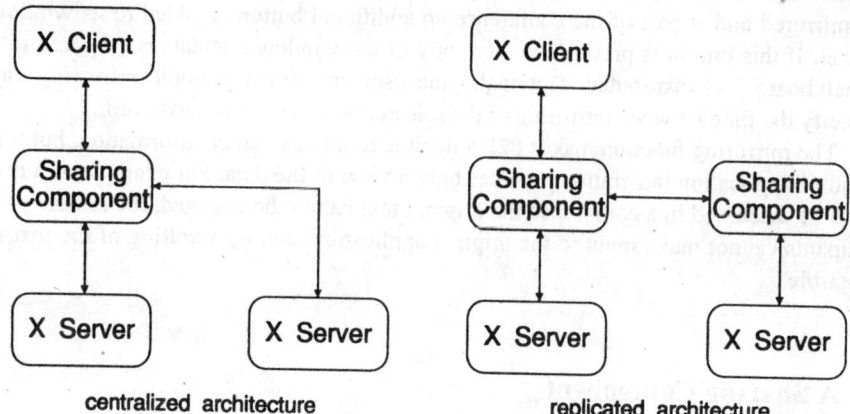

Fig. 1. X Sharing Architectures

There is another reason to favor the replicated approach. The communication between X client and server is based on TCP/IP. A centralized sharing component has to establish a TCP/IP connection to each X server. In the replicated approach, any protocol can be running between the sharing components. That way, connections already established by the conference system can be used by the sharing components. The sharing component has to build up a connection to the local X server. With regard to the integration of a sharing component into the Person to Person conference system the replicated architecture is more suitable. It allows larger conferences and avoids TCP/IP connections between a single sharing component and all involved X servers.

Besides the architecture of a sharing component its functionality is important for the choice of a certain sharing component. One design issue is the latecomer problem. In general, conferencing systems allow participants to join and leave the conference dynamically. Therefore, the sharing component should support dynamic participation in shared applications. Another important requirement is support of floor passing to control which user may provide input to a shared application. Only the user currently owning the floor is allowed to control an application. Moreover, the sharing component is to be integrated in a conference system that supports different window systems. Different X servers with different resources and capabilities may be used in a conference. Therefore the sharing component should support different server types.

The requirements on the sharing component to be integrated into a conferencing system can be summarized as follows:

- Replicated architecture
- Support of latecomers
- Providing floor passing mechanisms
- Handling of several server types

In literature several X sharing components have been introduced. A survey is provided in [3]. Table 1 provides an overview of the main requirements and the sharing components XMX, Xmux, X Teleconferencing and Viewing (XTV) and SharedX (ShX). The results in Table 1 are partially taken from [3].

	XMX	Xmux	XTV	ShX
Architecture	centralized	centralized	distributed	centralized
Floor Control	-	+	+	+
Late Connection	-	-	+	+
Server types	-	+	+/-	+

Table 1: X Sharing components and requirements for integration into conference systems

According to the requirements mentioned X Teleconferencing and Viewing (XTV) appears to be the best choice to be integrated into Person to Person. XTV was developed at Old Dominion University and at University of North Carolina, Chapel Hill. It has a distributed architecture. XTV provides solutions to the latecomer problem [7] and floor passing features. However, XTV cannot be used without changes in Person to Person. Modifications to XTV are described in chapter 5.

4.2 X Teleconferencing and Viewing

XTV has been developed as a conferencing system. It enables the conferees to invoke one or more X applications and to share them. Sharing means that applications may be joined by other conference participants. Each participant decides which applications to join. Applications joined are displayed on the participant's displays. The right to provide input to an application is associated with a token which is passed according to a chosen floor control policy. XTV distinguishes between the conference chairperson and other participants. The user who has initiated the conference becomes chairperson. All joining users get the role of participants.

XTV has a distributed architecture shown in Figure 2. An XTV component is started for each participating X server. One XTV component within the conference is running as packet switch process. This process is connected to all X applications. It is responsible to distribute the output of the shared applications to all of the other XTV components running as packet translator processes. A packet translator process modifies messages received from the packet switch process or X server. It translates resource identifiers and forwards the messages to the X server or the packet switch process.

The architecture of XTV reflects the different user roles. The conference chairperson is starting the packet switch process to which a packet translator process is connected for each participant. All applications to be shared within the conference have to be started under control of the packet switch process. A participant who wants to invoke an application has to start it against the XTV component running on the chairperson's site. I.e., a new TCP/IP connection has to be established over machine boundaries.

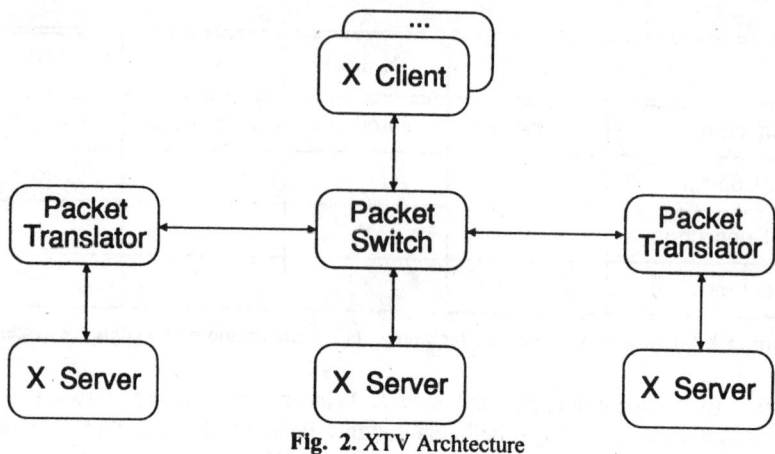

Fig. 2. XTV Archtecture

5 Modifying XTV to a Sharing Component

Both, Person to Person and XTV are conferencing systems. In order to use XTV as a sharing component within Person to Person, XTV has to provide a programming interface. Moreover, the architecture of XTV has to be converted from a distributed to a replicated one. Actually the following changes have been made:

- Changing XTV architecture
- Adding an application programming interface
- Using Person to Person connections for internal XTV communication
- Extending error handling

Person to Person has a replicated architecture. Each P2P node performs the same tasks and provides the same functionality to each conference participant. Moreover P2P has been designed to operate in a heterogeneous network environment. There is not necessarily a TCP connection between all nodes of a P2P conference. This architecture of a conference is very flexible but it also implies that an X application cannot communicate directly with the sharing component of another participant's site. As already mentioned, XTV has a distributed architecture consisting of a packet switch process and packet translator processes. Each application has to be started against the packet switch process. To use XTV without any changes would mean either to allow only one conferee to start and share applications or to guarantee that TCP/IP is running between all nodes. Both solutions would restrict the flexibility of the conference system considerably. That is why the architecture of XTV has been changed to a fully replicated one. That means, an XTV component must be able to run simultaneously as a packet switch process and as a packet translator process. In XTV a tool identifier is used to distinguish the applications in a conference. Since this identifier is created by the packet switch process it is always a conference wide unique identifier. With the replicated architecture more than one packet switch process would generate the same tool identifiers. To obtain a unique identifier per application a node identifier must be introduced. The node identifier is negotiated during connection setup of a Person to Person conference and handed over to the sharing component at start up time. If a new application is started in a conference the local sharing component generates a tool identifier and combines it with the local node identifier.

The design of the application programming interface to the sharing component was influenced by the objective to use management and control mechanisms already provided by Person to Person. Therefore all functions that are usually provided by a conference system like floor passing or group management have been removed from the sharing component. In the following we describe how the sharing tool is used during a conference. A participant can start an application and prepare some data without any notice by other conferees. Only if the application initiator makes the application explicitly available to all participants they will be informed by the conference system about a new shared application. Each conferee can decide whether and when to join the application. From the conferee's point of view an application can be in one of four states:

- Not existing
- Started: The conferee has started the application
- Joinable: A remote conferee has started and offered the application for sharing
- Joined: The conferee has started and shared the application or has joined the application that is shared by someone else.

Initially, the right to provide input to an application is assigned to the application initiator. To enable other participants to control applications Person to Person supports several floor passing policies. If the input right changes, the sharing component must change the input stream the application is listening to from one X server to another.

A remote pointing facility is an important feature provided by Person to Person. With a remote pointer, movements of one participant's cursor are visible to all conference participants. Since outside of the sharing component there is no knowledge about window identifiers and positions of shared applications a remote pointer can be displayed only at absolute screen coordinates. This is useful only if all conferees place all shared windows at the same position of the display. To display remote pointers relatively to a window's origin the sharing component must provide a routine to convert window identifiers. If a participant decides to use a remote pointer the remote pointer tool will determine the identifier of the window the cursor is pointing to. It requests the sharing component to convert this identifier into the identifier of the corresponding window at application initiator site. The initiator window identifier is distributed to all conferees. At each site this identifier is converted into the local corresponding window identifier by the sharing component. That way the remote pointer tool is able to display the remote pointer at the same position as it is displayed in the initiator's window.

The following application programming interface is deduced from the functionality described above:

XTV_Init (NodeIdentifier, XServer, ConnectionIdentifier)
The XTV component is initialized with a unique node identifier. It will connect to the specified X Server. The connection identifier is used for internal XTV communication.

XTV_StartApplication (ApplicationName)
The application specified with ApplicationName is started and displayed at the X Server specified during initialization of XTV. An application identifier will be generated to specify the application in following requests.

XTV_StopApplication (ApplicationIdentifier)
The specified application will be stopped by XTV.

XTV_JoinApplication (ApplicationIdentifier, InitiatorXTV)
The local XTV process will request the InitiatorXTV to join the specified application. The application's state will be displayed on the local X Server.

XTV_LeaveApplication (ApplicationIdentifier)
The local XTV process stops the presentation of the specified application.

XTV_ChangeControl (ApplicationIdentifier, UserIdentifier)
The input right is changed for the specified application to the specified user.

XTV_ConvertWindowID (ApplicationIdentifier, WindowIdentifier)
The window identifier of the specified application is converted into the initiator window identifier and vice versa.

XTV_Event ()
This callback function is called by XTV if it receives an event from a remote XTV process in the conference. The following events will be returned to the user of XTV:
    XTV_NewSubscriber: A user has joined a started application
    XTV_SubscriberLeft: A user has left a started application
    XTV_ControlChanged: The right to control an application has changed
    XTV_NewApplication: A new application has been started.
    XTV_ApplicationLeft: An application has been stopped.

The architectural changes and the implementation of an application programming interface enable XTV to profit from the network flexibility of Person to Person. The TCP/IP connections used for internal XTV data exchange has been replaced by communication connections already established between all conferees by the P2P conferencing system. In this version XTV has become a conference application under control of Person to Person.

Since Person to Person is running on different platforms the XTV sharing component should support a heterogeneous X environment, e.g., X servers of different platforms like AIX, OS/2 Presentation Manager or MS Windows may be used in a conference. In such an environment special problems can occure like different byte ordering or different resource names. Therefore, extended error handling has been added to the sharing component.

6 Conclusion and Outlook

In this paper we have shown how XTV, a public domain conferencing system to share existing applications, can be modified to become a versatile sharing component. After describing the benefits of the combination of unaware and aware application sharing we gave an overview of IBM's Person to Person conference system. A set of requirements was developed a sharing component has to meet for integration in a conference system that was designed to perform in a heterogeneous environment of network and window systems. XTV was compared to this list of requirements and changes to the architecture and functionality of XTV were described.

Currently we are working on a new sharing component supporting applications of other window systems. Local window systems like Presentation Manager and MS-Windows are taken into account. A new protocol for the communication between the sharing components is being developed.

References

[1] Hussein Abdel-Wahab and Mark Feit: *XTV: A Framework for Sharing X Window Clients in Remote Synchronous Collaboration.* Proceedings, IEEE TriComm '91: Communications for Distributed Applications & Systems, Chapel Hill, North Carolina, pp 159-167, April 1991.

[2] J.C.Lauwers, T.A.Joseph, K.A.Lantz and A.L.Romanow: *Replicated Architectures for Shared Window Systems: A Critique.* Conference on Office Information Systems, ACM, SIGOIS Bulletin Vol.11, Issues 2,3, April 1990.

[3] J.E.Baldeschwieler, T.Gutekunst, B.Plattner: *A Survey of X Protocol Multiplexers.* ACM SIGCOMM, Computer Communication Review, pp 16-24.

[4] M.Altenhofen: *Erweiterung eines Fenstersystems für Tutoring-Funktionen.* Diploma Thesis at Universität Karlsruhe, Karlsruhe, 1990.

[5] G. McFarlane: *Xmux - A system for computer supported collaborative work.* Proceedings, 1st Australian Multi-Media Communications, Applications & Technolog Workshop, Sydney, 1991.

[6] G.Dermler, K.Froitzheim: *JVTOS - A Reference Model for a New Multimedia Service.* 4th IFIP Conference on High Performance Networking (hpn '92), Liege 1992.

[7]) Goopeel Chung, Kevin Jeffay and Hussein Abdel-Wahab: *Dynamic Participation in Computer-based Conferencing System.* Journal of Computer Communications, 1993.

The CIO Multimedia Communication Platform

Andreas Rozek, Paul Christ

Stuttgart University Computer Center
Allmandring 30, D-70550 Stuttgart, Germany

Abstract. Within the context of an european networking project (RACE 2060 CIO) the University of Stuttgart is developing a common "Communication Platform" which can be used for (real-time multimedia) data transmission and runs on top of several operating and transport systems.

Primary intention is to allow for development of multimedia applications that make use of advanced networking features which will soon become available. While early implementations of the Communication Platform will have to simulate any features missing in current transport protocols, future advances in network research may be included without a need for changing the service, its programming interface, or any applications relying on the CIO Platform.

Beginning with a short overview of the whole project this paper briefly describes the Communication Service and its interface focusing on special characteristics of the CIO Communication Platform.

1 RACE 2060 CIO

CIO ("Coordination, Implementation and Operation of Multimedia Services") is a network research project in the context of RACE (Research and technology development in Advanced Communications technologies in Europe), situated in project line 8 ("Test, Infrastructure and Interworking") with a main emphasis on "interworking".

1.1 Project's Objectives

Main technical goal of CIO is the realization of *a common communication and service platform* based on *standard (programming) interfaces* and with implementations for a number of *different workstations*. Such a platform will allow the input and output of existing applications (that currently must be executed and controled on a single computer) to be distributed to multiple locations on a network.

1.2 Computer Supported Collaborative Work (CSCW)
by means of Teleservices

Following CIO's approach "computer supported collaborative work" becomes possible without the need for new or extended applications - all that needs to be done is to replace the module implementing one of the supported standard interfaces (e.g. XWindows) with the appropriate CIO *Teleservice* (see figure 1). Although the program itself remains completely unaware of being collaboratively used, any data passed down through that interface (e.g. an X-based GUI) may now be shared with other users which are also connected to the same teleservice.

Detail of the distribution process is hidden from the application. This includes any necessary communication control (e.g. initiating and terminating a conference) - the client gets a separate user interface for that purpose.

1.3 CIO Service Platform

To encourage a widespread use of teleservices an effort has been made to keep them independent of particular computer or operating systems and network technologies. Two *basic* teleservices have been developed into a *service platform* which is already suitable for a great variety of environments:

- a *Joint-Viewing and Tele-Operation Service (JVTOS)* and
- a *Multimedia Mail Messaging Service (MMMS)*

JVTOS is used to share input and output of X-Window applications. This allows distributed display of graphical data as well as *remote control* of programs (e.g. for the purpose of *joint editing*). Unambiguously pointing to objects within shared X windows is done by *telepointers*, and a *picturephone* provides audio and video communication.

The Multimedia Mail Messaging Service is built around X.400 and X.500 which have been extended with respect to "Multimedia Data" (such as Audio and Video). User agents are available for a number of workstations maintaining platform-specific "look-and-feel" - adding only those features necessary to achieve MMMS functionality.

1.4 The CIO Communication Platform

CIO teleservices rely on a common *Communication Platform* (CPf, see figure 1).

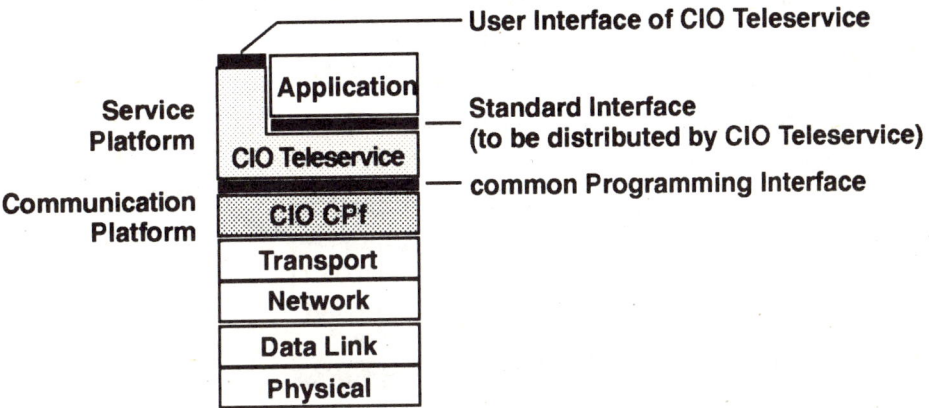

Fig. 1. CIO Communication and Service Architecture

In view of the real-time, bandwidth and connectivity requirements of modern multimedia applications, new transport systems are becoming necessary. Many research groups are currently working in this area, developing appropriate concepts and protocols. However, the variety of different approaches makes it difficult for an application programmer to apply these new protocols and exploit their benefits as doing so reduces the circulation area of product. In addition, individual solutions often focus on certain aspects (e.g. Qualities-of-Service, multicasting, etc.), while postponing other important issues (such as address management, routing, group integrity, etc.). Such an approach requires continuous updating to keep abreast of advances in networking.

In order to overcome this problem, CIO created a Communication Platform which combines and integrates several ideas and concepts of current network research arising from projects like OSI 95, MICE and Tenet. These include transport and routing protocols such as RTP, XTP(X), IP/Multicast, MTP, ST-II, or MOSPF, DVMRP and PIM.

The CPf - as seen from an Application (or Teleservice)

Applications use the Communication Service like an *enhanced transport service* (in the sense of OSI terminology) which is well suited to the needs of multimedia conferences and real-time data transmission. Its programming interface has been designed to be implemented on every computer and operating system foreseen in the project.

The CPf - as been supported by the Transport System

As the Communication System itself has to rely on an appropriate transport system, within CIO a new *Transport Service* has been defined [5] that keeps track of comparable developments in OSI 95 [7]. XTPX as the underlying protocol has been developed [6] to be an extension of XTP and cover transport and network layers of an OSI protocol stack.

CPf - the "Glue" between Application and Transport Systems

The first implementation of the Communication Platform runs on top of TCP/IP (for the sake of a widespread use) and XTPX (in order to exploit its Quality-of-Service capabilities). We will also integrate IP/Multicasting in order to connect to the MBone.

Every element of the Communication Service that is not already part of a particular transport system has to be simulated by the Communication Platform itself. However, this versatility offers the opportunity to introduce experimental protocols and test them under practical conditions without having to change any applications.

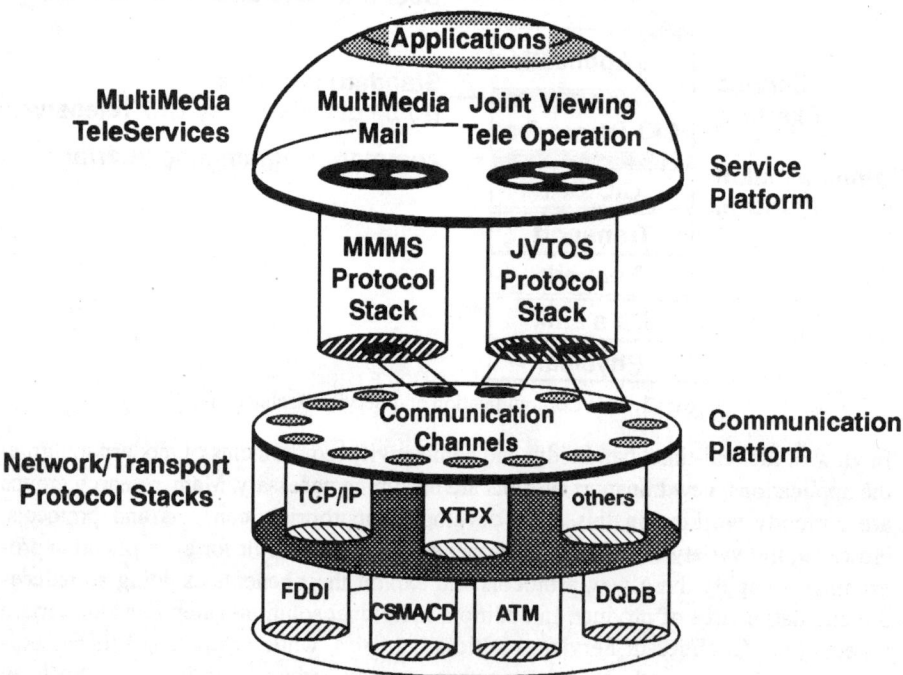

Fig. 2. CIO Platforms and Teleservices

1.5 Expected Results of CIO

CIO will establish prototypes for a complete protocol stack that supports transparent distribution of numerous standard applications using a few important workstations and network technologies. These prototypes will have been installed, tested and operated in local environments with interconnections through public networks.

2 The CIO Communication Service

The CIO Communication Service tries to offer a framework for various existing or upcoming transport protocols. Its specification resembles that of a classical transport system with a number of characteristic changes: e.g., the service provider characteristics have been extended in order to take into account its temporal behaviour and limited reliability. Special care has been taken to provide mechanisms which allow for allocation and optimization of network resources...

2.1 Basic Modes of Operation

The service provides for conversations between two, some or many partners according to the following basic modes of operation:

- **Peer-to-Peer**   (PP, two partners)
- **MultiPeer**   (MP, small number of partners)
- **Broadcast**   (BC, large number of participants)

2.2 Topologies and Transmission Directions

From the point of view of a communication "initiator" the "peer-aware" modes (PP and MP) allow for the following transmission topologies:

- **Peer-to-Peer**   (PP)
- **Peer-to-MultiPeer**   (P-MP)
- **MultiPeer-to-MultiPeer**   (MP-MP)

The P-MP topology connects a "central" peer with several participants (and, thus, fits well for lectures, etc.) while a MP-MP configuration provides for complete interconnection of all participants (which is well suited for conferences, etc.).

Transmission of user data can be

- **unidirectional**   (send-only or receive-only)
- **bidirectional**   (send and receive, half-duplex or full-duplex)

The initial choice of transmission direction(s) - to be done at connection setup - defines a set of *possible* data flows. In the course of a conversation, a participant may switch between different directions within these bounds of possibility - or "mute" to save bandwidth.

2.3 Topologies and Service Types

Instead of defining a small number of service types we distinguish between *Context Handling Attributes*, *Data Transfer Mechanisms* and *Data Transmission Attributes*.

Context Handling Attributes

With respect to the amount of work that has to be done inside of Interworking Units (IWUs, e.g. routers) the following Context Handling Attributes are distinguished:

- **Connectionless-Mode**   (no connection context needed)
- **Connection-Mode**   (simple connection context)
- **Broadcasting-Mode**   (special traffic handling required, see below)

The last mode has been introduced in order to provide for conversations between a large number of participants; it assumes that special *Broadcast Routers* take on a number of management tasks freeing the sender from that work.

Data Transfer Mechanisms

In advance to a communication, it is necessary to chose a single Context Handling Attribute which may be combined with one or multiple *Data Transfer Mechanisms*:

- **Datagram** (single message)
- **Transaction** (request-reply message pair)
- **Monitoring** (single request with multiple replies)

When creating a Communication Service Access Point (CSAP) it is necessary to specify a set containing all desired mechanisms; later on, they may be freely intermixed for data transfer on the same CSAP.

Data Transmission Attributes

One or multiple *Data Transmission Attributes* may be assigned to an individual message:

- **Acknowledgment** (asks for return of an acknowledgment)

  This kind of acknowledgments is intended to inform the sender about the actually achieved Quality-of-Service (see below);

- **Out-of-Band Transmission** (allows for expedited data transfer)
- **Retransmission** (to a subset of participants)
- **User-defined Notification** (allows to mix control information with normal data)

2.4 Multipeer Communication

Semantics and behaviour of multipeer communication primitives have been derived from their peer-to-peer counterparts according to a small set of rules like

- **Similarity to the corresponding PP Service** or

  A primary guideline was to issue a single primitive action for multiple recipients and - if need be - to get a single primitive event back;

- **Individual Peers**

  Logically, every single participant is still visible (e.g., in order to take care of individual capabilities) - although the underlying transport system may perform any traffic optimization that seems to be feasible (Multicast).

"Sum-up functions" are responsible for combining individual responses yielding a common result (e.g., the transmission of a single message may cause numerous acknowledgements. These have to be summed up in order to produce a single result). Three possible mechanisms are supported:

- **Pre-defined "Sum-Up" Strategies;**

  These are defined and performed by the communication system itself;

- **No "Sum-Up"** by the Communication System;

  Every single feedback message is passed on to the application which has to perform the sum-up itself;

- **User-defined "Sum-Up" Strategies;**

  The communication system calls a user-defined function which performs the sum-up. Data and feedback events are not intermixed; thus, it is easier for the application programmer to separate code for communication and sum-up;

2.5 Conversation Management

Communication between more than two peers requires special conversation management. In CIO this is based on *roles* which may be assigned to individual participants [4].

- an *organizer* is responsible for conference planning and establishment, it also becomes the first manager of that conference;
- a *manager* is designated to control a conversation - e.g., to add or remove participants, to assign roles, manage tokens or to terminate a conference;
- a *talker* is allowed to send data;
- a *listener* actually receives data (this role may be temporarily given up due to a local mute for the sake of saving bandwidth);

Role assignment may change in the course of a conversation. A channel with n participants may have 1 organizer, a few managers (usually less than n), a few simultaneous talkers (usually less than n) and up to n listeners.

The protocol used for implementing conversation management has been designed to be robust, i.e. to behave well even in case of control message losses.

2.6 Access Control

Conversations between multiple participants are principally insecure, CIO therefore provides two levels of communication control in order to get rid of that problem:

- **Role Assignment**

  The abovementioned roles are always associated with the access rights which are required in order to perform a related operation.

- **Access Control, Voting, Notification**

  The right to perform certain communication operations (like adding/removing participants, sending data, ...) can be assigned on an individual basis. It might be granted unconditionally or depending on the "vote" of other attendees. Other participants might have to be informed about (im)proper completion of an operation.

2.7 Broadcasting

Being aware of every peer in a multi-party configuration may become infeasible for a large number of participants. However, even today there is already some demand for "TV" or "radio broadcasting" over the internet.

A special broadcasting mode takes care of this problem as it assumes that special Broadcast Routers (BCRs) take on some management tasks: e.g., a new participant going to join an existing broadcast conversation no longer has to contact the sender itself. Instead, when constructing a path to the sender, the first BCR encountered which already processes traffic from that conversation may complete the join. The sender is still unaware of this new recipient as the "join" request is not further propagated - the broadcast mode spreads the cost of conversation management (see above) over the whole topology.

From then on, outgoing messages are branched off at that BCR (and additionally sent to the new recipient) while incoming messages (feedback) are summed-up at the same site in order to provide for some traffic in the opposite direction ("interactive TV").

Topologies and Transmission Directions

Depending on how many senders are foreseen the following topologies are supported:
- **Peer-to-World** (P-W)
- **MultiPeer-to-World** (MP-W)
- **World-to-World** (W-W)

The first two topologies assume a peer belonging to the "world" to be a receiver only while the latter one allows for bidirectional data transmission between all peers. Knowledge of a particular topology may help routers in constructing their distribution trees.

From Peer-to-Peer to Broadcast

Again, a small set of rules is used to derive broadcast service primitives from their peer-to-peer counterparts. Compared to multipeer configurations, important consequences are:
- **Pre-defined "Sum-Up" Strategies** only;
  As BCRs have to perform the sum-up, user-defined strategies are infeasible;
- **Restricted Form of Replies for Transactions and Monitors**;
  As BCRs have to combine transaction and monitor replies, their contents are restricted to a few pre-defined data types for which sum-up functions can be defined;
- **Timeout Definitions necessary**;
  When collecting feedback or replies, a BCR waits for a contribution from all "neighbours" - or until a timeout occurs.

2.8 Arrangement: Bundles and Channels

Multimedia applications often handle a number of different data streams (e.g. audio and video) which - although they belong together (e.g. because of the same set of communication participants) - have different transmission characteristics.

Channels and Bundles

Actual data transmission is performed using *Channels* which model single (bidirectional) data streams between two or more participants. *Bundle*s are used for grouping multiple channels (and further bundles) together in order to handle possible relationships (for the purpose of multiplexing, synchronization a.s.o.) between all bundled objects which have to be handled by the local and remote communication system entities.

Addressing

The CIO Communication System allows for two different ways of addressing:
- **CSAP Addressing**
  A certain participant may be addressed "conventionally" using the port number of the CSAP that is used to access a given channel;
- **Channel Addressing**
  If a new participant is going to join an existing channel it is not necessary to contact the current attendees individually (in a broadcasting environment this is even impossible as the complete set of participants is unknown). Instead, the application only has to supply the name of a channel (which is guaranteed to be unique) and the communication system performs the contact. Subaddressing (i.e., sending to a subset of participants) still remains possible;

2.9 Qualities-of-Service

Quality-of-Service Parameters

The CIO Communication System defines QoS parameters with user-level semantics and maps them to the appropriate transport system counterparts, if necessary. In addition to attributes describing traffic characteristics a.s.o. a few parameters are available that can be used for special purposes (like resource optimization etc.):

- **Selective Forwarding**

  A router (or the communication system entity itself) may be instructed to selectively dismiss certain messages according to a priority scheme in order to send the same data stream to more powerful and less powerful recipients;

- **Number of Simultaneous Talkers**

  Usually, in a multipeer configuration only a limited number of talkers are sending simultaneously - this limits the bandwidth which has to be allocated;

- **Half-Duplex Transmission**

  If a certain peer is able either to send or to receive data bandwidth requirements can be further reduced;

Separate QoS parameters exist for channels and bundles. By specifying appropriate values it is possible to optimize resource requirements, to establish (and to negotiate) multiple channels at the same time, to specify which channel should be preferred in case of resource bottlenecks, etc.

Initial QoS Negotiation vs. QoS Renegotiation

Establishment of connection-mode conversations may be combined with an initial QoS negotiation: the path between sender and receiver is chosen according to the requested QoS parameters, a negotiation failure will prevent connection establishment. Later, the initial QoS values may be renegotiated, but the previously chosen path remains the same and negotiation failures don't affect the communication. If certain QoS parameters are unknown (e.g., because the initiator of a QoS negotiation is different from the sender) they may be left open and filled in by the responder.

Resource Reservation, QoS Announcement and Application

If need be, the transport provider is wanted to reserve resources according to the actually known QoS settings, at least for a certain amount of time. The application is then asked to announce the final negotiation results (which may differ due to multipeer-negotiations) and/or to inform the communication system when these settings are to be used.

Quality-of-Service Probing

A special service primitive informs about theoretically and practically possible QoS values. The service provider is wanted to behave similar to a QoS negotiation (e.g., to chose the proper path) but not to reserve any resources.

QoS Handling

While Quality-of-Service *provision* may depend on the capabilities of the underlying transport system, it is always possible to monitor the achieved qualities. Up to five QoS values may be specified for negotiation in order to

- ask for provision of a certain quality,
- specify a narrow range for Quality-of-Service control,
- define absolute outer QoS limits from where on messages should be dismissed.

Measurement of QoS Values

Every QoS parameter measurement is smoothed using a floating average technique which may be adjusted by the application. It is possible to specify when to start with QoS control (see figure 3) and whether to start at zero or with the first measured value.

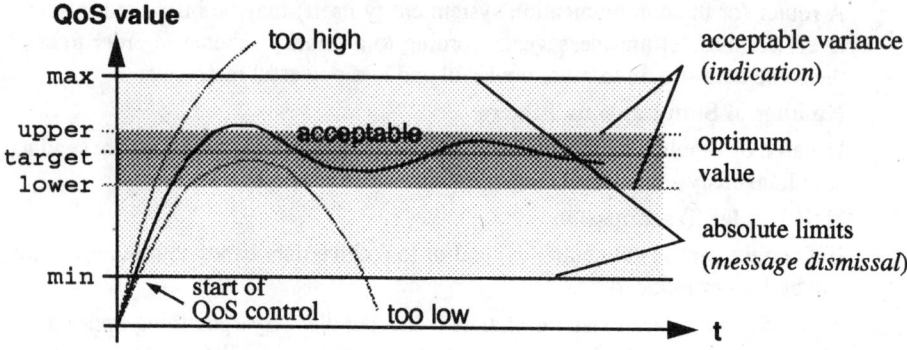

Fig. 3. Illustration of Model for QoS Monitoring

2.10 Pathological Situations

As a consequence of multipeer communication with access control a conference may encounter "pathological situations" which require a special handling in order to continue with normal conversation. Examples of pathological situations are:

- **Talker Loss;**

    The only talker in a conference might crash or get differently disconnected from all the other participants without having had the chance to pass on the right to send data.

    A special primitive allows to find out which participant actually owns certain access rights - it is up to a "manager" to detect this situation and to commit a new talker.

- **Manager Loss;**

    Similarly, the manager(s) of a conference might become unavailable.

    The abovementioned primitive can be used (by any participant) to detect this situation and to initiate a three-step approach leading to the commitment of new managers - always trying to respect the previously fixed access control settings.

- **Conference Decomposition;**

    If the attendees of a conference get cut into multiple groups (perhaps due to the failure of an intermediate router) several fragments of that conference may continue to coexist independently until the broken link becomes available again. Then, a special algorithm starts to "recompose" the original conference - however, this procedure may fail for many reasons (f.e., because the requested Qualities-of-Service might not be available any longer).

These facilities are intended to work hand in hand with an application (e.g., a "session manager") as it might not always be possible for the communication system to detect a pathological situation whereas it is necessary for it to control any repair in order to keep internal communication contexts up-to-date.

2.11 Assisting Services

Apart from the abovementioned services (which are related to communication itself) there are a few additional ones performing assisting tasks.

User and Group Management

Individual users or user groups can be defined together with their (initial) access rights outside the context of a conversation. These definitions may be independent of or bound to a concrete channel or bundle and may be overridden during a conference.

Scheduling

In spite of Quality-of-Service negotiation it may happen that establishment of a conference fails due to missing resources (perhaps because of the actual trafic load).

In order to get rid of that problem it is possible to plan a conference (with channels and bundles and the foreseen participants) in advance and to announce it to the transport system. Provided that PNOs support this kind of scheduling it is then possible to coordinate the conference with conversations announced by other parties and to reserve appropriate resources. Later on, when this conference is going to be established a simple reference to the previous reservation is sufficient in order to get access to the related resources.

3 The Communication Platform Programming Interface

A primary design goal of the CIO Programming Interface (API) has always been to offer a *common* (similar calls for every CIO hardware platform), *uniform* (similar calls and data structures for every kind of communication service) and *application oriented* (hiding details of the underlying communication system) interface to the Communication Platform. Consideration of this design goal had some important consequences:

- **Blocking and Non-Blocking IO Calls**

  Macintosh computers have a strong need for non-blocking IO calls which return to the calling program before having completed their operation;

- **Event-Oriented Control Mechanisms**

  The Macintosh Operating System is based on *Event*s and requires appropriate *Event Handling* mechanisms for processing of incoming events;

- **Efficient Data Transfer Mechanisms**

  With regard to smaller computers (like apple Macintosh and PC/AT systems) the API provides mechanisms for efficient data movement between application and network;

4 Work in Progress

A first prototype of the Communication Platform (based on TCP/IP and XTPX) described herein is currently being tested on a Sun SPARCstation (SunOS 4.1.x), an IRIS Indigo (IRIX 4.0.5f), an apple Macintosh (MacOS 7.x) and an PC compatible (WfW 3.11). It offers the complete interface but limited functionality (e.g., User and Group Management, and Scheduling use a local database instead of a distributed one).

5 Related Work

The CIO Communication Platform tries to combine a number of trends in real-time multi-party and multi-media networking yielding a common service which is prepared to integrate foreseen advances in the near future without having to change its interface or principal behaviour. F.e., the following projects and protocols have been taken into account:

- **OSI 95 Enhanced Transport Services (ETS)**
  ETS provides a number of peer-to-peer transport services with good QoS support [7]. It is subject to be standardized by ISO;
- **IP/Multicast**
  Practical experience with the MBone has shown some deficiencies with respect to resource allocation and admission control;
- **Multicast Transport Protocol (MTP)**
  MTP provides a reliable transport service on top of network layers with multicast capability. A conference is controlled by a central "master" which is responsible for group management and token handling;
- **ST-II**
  ST-II is a network protocol providing guaranteed end-to-end bandwidth and delay together with some multicast support;

In addition, it has been tried to consider the aspects of routing - not to jeopardize the basic mechanisms of routing protocols:

- **Distance Vector Multicast Routing Protocol (DVMRP)**
  DVMRP is an experimental routing protocol for internetwork multicasting which has become very popular with the Mbone. Problems concerning efficiency ("truncated broadcasting" with "multicast pruning") have lead to development of PIM;
- **Multicast OSPF (MOSPF)**
  MOSPF provides Multicast Extensions to the OSPF (Open Shortest Path First) routing protocol. A "source/destination routing" approach leads to distribution trees with "least cost paths", path commonalities may be used in order to reduce the number of datagram replications at tree branches;
- **Protocol Independent Multicast (PIM)**
  PIM is a very new approach to multicast routing: two different modes (dense and sparse mode) ensure efficiency both when members of a multicast group are situated close together as well as when they are distributed sparsely across a wide area.

6 Future Plans

Three major topics outline our future plans in that area:

- **MBone**
  Apart from just extending the scope of CIO a contact to the MBone could offer the opportunity to use the scheduling service for announcing conferences (and, perhaps, to implement a method for multicast address assignment);

- **Transport Protocols with Quality-of-Service Support**

  The proposed QoS concepts are not yet part of any transport protocol - however, it is intended to integrate (some of) them into XTPX (incl. bundles);

- **Routing with proper Quality-of-Service Handling**

  While there are already some protocols with QoS support running on a single LAN proper routing algorithms still have to be developed. PIM seems to be an interesting approach - in addition, it offers mechanisms that fit well to CIO's view of multicasting and broadcasting.

In all cases, applications relying on the Communication Platform may be used in order to test and to compare different transport and routing protocols under practical conditions.

7 Acknowledgments

This work was founded by the European Commission in context of the RACE program. I would like to thank all project partners for their ideas, their cooperation and for their patience when I tried to design the CIO Communication Service.

8 References

[1] Edgar Ostrowski et al.
2nd Version of Requirements to the Transport Infrastructure
RACE Project 2060, Internal Report, 30. June 1992

[2] R. Braudes, S. Zabele
Requirements for Multicast Protocols
RFC 1458, May 1993

[3] D. Ferrari
Client Requirements for Real-Time Communication Services
RFC 1193, November 1990

[4] Clemens Szyperski, Giorgio Ventre
Efficient Multicasting for Interactive Multimedia Applications
Tenet Group, Computer Science Division, Department of EECS, University of California and International Computer Science Institute, Berkeley

[5] Lutz Henckel, Spiridon Damaskos
Multimedia Communication Platform:
Specification of the Broadband Transport Service
RACE Project 2060, Deliverable No. 14a, 15. December 1992

[6] Bernhard Metzler, Ilka Miloucheva, Klaus Rebensburg
Multimedia Communication Platform:
Specification of the Broadband Transport Protocol XTPX
RACE Project 2060, Deliverable No. 14b, 30. September 1992

[7] Yves Baguette, Luc Leonard, Guy Leduc, Andre Danthine
The OSI95 Enhanced Transport Services
RACE Project 2060, January 27, 1993

[8] P. Jones
Resource Allocation, Control and Accounting for the Use of Network Resources
RFC 1346, June 1994

BROADBAND MULTIMEDIA AND COLLABORATION TOOLS. IDEA PROJECT

by Miguel A. Blanco and Ramón Montero (TELEFONICA I+D), Franco Almerico (TECNATION), Giovanni Venuti (CSELT), and Piergiorgio Cremonese (FINSIEL).

Abstract: *This paper will address the deployment of multimedia technologies, services and tools, focusing on CSCW, over the broadband networks that will be available in Europe from this year, although in an experimental phase, and will analyze their impact over new services for end-users. A particular case, the IDEA project within the TEN-IBC program, in the automotive design area will be presented.*

1. Introduction. Links between multimedia services and broadband communications.

Although multimedia is receiving considerable attention from researchers, service providers, telecommunications carriers and end users, there are significant issues to consider for users implementing multimedia technology over wide area distances. As new high-speed local area network technology begins to be deployed, more users are gaining the capability to run multimedia applications.

The recent directions taken by the B-ISDN are influenced by a number of parameters, the most important being the emergence of a large number of teleservices with different, sometimes yet unknown requirements. In this information age, customers are requesting an ever increasing number of new services. The most exciting teleservices to appear in the next future are HDTV (High Definition TV), videoconference, high speed data transfer, videotelephony, multimedia messaging, videolibrary, home education and video on demand (VoD). The key to effective multimedia services over wide area connections is bandwidth, due to the fact that real-time applications supporting quality audio and quality video require significantly high transmission speeds.

Both the need for a flexible network and the progress in technology and system concepts led to the definition of the *Asynchronous Transfer Mode* (ATM) principle. This ATM concept is now accepted as the ultimate solution for the B-ISDN by ITU, and plans are being made by different entities to realize experimental ATM pilots. A few examples of these experiments are the RACE, ESPRIT and TEN-IBC programs in Europe, the Multigigabit project and BAGNet (Bay Area Gigabit Network) in the US, and an Australian experiment transporting ATM over satellite.

With respect to the market expectation, two classes of subscriber for B-ISDN can be distinguished: the residential one and the business one. Both have their own service requirements.

- Residential subscriber expectations can be found in videotelephony and entertainment services, as TV distribution (in all sorts of quality and accessibility), teleshopping, home education, travelling and house-rental information, ... It is clear that once the B-ISDN will become available to every residential subscriber, a large number of new possibilities, which are currently not considered or are even unknown, will emerge.
- Business subscribers require services resulting in increased productivity. One of the main services that will become generally available will be videoconferencing, allowing multiparty videotelephony. LAN interconnectivity will offer distributed database access. Even with the ever increasing capabilities of PCs and workstations in terms of processing speed and disk storage, more and more software applications run on different machines in a distributed environment. Home office workers will benefit from these new possibilities. Applications like high quality medical image transfer, corporate education, multimedia electronic mail and desktop multimedia teleconferencing in the office environment are also expected. In the manufacturing environment, cooperative work, in form of multi-site CAD/CAM/CAE and remote visual inspection, will spread more and more.

All the above indicates that broadband communications and extensive use of multimedia applications/services are now, and will be in the future, issues tightly coupled. In this paper we will present an overview over the emerging multimedia services and applications, focusing on Cooperative Work, and its use within the IDEA Project; the initiatives that are being taken in the European frame for the deployment of trans-national broadband communications; and, finally, a particular application in the automotive design field (the TEN-IBC IDEA Project) of these pilot experiences.

2. Multimedia services and tools. Cooperative Working.

Multimedia applications are considered as the driving force that will make broadband communications a reality at home, at the desktop, and at the office. We will present a simple clasification of broadband services, and indicate some of the multimedia applications that these services will allow. Afterwards, we will focus on one of these generic applications: Computer Supported Cooperative Working (CSCW).

2.1 Overview of Multimedia services and applications.

According to ITU recommendations, services are classified into *interactive* and *distribution* services. Interactive services comprise *conversational, messaging* and *retrieval* services. The most representative services in each category currently are:

- *Conversational services: Computer Supported Cooperative Working (CSCW).* This paper will mainly focus on this service, and will be treated in deep in the following sections.
- *Messaging services: Multimedia Electronic Mail.* There are currently two main lines of work in the Multimedia Mail area:
  1) The toolkits and applications based on MIME (Multipurpose Internet Mail Extensions), an emerging standard based on the Internet SMTP (Simple Mail Transfer Protocol). Several commercial tools based on MIME exist today.
  2) An X.400 compliant (but enhanced) message transfer service, extended to convey multimedia information and to handle external references which represent part of the information not included directly into the message, but in a so-called global store. Efforts are being made in several R&D European projects in this direction.
- *Retrieval services: Video on Demand (VoD).* This new service offers the possibility to bring films to the home, in an interactive way. Users are able to select and fully control audiovisual documents from a remote database. Operators show currently a high interest in this service, mainly due to the possibility of offering it through public networks, using ADSL technology and installing TV set-tops (MPEG decoders able to work with standard TVs) at the subscriber's home.
- *Distribution services: TV Distribution.* This service allows the reception of multiple TV channels via cable or satellite. In the next future, High Definition TV (HDTV) channels will be available to users at home.

TYPES OF SERVICES	EXAMPLES OF SERVICIES
Conversational services	Broadband telephony Broadband videoconference Video-surveillance High speed digital transmission High volume file transfer service High speed teleaction High speed telefax Multimedia document transmission Telegames and remote Virtual Reality
Retrieval services	Broadband multimedia videotex Video retrieval service - Video on Demand (VoD) Multimedia document retrieval (training, medical) Data retrieval service - Telesoftware
Messaging services	Multimedia mail service
Distribution services	High speed digital distribution service News distribution TV and HDTV distribution

2.2 Computer Supported Cooperative Working.

Computer Supported Cooperative Working (CSCW) is a set of facilities, that allows users to engage in multimedia conference. The IDEA Project will use these CSCW facilities, as we will see in the next sections. During a conference users can share applications and communicate audiovisually with each other.

A CSCW tool must allow audiovisual conferencing, sharing of applications, and should also support other facilities as electronic whiteboard, shared CAD/CAM/CAE applications, file transfer and terminal emulation on remote host.

The purpose of CSCW is to ease cooperative working for projects inside and between companies. In the past, videoconferencing has been seen as a horizontal application with sales to diverse customers. However, the demand for pure videoconferencing - meetings between high-level executives - is limited. In future, products must also include a range of multimedia and groupware features designed for vertical markets such as the ones previously mentioned. Some example applications areas where these capabilities are needed are given: automotive and general product design, architecture graphic design, medical treatment planning, scientific modeling, ...

The requirement for high definition images gives rise to a need to transfer very large amounts of information (since single images can require tens or hundreds of megabytes of data). The processing required to generate the images is frequently only available from top of the range supercomputers - this is the common experience in the aerospace and automotive industry. It requires the use of high-speed networks to interconnect the supercomputers that generate the images with the workstations used to view and manipulate them. Broadband networks providing tens of Mbit/s capacity are needed. This demand can best be met both on the premises and in the wide area through ATM.

Virtual reality is an extension of computer visualization techniques that have been found beneficial in some problem-solving applications, through more deeply immersing the users in the solution process. Its use in "walking through" completed building designs is also well known. Thus, virtual reality is a tecnique that can be very useful for developing CSCW applicacions.

2.3 Multicast capability requirements in CSCW tools.

Multicast comes up as a key issue in any conferencing and CSCW system, to optimize usage of network resources. On the user side there are also specific aspects of these environments like token management and one-to-many communication, which must be addressed. Official and de facto standards are emerging in the current moment. True standard multipoint communication is performed via the Multipoint Communication Provider Module, specified according ITU recommendations. It is assumed in this context that a Multipoint Communication Service (MCS) is in charge of multipoint connection management. This module performs services such as: multicast addressing: one-one, one-all and one-subgroup, multipoint routing, and token management for the resource contention resolution

On the other hand, the solution currently adopted by available products is to use multicasting at network layer. The Internet Protocol (IP) provides a mechanism supporting multicasting, supplying and end-to-end network one-to-many service. It provides for the transmission of datagrams to a "hosts group", that is to say, to a set of hosts identified by a single destination.

The transport protocol between the network layer and applications can be connectionless or connection-oriented. The available products use IP multicasting by means of UDP for unreliable communications (audio/video), and multiple point-to-point connections for reliable communications (shared application, ...).

2.4 Commercial Tools for CSCW.

What commercial CSCW packages have usually in common is the possibility of establishing audio/video conferences. Audio/video synchronization is an important issue to handle for realistic communication. Images can be compressed by means of H.261, MPEG or JPEG codecs, resulting in constant or variable bit rates. The solution currently implemented in the most of CSCW commercial packages is the VBR (Variable Bit Rate) transfer of compressed audio and video over IP, due to the fact that bandwidth is not guaranteed, sacrificing sometimes the quality of audio and video. This is consistent with the availability of AAL 3/4 and 5 host interfaces for leading computer platforms (SUN, Silicon Graphics, and Hewlett Packard), whereas AAL1 (CBR transfer) hardware is still not widely available. Currently, there are four key platforms that deliver desktop multimedia:

- Apple's Macintosh with QuickTime
- Intel-based PCs with Microsoft's Video for Windows.
- IBM's platforms with OS/2 Multimedia Presentation Manager/2.
- UNIX-based workstations with the X Window System.

Talking about UNIX workstations, there exist several commercial packages that could be considered as CSCW tools, for the most popular platforms:

- *Communique!* from Insoft Inc., and *ShowMe* from Sunsoft, for SUN Sparc workstations.
- *In Person*, for SGI platforms.
- *M-Power*, for HP machines.

All of them provide, with a major or minor degree of success, the basic tools for CSCW that we have previously mentioned. A deep analysis of these products is out of the scope of this paper.

3. The Pan-European Pilot ATM Network.

Finally, commercial broadband services are at last about to become a reality in Europe: the approach to transport infrastructure comprises a sequence of steps towards B-ISDN to be developed by European PNOs on a progressive schema. One of them is the Pan-European ATM Pilot Network. The IDEA Project will make use of the ATM Pilot Network infraestructure. The European ATM Pilot Network, planned to be operational by July 1994 for a period of one year with a possible extension of another six months, will provide the opportunity to check proposed international ITU recommendations and ETSI standards as well as relevant EURESCOM specifications. This network, which will be the first B-ISDN based on ATM at European level, will enable the standards to be interpreted in an unambiguous manner, which is essential to the achievement of international interoperability.

Multimedia applications are considered to be one of the driving forces behind future services and network evolution, and an important revenue generator for the PNOs. A focussing on experiments on selected multimedia services will offer a feedback to the PNOs both on the services they need to offer in the future, as well as on the traffic performance required for such services. It is essential that applications are mapped against the important parameters defining the European networking requirements, i.e.: bitrate, routing, traffic characteristics (bursty, continuous, asymmetric, etc.), QoS, connection duration, call frequency, etc.

3.1 Objectives of the Pilot.

The ATM Pilot will provide the opportunity to verify proposed international standards and recommendations as well as the relevant EURESCOM specifications (P105). Mature standards are expected to foster the development of ATM equipment by providing a clear focus for suppliers' developments. The ATM Pilot will be used to validate the technical and standards issues involved in the support of benchmark services internationally. The following have been identified as benchmark services:

- FMBS/Frame Relay
- CBDS/SMDS
- CBR Circuit emulation
- Virtual Path Bearer Service

The network concept relies on a logical ATM Virtual Path network overlaying current infrastructures with standardised interfaces to both terminals and service providers. This network will be initially a backbone one based on ATM cross-connects supporting point-to-point VP connections. Principal services to be offered would be LAN and WAN interconnection. Traditional networks will coexist with this ATM infrastructure and SDH transmission systems will be deployed.

Figure 1. Pan-European ATM Pilot Network links.

4. Industrial Design across Europe through ATM: IDEA.

Industrial Design across Europe through ATM (IDEA) is a project framed within the TEN-IBC Programme that makes use of the ATM Pilot Network infrastructure and currently available multimedia collaboration tools to establish a CSCW environment in some particular automotive design centres.

4.1 The TEN-IBC Programme: Trans-European Networks - Integrated Broadband Communications.

Consequently with current efforts adopted by the European PNOs for Trans-European Networks implementation, guidelines for the introduction of broadband networks in Europe have been established. As one of the first steps, the European Commission issued a Call for Proposals for preparatory actions within the TEN-IBC Programme (Trans-European Networks - Integrated Broadband Communications). This programme intends to match end-user broadband communications needs with the technology and networking infrastructure available or nearly available. Common Interest Groups have been established to specify and implement IBC trials during 1994.

A two-phase schema has been adopted in TEN-IBC. The objective of Phase I is to develop specifications for IBC trials. The following areas are covered: citizen network, scientific network, industry network, business network, administrative network, and media (video and multimedia) network.

In a second stage these specifications will be implemented in real trials whose results will provide guidelines for deployment of advanced communications services supporting Europe-wide applications. Examples of these applications are remote access to databases including video and interactive functionalities, joint editing, and video on demand.

One of the projects currently facing its 2nd phase is IDEA, the project that we are considering in this paper.

4.2 Objectives of IDEA project.

The acronym IDEA stands for Industrial Design across Europe through ATM. The main objective of this TEN-IBC project is the specification of a CSCW trial involving industrial design centres operating in the automotive field, with possible extension to other industries with similar requirements of remote cooperation among designers. Transport infrastructure will be provided by the Pan-European ATM Network.

The principal requirement of the users is to offer to the technicians remotely located a cooperative environment where they can share in real time the documents they are working on, allowing the modification and later diffusion or storage of those documents.

Other objectives to be issued are the identification of functional requirements in the industrial design, the identification of suitable applications and tools for cooperative work, the study of the communication requirements and the selection of the network service that will support the application. The final result, in the first phase, is the specification of an IBC trial, in order to tests the obtained results.

The identified activities to be performed are:

- User requirements identification
- Bearer and telecommunication services identification
- HW/SW platform identification
- Application requirements and identification
- Trial plan

Those activities were distributed among the partners, and mapped into a number of Deliverables, showing the obtained results.

4.3 Participants in the IDEA Consortium.

A Common Interest Group (CIG) that develops a TEN-IBC project is usually established by joining end users, service providers, applications providers and telecommunications consulting companies in order to ensure correlation among user requirements, expertise and network/services capabilities to perform an IBC trial. Under a legal scope, the leader of a TEN-IBC project is the prime and unique contractor with the E.C. and the coordinator of the other partners, that sign a subcontract with the leader to assume part of the planned activities. Following, the IDEA partners and their roles are described:

- *TECNATION.* An Italian company that operates in the areas of telecommunications, industrial automation, quality systems, energy and logistics. It is the coordinator of the project, and consequently is in charge of project management, and performs the planning, organization and monitoring of the Consortium's work. It is as well responsible for the relationship with the European Commission.

- *ITALDESIGN.* It is an Italian automotive design company located in Torino. Its activities comprise all interaction among the styling centre, engineering, production, prototype building and innovative computerized techniques. As the main user, ItalDesign has been involved in the users requirements identification and the definition of scenarios.

- *DISEÑO INDUSTRIAL ITALDESIGN.* It is the Spanish branch of ItalDesign, located in Barcelona, and has been collaborating in the same activities performed by its mother-company.

- *FINSIEL.* It is a R&D company working in the telecommunication and information areas, and collaborates with users and computer manufacturers in development of communication systems. It is in charge of studying the application requirements and identifying the most suitable software tools.

- *CSELT.* It is the R&D department of the Italian PNO. It performs activities in the areas of feasibility studies, tests, experimentation, standardization and advanced research. Within the project, CSELT is dealing with network aspects.

- *TELEFONICA I+D.* It is the R&D department of the Spanish PNO and performs activities similar to those performed by the Italian counterpart.

- *SEAT and BUGATTI* .They are the car manufacturers on the Spanish side and on the Italian side, respectively, and will be connected with ITALDESIGN adn DISEÑO INDUSTRIAL.

4.4 Definition of scenarios, user requirements, application requirements and communications aspects.

a) Scenarios and user requirements.

Technical activities began with the current scenarios and users requirements analysis in order to determine the foreseeable necessities. The users in the IDEA project are design centres working in the automotive field. Manufacturers are currently outsourcing design and prototyping activities to specialized companies. These activities are carried out with the aid of CAD/CAM/CAS systems running on high-performance graphic workstations. Interactive communications among designers, engineers, managers and suppliers remotely located are required to allow cooperative work.

Frequent meetings are established during all the phases in a design project involving persons at different staff levels. A videoconference system would avoid a lot of travels and related costs and would increase productivity. Such a system must provide multipoint capabilities in order to allow audio mixing - to support users' voices - and video switching - to display the images of one or more than one of the participants - dynamically.

Work is currently done by using a great number of applications and platforms, usually running UNIX and CAD/CAM/CAE tools, on top of the X Windows System. So, high resolution and 3-D capabilities are required. The Customer Premises Networks consist on several LANs, most of them Ethernet and Token Ring. In some of the cases they are linked to X.25 public networks, while in other cases they only carry internal traffic. In all cases, LAN interconnection is required, because of the need of sharing remote resources among designers, engineers, managers, etc.

The figure below illustrates a client network:

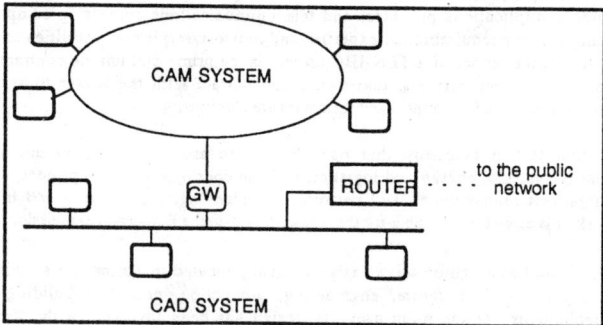

Figure 2. LAN configuration in the client centre.

Work will be optimized by providing a desktop multimedia conferencing system with one application offering hearing, viewing and joint editing capabilities, all on a multipoint configuration. Technicians might join or leave dynamically the conference, or remove one of the active media components. It is needed a *whiteboard* functionality to support joint drawing or editing of a document during the conference in a cooperative way, that is to say, the possibility of being interactively modified and manipulated by the participants in the conference. In addition, CAD/CAE/CAS applications and office automation tools must be shared.

b) CSCW Application requirements.

The CSCW application must integrate other generic facilities in order to offer efficient cooperative work (groupware), and to avoid having to leave the active facility when using a different one. These facilities are:

- **File transfer.** When a multiconference is finished, the output data must be sent to the participants and properly stored. Another particular case is the transfer of model parts of a car from the design centre to

the car manufacturer, that are included in files ranging from 10 to 20 MBytes. The total average in a project comprises GBytes figures. The transaction must assure data integrity in spite of the data volume.
- **Terminal emulation,** offering individual access to remote hosts. In a UNIX context the Telnet protocol is provided by the system and allows access via Internet to remote hosts over TCP/IP protocols.
- **Electronic mail,** providing document transfer and messaging facilities.

Such application will demand considerable bandwidth from the network, not in continuous but in burst mode. This bandwidth demand could collapse the LAN throughput, so a network configuration with traffic segmentation should be considered. Another issue to be considered at the time of choosing the CSCW application is that existing platforms at the design centres should be maintained, in order to avoid new investments in HW and SW. The user interface should be as friendly as possible. Taking into considerations all these requirements, some points about the CSCW application could be concluded:

- The application should be as independent as possible of the underlaying network, because different network services are considered, as VP service, SMDS/CBDS, FR and others.
- Tools such as file transfer, electronic mail and terminal emulation on remote host should be supported in an unique session, thus easing cooperative working.
- Available commercial products for CSCW make use of UDP and TCP protocols over IP, so they will be used as transport protocols, as no developments are considered in the TEN-IBC projects.
- Bandwidth requirements strongly depend on the kind and size of the application and data to be shared; it is usual to manipulate sketches by rotating or making zoom on pieces or parts of a car. It can generate traffic peaks between server and client around 8 Mbit/s, so this traffic average must be supported in the private and public domains: customer networks, access equipment and public network.

c) Communication aspects.

Possible network services supporting the CSCW application could be VP, SMDS/CBDS, FR or CBR services over the ATM Pilot Network. Initially, two different configurations are considered as feasible:

- The first one supposes that a Virtual Path Service is provided. This service currently can be provided by means of an ATM switch or a DSU (Data Service Unit). The following picture shows this configuration:

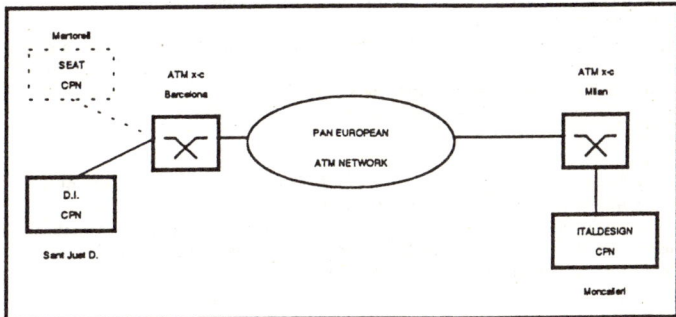

Figure 3. Virtual Path service configuration.

- The second one considers the use of a SMDS/CBDS service, provided by a DQDB MAN located in Torino, using an Interworking Unit (IWU) in order to reach the ATM network. The CPN (Customer Premises Network) of the users in the Spanish side will include a router and a DSU, will offer SMDS point-to-point service, and will be acting as remote terminals of the Italian MAN. Delays in the availability of the IWUs could affect this solution. The following picture depicts the described scenario:

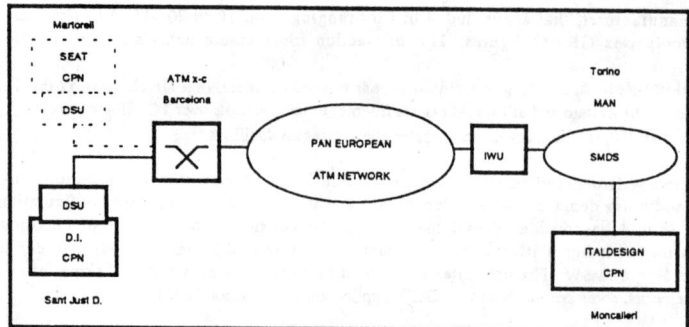

Figure 4. SMDS service configuration.

In both cases, LANs at the users centres will be connected to the ATM Pilot through the respective national segments. The ATM cross-connects involved in the trial of the IDEA project will be those located in Milano, Paris, Madrid and Barcelona.

4.5 Trial Definition and Expected Results.

It is envisaged by the PNOs, in the ATM Pilot Network environment, to start with providing Virtual Path Bearer service on July 1994. Benchmark services will be operational on a step-by-step schema adopted by every operator. As PNOs are not due to offer the same services at the same time, projects must assume some uncertainties about exact dates. The same applies to billing, that are not officially closed at this time. Keeping all this in mind, the IDEA trial is scheduled in a realistic way in order to set up international interconnection when network services have been tested and are stable. The following activities are planned to be performed:

- Installation, testing and integration start-up. This phase will include contract agreements, acquisition and installation of the platforms and access equipment as the first step, and performance of a series of local, national and transnational tests as the second step.
- Trial Planning, including test cases, evaluation criteria and time/personal assignment.
- De-facto Trial, including data collection.
- Evaluation, covering aspects as cost and time reduction, quality improvements, impact on user activities and design process, and statistics.
- Dissemination of results to users, manufacturers and operators.

Expected results are:

- Cost and time reduction: telecommunications weight is estimated around 11.4% of the total costs of a project and they will be decreased by bandwidth efficient use.
- Quality and cooperation improvements by avoiding human errors, travels or isolated applications.
- Impact on user activities and design processes, due to the introduction of a CSCW environment.
- Extension to other environments where cooperative work is feasible.

5. Conclusions.

Broadband communications and multimedia services deployment are interrelated issues. Both will cause changes in the behaviours of users at their work and home environments: cooperative working, home office and new entertainment alternatives are some examples. Cooperative working tools, making use of broadband networks are likely to become widely used, specially in environments such as the automotive and aerospatial design industries.

The results obtained in projects such as IDEA, using the ATM trans-national infrastructure deployed by PNOs, are expected to contribute in the extensive use of multimedia technologies, services and tools by end-users.

6. Acknowledgements.

The authors want to thank the collaboration and help received from the members in the IDEA Consortium, specially G. Savino (ITALDESIGN), M. Scamuzzi (DISEÑO INDUSTRIAL), and M. Re (FINSIEL).

7. Glossary.

AAL	ATM Adaptation Layer.
ADSL	Asymmetric Digital Subscriber Line.
CBDS	Connectionless Broadband Data Service.
CBR	Constant Bit Rate.
CLSF	ConnectionLess Service Functions.
CSCW	Computer Supported Cooperative Working.
ITU	International Telecommunications Union.
MCS	Multipoint Communication Service.
MIME	Multipurpose Internet Mail Extensions.
NNI	Network to Network Interface.
PDH	Plesiochronous Digital Hierarchy.
PNO	Public Network Operator.
QoS	Quality of Service.
SDH	Synchronous Digital Hierarchy.
SMDS	Switched Multi-megabit Data Service.
SMTP	Simple Mail Transfer Protocol.
UNI	User to Network Interface.
VBR	Variable Bit Rate.
VoD	Video on Demand.
VP	Virtual Path.
X.400	Mail Service ITU Recommendation.

8. References.

[1] HANDEL, R., HUBER, M.N.: Integrated Broadband Networks. An Introduction to ATM-based Networks. Addison-Wesley Publishing Company, Munich, 1991.

[2] DE PRYCKER, M.: Asynchronous Transfer Mode. Ellis Horwood Limited, 1991.

[3] STEVENSON, I., HOWARD-HEALY, M.: ATM: Market Strategies. Ovum Limited, London 1993.

[4] JEFFCOATE, J., MAN-SZE, L., TIMMS, S.: Networked Multimedia. Ovum Limited, London 1993.

[5] SAVINO, G., SCAMUZZI, M., ALMERICO, F.: Report on user scenario and requirements. IDEA CONSORTIUM - TEN-IBC IDEA Project, 1994

[6] MONTERO, R., VENUTI, G., CREMONESE, P.: Specification of the protocol stack and the selected network services. IDEA CONSORTIUM - TEN-IBC IDEA Project, 1994

[7] CREMONESE, P., RE, M.: Specification of an user application for cooperative working in the industry of car design. IDEA CONSORTIUM - TEN-IBC IDEA Project, 1994

[8] MULTIMEDIA COMMUNICATIOS FORUM: White Paper on Standards Organizations and Industry Interest Groups with ongoing or planned activities in the area of Multimedia Communications. MMCF, Atlanta 1994.

Authors Index

Akyildiz I. F., Georgia Insitute of Technology (USA)	168
Almerico F., TECNATION	368
Alves A.P., INESC (P)	310
Andrade T., INESC (P)	310
Bhonsle S.K., National University of Singapore	103
Blair G.S., Lancaster University (UK)	145
Blair L., Lancaster University (UK)	145
Blanco M.A., TELEFONICA I+D (E)	368
Blum C., Fraunhofer Institute for Computer Graphics (D)	35
Bonaventure O., University of Liege (B)	199
Bowman H., University of Kent (UK)	145
Bozios T., INTRACOM S.A. (GR)	340
Carle G., University of Karlsruhe (D)	219
Chetwynd A.G., University of Kent (UK)	145
Christ P., Stuttgart University Computer Center (D)	356
Clark W.J., BT Labs (UK)	322
Cremonese P., FINSIEL	368
Dermler G., University of Stuttgart (D)	334
Duboc Y., ENST de Bretagne (F)	294
Fedaoui L., University of Paris (F)	124
Ferry M., ENST de Bretagne (F)	294
Frimpong-Ansah K., Alcatel Austria AG (A)	47
Gay V., University of Paris (F)	1, 89
Gutekunst T., Swiss Federal Institute of Technology (CH)	334
Hamdi M., ENST de Bretagne (F)	294
Horlait E., University of Paris (F)	1, 89, 124
Kervella B., University of Paris (F)	1
Lazar A., Columbia University (USA)	103
Leopold H., Alcatel Austria AG (A)	47
Lim K.S., National University of Singapore	103
Loge C., University of Paris (F)	89

Mathy L., University of Liege (B)	199
Montero R., TELEFONICA I+D (E)	368
Neumann L., Fraunhofer Institute for Computer Graphics (D)	35
Ostrowski E., Technical University of Berlin (D)	334
Pires N.B., INTERSIS Automacao	334
Popescu-Zeletin R., Technical University of Berlin (D)	241
Pronios N. B., INTRACOM S.A. (GR)	340
Rakow Th.C., GMD-IPSI (D)	14
Riexinger D., IBM ENC (D)	346
Röhr K., GMD-Fokus (D)	14
Rolin P., ENST de Bretagne (F)	294
Rozek A., Stuttgart University Computer Center (D)	356
Sakatani T., NTT Human Interface Labs(J)	256
Sandvoss J., IBM ENC (D)	274
Schatzmayr R., Technical University of Berlin (D)	241
Schiller J., University of Karlsruhe (D)	219
Schmidt C., University of Karlsruhe (D)	219
Schmidt T., Siemens AG (D)	334
Schooler E., USC/ISI (USA)	69
Seneviratne A., University of Technology Sydney (AUS)	124
Shenker S., XEROX (USA)	69
Singer N., Alcatel Austria AG (A)	47
Szyperski C., International Computer Sciences Institute, Berkley (USA)	185
Thimm H., GMD-IPSI (D)	14
Ventre G., University of Napoli (I)	185
Venuti G., CSELT (I)	368
Weber M., Siemens AG (D)	334
Weinrib A., Bellcore (USA)	69
Werner K., IBM ENC (D)	346
Winckler J., IBM ENC (D)	274
Wittig H., IBM ENC (D)	274
Wolf H., University of Ulm (D)	334
Yen W., Georgia Insitute of Technology (USA)	168

Springer-Verlag and the Environment

We at Springer-Verlag firmly believe that an international science publisher has a special obligation to the environment, and our corporate policies consistently reflect this conviction.

We also expect our business partners – paper mills, printers, packaging manufacturers, etc. – to commit themselves to using environmentally friendly materials and production processes.

The paper in this book is made from low- or no-chlorine pulp and is acid free, in conformance with international standards for paper permanency.

Lecture Notes in Computer Science

For information about Vols. 1–801
please contact your bookseller or Springer-Verlag

Vol. 802: S. Brookes, M. Main, A. Melton, M. Mislove, D. Schmidt (Eds.), Mathematical Foundations of Programming Semantics. Proceedings, 1993. IX, 647 pages. 1994.

Vol. 803: J. W. de Bakker, W.-P. de Roever, G. Rozenberg (Eds.), A Decade of Concurrency. Proceedings, 1993. VII, 683 pages. 1994.

Vol. 804: D. Hernández, Qualitative Representation of Spatial Knowledge. IX, 202 pages. 1994. (Subseries LNAI).

Vol. 805: M. Cosnard, A. Ferreira, J. Peters (Eds.), Parallel and Distributed Computing. Proceedings, 1994. X, 280 pages. 1994.

Vol. 806: H. Barendregt, T. Nipkow (Eds.), Types for Proofs and Programs. VIII, 383 pages. 1994.

Vol. 807: M. Crochemore, D. Gusfield (Eds.), Combinatorial Pattern Matching. Proceedings, 1994. VIII, 326 pages. 1994.

Vol. 808: M. Masuch, L. Pólos (Eds.), Knowledge Representation and Reasoning Under Uncertainty. VII, 237 pages. 1994. (Subseries LNAI).

Vol. 809: R. Anderson (Ed.), Fast Software Encryption. Proceedings, 1993. IX, 223 pages. 1994.

Vol. 810: G. Lakemeyer, B. Nebel (Eds.), Foundations of Knowledge Representation and Reasoning. VIII, 355 pages. 1994. (Subseries LNAI).

Vol. 811: G. Wijers, S. Brinkkemper, T. Wasserman (Eds.), Advanced Information Systems Engineering. Proceedings, 1994. XI, 420 pages. 1994.

Vol. 812: J. Karhumäki, H. Maurer, G. Rozenberg (Eds.), Results and Trends in Theoretical Computer Science. Proceedings, 1994. X, 445 pages. 1994.

Vol. 813: A. Nerode, Yu. N. Matiyasevich (Eds.), Logical Foundations of Computer Science. Proceedings, 1994. IX, 392 pages. 1994.

Vol. 814: A. Bundy (Ed.), Automated Deduction—CADE-12. Proceedings, 1994. XVI, 848 pages. 1994. (Subseries LNAI).

Vol. 815: R. Valette (Ed.), Application and Theory of Petri Nets 1994. Proceedings. IX, 587 pages. 1994.

Vol. 816: J. Heering, K. Meinke, B. Möller, T. Nipkow (Eds.), Higher-Order Algebra, Logic, and Term Rewriting. Proceedings, 1993. VII, 344 pages. 1994.

Vol. 817: C. Halatsis, D. Maritsas, G. Philokyprou, S. Theodoridis (Eds.), PARLE '94. Parallel Architectures and Languages Europe. Proceedings, 1994. XV, 837 pages. 1994.

Vol. 818: D. L. Dill (Ed.), Computer Aided Verification. Proceedings, 1994. IX, 480 pages. 1994.

Vol. 819: W. Litwin, T. Risch (Eds.), Applications of Databases. Proceedings, 1994. XII, 471 pages. 1994.

Vol. 820: S. Abiteboul, E. Shamir (Eds.), Automata, Languages and Programming. Proceedings, 1994. XIII, 644 pages. 1994.

Vol. 821: M. Tokoro, R. Pareschi (Eds.), Object-Oriented Programming. Proceedings, 1994. XI, 535 pages. 1994.

Vol. 822: F. Pfenning (Ed.), Logic Programming and Automated Reasoning. Proceedings, 1994. X, 345 pages. 1994. (Subseries LNAI).

Vol. 823: R. A. Elmasri, V. Kouramajian, B. Thalheim (Eds.), Entity-Relationship Approach — ER '93. Proceedings, 1993. X, 531 pages. 1994.

Vol. 824: E. M. Schmidt, S. Skyum (Eds.), Algorithm Theory – SWAT '94. Proceedings. IX, 383 pages. 1994.

Vol. 825: J. L. Mundy, A. Zisserman, D. Forsyth (Eds.), Applications of Invariance in Computer Vision. Proceedings, 1993. IX, 510 pages. 1994.

Vol. 826: D. S. Bowers (Ed.), Directions in Databases. Proceedings, 1994. X, 234 pages. 1994.

Vol. 827: D. M. Gabbay, H. J. Ohlbach (Eds.), Temporal Logic. Proceedings, 1994. XI, 546 pages. 1994. (Subseries LNAI).

Vol. 828: L. C. Paulson, Isabelle. XVII, 321 pages. 1994.

Vol. 829: A. Chmora, S. B. Wicker (Eds.), Error Control, Cryptology, and Speech Compression. Proceedings, 1993. VIII, 121 pages. 1994.

Vol. 830: C. Castelfranchi, E. Werner (Eds.), Artificial Social Systems. Proceedings, 1992. XVIII, 337 pages. 1994. (Subseries LNAI).

Vol. 831: V. Bouchitté, M. Morvan (Eds.), Orders, Algorithms, and Applications. Proceedings, 1994. IX, 204 pages. 1994.

Vol. 832: E. Börger, Y. Gurevich, K. Meinke (Eds.), Computer Science Logic. Proceedings, 1993. VIII, 336 pages. 1994.

Vol. 833: D. Driankov, P. W. Eklund, A. Ralescu (Eds.), Fuzzy Logic and Fuzzy Control. Proceedings, 1991. XII, 157 pages. 1994. (Subseries LNAI).

Vol. 834: D.-Z. Du, X.-S. Zhang (Eds.), Algorithms and Computation. Proceedings, 1994. XIII, 687 pages. 1994.

Vol. 835: W. M. Tepfenhart, J. P. Dick, J. F. Sowa (Eds.), Conceptual Structures: Current Practices. Proceedings, 1994. VIII, 331 pages. 1994. (Subseries LNAI).

Vol. 836: B. Jonsson, J. Parrow (Eds.), CONCUR '94: Concurrency Theory. Proceedings, 1994. IX, 529 pages. 1994.

Vol. 837: S. Wess, K.-D. Althoff, M. M. Richter (Eds.), Topics in Case-Based Reasoning. Proceedings, 1993. IX, 471 pages. 1994. (Subseries LNAI).

Vol. 838: C. MacNish, D. Pearce, L. Moniz Pereira (Eds.), Logics in Artificial Intelligence. Proceedings, 1994. IX, 413 pages. 1994. (Subseries LNAI).

Vol. 839: Y. G. Desmedt (Ed.), Advances in Cryptology - CRYPTO '94. Proceedings, 1994. XII, 439 pages. 1994.

Vol. 840: G. Reinelt, The Traveling Salesman. VIII, 223 pages. 1994.

Vol. 841: I. Prívara, B. Rovan, P. Ružička (Eds.), Mathematical Foundations of Computer Science 1994. Proceedings, 1994. X, 628 pages. 1994.

Vol. 842: T. Kloks, Treewidth. IX, 209 pages. 1994.

Vol. 843: A. Szepietowski, Turing Machines with Sublogarithmic Space. VIII, 115 pages. 1994.

Vol. 844: M. Hermenegildo, J. Penjam (Eds.), Programming Language Implementation and Logic Programming. Proceedings, 1994. XII, 469 pages. 1994.

Vol. 845: J.-P. Jouannaud (Ed.), Constraints in Computational Logics. Proceedings, 1994. VIII, 367 pages. 1994.

Vol. 846: D. Shepherd, G. Blair, G. Coulson, N. Davies, F. Garcia (Eds.), Network and Operating System Support for Digital Audio and Video. Proceedings, 1993. VIII, 269 pages. 1994.

Vol. 847: A. L. Ralescu (Ed.) Fuzzy Logic in Artificial Intelligence. Proceedings, 1993. VII, 128 pages. 1994. (Subseries LNAI).

Vol. 848: A. R. Krommer, C. W. Ueberhuber, Numerical Integration on Advanced Computer Systems. XIII, 341 pages. 1994.

Vol. 849: R. W. Hartenstein, M. Z. Servít (Eds.), Field-Programmable Logic. Proceedings, 1994. XI, 434 pages. 1994.

Vol. 850: G. Levi, M. Rodríguez-Artalejo (Eds.), Algebraic and Logic Programming. Proceedings, 1994. VIII, 304 pages. 1994.

Vol. 851: H.-J. Kugler, A. Mullery, N. Niebert (Eds.), Towards a Pan-European Telecommunication Service Infrastructure. Proceedings, 1994. XIII, 582 pages. 1994.

Vol. 852: K. Echtle, D. Hammer, D. Powell (Eds.), Dependable Computing – EDCC-1. Proceedings, 1994. XVII, 618 pages. 1994.

Vol. 853: K. Bolding, L. Snyder (Eds.), Parallel Computer Routing and Communication. Proceedings, 1994. IX, 317 pages. 1994.

Vol. 854: B. Buchberger, J. Volkert (Eds.), Parallel Processing: CONPAR 94 – VAPP VI. Proceedings, 1994. XVI, 893 pages. 1994.

Vol. 855: J. van Leeuwen (Ed.), Algorithms – ESA '94. Proceedings, 1994. X, 510 pages.1994.

Vol. 856: D. Karagiannis (Ed.), Database and Expert Systems Applications. Proceedings, 1994. XVII, 807 pages. 1994.

Vol. 857: G. Tel, P. Vitányi (Eds.), Distributed Algorithms. Proceedings, 1994. X, 370 pages. 1994.

Vol. 858: E. Bertino, S. Urban (Eds.), Object-Oriented Methodologies and Systems. Proceedings, 1994. X, 386 pages. 1994.

Vol. 859: T. F. Melham, J. Camilleri (Eds.), Higher Order Logic Theorem Proving and Its Applications. Proceedings, 1994. IX, 470 pages. 1994.

Vol. 860: W. L. Zagler, G. Busby, R. R. Wagner (Eds.), Computers for Handicapped Persons. Proceedings, 1994. XX, 625 pages. 1994.

Vol: 861: B. Nebel, L. Dreschler-Fischer (Eds.), KI-94: Advances in Artificial Intelligence. Proceedings, 1994. IX, 401 pages. 1994. (Subseries LNAI).

Vol. 862: R. C. Carrasco, J. Oncina (Eds.), Grammatical Inference and Applications. Proceedings, 1994. VIII, 290 pages. 1994. (Subseries LNAI).

Vol. 863: H. Langmaack, W.-P. de Roever, J. Vytopil (Eds.), Formal Techniques in Real-Time and Fault-Tolerant Systems. Proceedings, 1994. XIV, 787 pages. 1994.

Vol. 864: B. Le Charlier (Ed.), Static Analysis. Proceedings, 1994. XII, 465 pages. 1994.

Vol. 865: T. C. Fogarty (Ed.), Evolutionary Computing. Proceedings, 1994. XII, 332 pages. 1994.

Vol. 866: Y. Davidor, H.-P. Schwefel, R. Männer (Eds.), Parallel Problem Solving from Nature - PPSN III. Proceedings, 1994. XV, 642 pages. 1994.

Vol 867: L. Steels, G. Schreiber, W. Van de Velde (Eds.), A Future for Knowledge Acquisition. Proceedings, 1994. XII, 414 pages. 1994. (Subseries LNAI).

Vol. 868: R. Steinmetz (Ed.), Multimedia: Advanced Teleservices and High-Speed Communication Architectures. Proceedings, 1994. IX, 451 pages. 1994.

Vol. 869: Z. W. Raś, Zemankova (Eds.), Methodologies for Intelligent Systems. Proceedings, 1994. X, 613 pages. 1994. (Subseries LNAI).

Vol. 870: J. S. Greenfield, Distributed Programming Paradigms with Cryptography Applications. XI, 182 pages. 1994.

Vol. 871: J. P. Lee, G. G. Grinstein (Eds.), Database Issues for Data Visualization. Proceedings, 1993. XIV, 229 pages. 1994.

Vol. 873: M. Naftalin, T. Denvir, M. Bertran (Eds.), FME '94: Industrial Benefit of Formal Methods. Proceedings, 1994. XI, 723 pages. 1994.

Vol. 874: A. Borning (Ed.), Principles and Practice of Constraint Programming. Proceedings, 1994. IX, 361 pages. 1994.

Vol. 875: D. Gollmann (Ed.), Computer Security – ESORICS 94. Proceedings, 1994. XI, 469 pages. 1994.

Vol. 876: B. Blumenthal, J. Gornostaev, C. Unger (Eds.), Human-Computer Interaction. Proceedings, 1994. IX, 239 pages. 1994.

Vol. 877: L. M. Adleman, M.-D. Huang (Eds.), Algorithmic Number Theory. Proceedings, 1994. IX, 323 pages. 1994.

Vol. 878: T. Ishida; Parallel, Distributed and Multiagent Production Systems. XVII, 166 pages. 1994. (Subseries LNAI).

Vol. 879: J. Dongarra, J. Waśniewski (Eds.), Parallel Scientific Computing. Proceedings, 1994. XI, 566 pages. 1994.

Vol. 880: P. S. Thiagarajan (Ed.), Foundations of Software Technology and Theoretical Computer Science. Proceedings, 1994. XI, 451 pages. 1994.

Vol. 882: D. Hutchison, A. Danthine, H. Leopold, G. Coulson (Eds.), Multimedia Transport and Teleservices. Proceedings, 1994. XI, 380 pages. 1994.